B

Central Cholinergic Synaptic Transmission

Edited by Michael Frotscher
Ulrich Misgeld

1989
Birkhäuser Verlag
Basel · Boston · Berlin

Editors:

Prof. Dr. Michael Frotscher
Anatomisches Institut
J. W. Goethe-Universität
Theodor-Stern-Kai 7
6000 Frankfurt a.M. 70
Federal Republic of Germany

Priv.-Doz. Dr. Ulrich Misgeld
Max-Planck-Institut für
Psychiatrie
Am Klopferspitz 18A
8033 Planegg-Martinsried
Federal Republic of Germany

Library of Congress Cataloging in Publication Data

Central cholinergic synaptic transmission/edited by Michael
Frotscher, Ulrich Misgeld.
    p. cm.—(Experientia. Supplementum: vol. 57
    Includes index.
    ISBN 0-8176-2267-5 (U.S.: alk. paper: 98.00F (Switzerland:
est.))
    1. Neural transmission. 2. Acetylcholine—Receptors.
    3. Cholinergic mechanisms. 4. Central nervous system.
    I. Frotscher, Michael, 1947–   . II. Misgeld, Ulrich, 1943–
    III. Series: Experientia. Supplementum: v. 57.
    [DNLM: 1. Neural Transmission. 2. Neuroregulators. 3. Receptors,
Cholinergic. 4. Synapses—physiology. W1 EX23 v. 57/WL 102.8
C3975]
    OP364.7.C46 1989
    599.0188—cd19
    DNLM/DLC

CIP-Titelaufnahme der Deutschen Bibliothek

Central cholinergic synaptic transmission/ed. by Michael
Frotscher; Ulrich Misgeld.–Basel; Boston; Berlin:
Birkhäuser, 1989
    (Experientia: Supplementum; Vol. 57)
    ISBN 3-7643-2267-5 (Basel ...) Pb.
    ISBN 0-8176-2267-5 (Boston) Pb.
NE: Frotscher, Michael [Hrsg.]; Experientia/Supplementum

© 1989 Birkhäuser Verlag
      P.O. Box 133
      4010 Basel
      Switzerland

Printed in Germany on acid-free paper
ISBN 3-7643-2267-5
ISBN 0-8176-2267-5

Contents

vi

Cholinergic receptor types and their modulation

Acetylcholine and plasticity of the CNS

Preface

Our knowledge of cholinergic synaptic transmission in the peripheral nervous system (PNS) has expanded enormously since the early 1930's, at which time O. Loewi discovered chemical neurotransmission through acetylcholine (ACh) and the pharmacological actions of ACh were described by H. Dale and his colleagues. Description of ACh's actions and receptors in various parts of the brain was followed by a detailed analysis of ACh's synthesis, release mechanism, removal from the synaptic cleft, modes of agonist-receptor interactions, properties of regulated conductances and of the pre- and postsynaptic modulation of cholinergic synapses. Our knowledge has been increasingly consolidating, leading us to the recent clarification of the structure of the peripheral nicotinic receptor-ion channel and its functional properties. It is appropriate to maintain the claim that the fundamentals of our understanding of synaptic transmission come from studies on cholinergic synapses in the PNS.

Contrastingly, views held on the role of cholinergic synapses in the mammalian central nervous system (CNS) are extremely controversial, although it has been clear for some time that ACh is present in the brain. Illustrating this, no unanimous view is held on the role of nicotinic receptors in the CNS. There is one notable exception to this confusion: Studies begun by J. C. Eccles and associates in the 1960's and completed by D. R. Curtis and R. W. Ryall in the 1970's on the motor axon collateral-Renshaw cell pathway in cats revealed the same principles of cholinergic synaptic transmission in the spinal cord as those revealed in studies on neuro-neuronal synapses in autonomic ganglia. In both autonomic ganglia and brain, however, the functional role of 'slow' muscarinic transmission is unknown.

The main obstacle to a better understanding of central cholinergic synaptic transmission is nothing less than the complexity of the brain itself. However, fundamental information provided by studies on *peripheral* cholinergic synaptic transmission strongly suggests that *central* cholinergic synaptic transmission is a key to understanding central synaptic transmission in general. The significance of cholinergic synaptic mechanisms is also weightened by the possibility that they may have a pivotal role in CNS disorders such as Alzheimer's disease.

New techniques and new tools, particularly those developed in the last decade, facilitated the discovery of many new facts concerning cholinergic synaptic transmission in the CNS. Choline acetyltransferase immunocytochemistry and various *in vitro* preparations of the mammalian CNS, with the aid of refined electrophysiological techniques, are just a few examples. Other techniques, single channel recording and molecular biology approaches, are beginning to show their potential usefulness. Given this state of methodological advancement, we considered it worthwhile to collect current views held on central cholinergic synaptic transmission.

This book is the manifestation of that collection of views. Most of the articles were first presented at a meeting with the same title in September, 1988 at Ringberg Castle in Bavaria. The book starts with a series of morphological studies on the distribution of cholinergic neurons, the course of cholinergic fibers and the ultrastructural features of cholinergic synapses in the CNS. Following this, a series of electrophysiological studies are presented centering around synaptic responses, involvement of various receptor types and their coupling to second messengers. The third group of articles introduces molecular biology techniques and pharmacological approaches for the study of receptors. Finally, there is a series of studies dealing with the influence of environmental cues on growth and regrowth of cholinergic fibers and the involvement of cholinergic systems in the plasticity of the brain.

This book does not aim to resolve all matters of controversy on the topic. On the contrary, the reader will find a number of controversial and even contradicting statements in the different articles. The book does, however, provide a comprehensive account of 'state of the art' research on central cholinergic synaptic transmission, which we hope will prove provoking and stimulating for future research.

The editors wish to thank the Stiftung Volkswagenwerk, the University of Frankfurt, the Sonderforschungsbereich 45 and 220 of the Deutsche Forschungsgemeinschaft, the Dr. Ernst-Rudolf Schloeßmann Stiftung, the Max-Planck-Gesellschaft, CIBA-Geigy GmbH, Hoffmann-La Roche AG, E. Merck, Sandoz AG, and Boehringer Ingelheim KG for making the meeting previous to the preparation of this book possible, and Birkhäuser Verlag for its publication. Special thanks to E. Schroeder for her organizational and editorial assistance.

M. Frotscher and U. Misgeld

Behavioral neuroanatomy of cholinergic innervation in the primate cerebral cortex

M.-Marsel Mesulam

Department of Neurology, Beth Israel Hospital, 330 Brookline Avenue, Boston, MA 02215, USA

Summary. This brief review summarizes the anatomy and behavioral affiliations of cortical cholinergic innervation. A depletion of this innervation has been reported in a number of human neurodegenerative conditions (including Alzheimer's disease) and may contribute to the genesis of the associated behavioral disturbances.

Central cholinergic pathways have been implicated in the regulation of extrapyramidal motor function, arousal, sleep, mood and especially memory (Karczmar, 1975). In keeping with the diversity of these behavioral affiliations, cholinergic pathways reach all levels of the neuraxis. Although ubiquitous, these pathways also display a strict topographical arrangement. Modern methods for axonal tracing and immunohistochemistry have led to the accumulation of additional information on the anatomical organization of these pathways.

The cholinergic innervation of the striatal complex is predominantly intrinsic and originates from cholinergic interneurons (Woolfe and Butcher, 1981). In contrast, the cholinergic innervation of the cerebral cortex, amygdaloid nuclei, olfactory bulb, thalamus and mesencephalic tectum is almost exclusively extrinsic, at least in the adult primate. The hippocampal formation receives its major cholinergic innervation from the medial septal nucleus (the cholinergic neurons of which are also designated as Ch1) and the vertical limb nucleus of the diagonal band (Ch2); the olfactory bulb from the horizontal limb nucleus of the diagonal band (Ch3); the cerebral cortex and amygdaloid nuclei from the nucleus basalis (Ch4); the thalamus from the pedunculopontine (Ch5) and laterodorsal tegmental nuclei (Ch6); the interpeduncular nucleus from the medial habenula (Ch7) and the superior colliculus from the parabigeminal nucleus (Ch8) (Mesulam et al., 1983a, b; Mufson et al., 1986).

The cholinergic nature of the nucleus basalis (NB) neurons has been demonstrated in a large number of animal species by showing the presence of an AChE-rich and ChAT-positive staining pattern (Wainer et al., 1984). In both man and monkey, more than 90% of the magnocellular NB neurons are also ChAT-positive (Mesulam et al., 1983a; Mesulam and Geula, 1988). These observations justify the designation

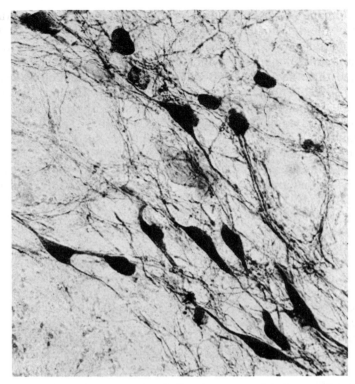

Choline acetyltransferase immunoreactive cholinergic neurons of the human Ch4 complex.

of this cell group, in the man and monkey, as the NB-Ch4 complex. We use the term Ch4 (sensu strictu) when referring exclusively to the ChAT-positive (cholinergic) neurons and the term NB-Ch4 when referring to the entire NB complex. The neurochemical properties of the ChAT-negative NB neurons remain to be determined. Some of these are likely to be NADPH and/or somatostatin-positive (Candy et al., 1985; Ellison et al., 1987).

That Ch4 is the major source of cortical cholinergic innervation has been demonstrated unequivocally by two types of observations in experimental animals. First, most of the cholinergic neurons projecting to cortex belong to the Ch4 complex (Mesulam et al., 1983a, b). Second, lesions involving Ch4 cause a major reduction of cortical ChAT levels and cholinegic fiber staining (Johnston et al., 1979; Mufson et al., 1987). This type of evidence is not available for the human brain. However, indirect support has been gathered from patients with Alzheimer's disease in whom cell loss of the NB-Ch4 complex is almost always associated with a corresponding depletion of cortical ChAT (Etienne et al., 1986).

The human NB-Ch4 extends from the level of the olfactory tubercle to that of the anterior hippocampus, spanning a distance of 13–14 mm in the sagittal plane. It attains its greatest width of 18 mm within the substantia innominata (subcommissural gray). Arendt et al. (1985) have estimated that the human NB-Ch4 complex contains 200,000 neurons in each hemisphere. Thus, the NB-Ch4 is about 10 times larger than the nucleus locus coeruleus which contains approximately 15,000 neurons in the adult human brain (Vijayashankar and Brody, 1979). The human NB-Ch4 can be subdivided into six subsectors that occupy its anteromedial (Ch4am), anterolateral (NB-Ch4al), anterointermediate (NB-Ch4ai), intermediodorsal (NB-Ch4id), intermedioventral (NB-Ch4iv) and posterior (NB-Ch4p) regions (Mesulam and Geula, 1988).

Gorry (1963) has pointed out that the NB displays a progressive evolutionary trend, becoming more and more extensive and differentiated in more highly evolved species, especially in primates and cetacea. Our observations in the brains of turtles, mice, rats, squirrel monkeys, rhesus monkeys and humans are consistent with this general view and show that the primate NB-Ch4 is a very complex, highly differentiated and relatively large structure. Although many morphological features of the human NB-Ch4 are similar to those described for the rhesus monkey, there is also a sense of increased complexity and differentiation.

The NB-Ch4 complex has been designated as an 'open' nucleus. There is a certain overlap with surrounding cell groups such as the olfactory tubercle, preoptic area, hypothalamic nuclei, striatal structures, nuclei of the diagonal band, amygdaloid nuclei and globus pallidus. An evolutionary trend can also be discerned in that the overlap is less in the monkey than in the rodent and somewhat less in the human than in the monkey. There is also no strict delineation between nuclear aggregates and passing fiber tracts. A substantial number of NB-Ch4 neurons, for example, are embedded within (or abut upon) the internal capsule, the diagonal bands of Broca, the anterior commissure, the ansa peduncularis (inferior thalamic peduncle) and the ansa lenticularis. In fact, previous designations for the nucleus basalis used terms such as the 'nucleus of the ansa peduncularis' and the 'nucleus of the ansa lenticularis'. The physiological implication of this intimate association with fiber bundles is unknown. Conceivably, the NB-Ch4 complex could monitor and perhaps influence the electrical activity along these fiber tracts. In addition to this open nuclear structure, the neurons of NB-Ch4 are heteromorphic in shape and have an isodendritic morphology with overlapping dendritic fields. These characteristics are also present in the nuclei of the brainstem reticular formation and has led to the suggestion that the NB-Ch4 complex could be conceptualized as a telencephalic extension of the brainstem reticular core (Ramon-Moliner and Nauta, 1966).

Experimental neuroanatomical methods in the monkey brain have shown that different cortical areas receive their major cholinergic input from different subsectors of the NB-Ch4 complex. Thus, Ch4am provides the major source of cholinergic input to medial cortical areas including the cingulate gyrus; Ch4al to frontal and parietal opercular regions and the amygdaloid nuclei; Ch4i to lateral frontal, parietal, peristriate and temporal regions; and Ch4p to superior temporal and temporopolar areas (Mesulam et al., 1983a, 1986a). The experimental methods that have been used to reveal this topographic arrangement cannot be used in the human brain. However, indirect evidence for the existence of a topographical arrangement can be gathered from disease states that selectively affect individual sectors of Ch4 or only parts of the cortical cholinergic innervation. For example, we described two cases with Alzheimer's disease in whom extensive loss of cholinergic fibers in temporopolar but not frontal opercular cortex was associated with marked cell loss in Ch4p but not Ch4a. This relationship is consistent with the topography of these projections in the monkey brain (Mesulam and Geula, 1988). Many additional cases will be necessary in order to unravel the complete topography of the corticopetal cholinergic projections that arise from the NB-Ch4 complex. The two cases that we have reported indicate that this projection is not diffuse and that it obeys a pattern that may well parallel the overall organization described in the rhesus monkey.

The AChE-rich fibers of the human cortex can be revealed in great detail with the help of recently developed and highly sensitive modifications of the Karnovsky-Roots procedure (Tago et al., 1986; Mesulam et al., 1987). Several lines of evidence indicate that these fibers are the cholinergic axons originating from Ch4 (see Mesulam et al., 1984; Mufson et al., 1987 for discussion). The pattern of laminar organization and regional density of cortical cholinergic fibers varies from one cortical area to another. Histochemical and neurochemical studies in the rhesus monkey have shown that limbic and paralimbic areas receive the most intense cholinergic input whereas the primary sensory-motor and isocortical association areas have a much less intense input. These differences are quite marked and may be as high as seven-fold (Mesulam et al., 1984, 1986b). Observations in the human brain show an almost identical pattern, including a close adherence to cytoarchitectonic boundaries (Mesulam and Geula, 1988). Thus, less well differentiated and non-isocortical (agranular and dysgranular) regions contain a greater cholinergic innervation than the immediately adjacent but more differentiated granular components of the same paralimbic area.

An even closer analysis of the regional variations in the cholinergic innervation of the cerebral cortex suggests that sensory information about extrapersonal events (especially those that are motivationally relevant) is likely to come under progressively greater cholinergic influence as it is conveyed along multisynaptic pathways leading to limbic structures such as the amygdala (Mesulam et al., 1986b). We speculated that these cholinergic pathways may provide a gating mechanism for regulating the access of sensory information into the limbic system. The disruption of this gating by cholinergic antagonists (e.g. scopolamine) or by disease states (e.g. Alzheimer's disease) may underlie the associated memory deficits. Thus, the behavioral affiliations of a transmitter that is ubiquitous in cortex can have a relative selectivity that reflects its differential anatomical distribution.

Neuroanatomical experiments in the monkey show that the reciprocal connectivity between the NB-Ch4 complex and the cerebral cortex is highly skewed (Mesulam and Mufson, 1984; Russchen et al., 1985). Although this nucleus projects to all cortical areas, it receives input only from a handful of regions all of which belong to the limbic and paralimbic parts of the brain (piriform, orbitofrontal, insular, temporopolar, parahippocampal and probably cingulate regions). Additional input comes from other limbic structures such as the amygdala, hypothalamus, septum and nucleus accumbens (or ventral globus pallidus). In keeping with this connectivity pattern in the rhesus monkey, our preliminary observations based on silver staining of 2 brains with circumscribed lesions suggests that the cingulate gyrus and amygdaloid nuclei in the human brain send projections to the NB-Ch4 complex. Whether the skewed connectivity pattern described in the monkey will also be found to exist in the human brain must await many additional cases that lend themselves to neuroanatomical investigation. In the human brain, catecholaminergic fibers (probably originating from the nucleus locus coeruleus) have also been shown to reach the region of the NB-Ch4 (Gaspar et al., 1985).

The skewed connectivity between NB-Ch4 and the cerebral cortex suggests that most cortical areas (primary sensory, motor and high order association areas) have no direct feedback control upon the cholinergic input that they receive whereas a handful of limbic and paralimbic areas can have direct control over the cholinergic input that they receive as well as over the cholinergic input that reaches all other parts of the cerebral cortex. Thus, the NB-Ch4 complex is in a position to act as a cholinergic relay for rapidly modulating the activity of the entire cortical surface according to the prevailing motivational conditions as reflected by the limbic system. This skewed organization, which is probably also found in the nucleus locus coeruleus, the brainstem raphe nuclei and the substantia nigra is well suited for modulating behavioral states (Mesulam, 1987). We had suggested that all complex behaviors can be conceptualized in the form of neural networks consisting of channel and state functions (Mesulam, 1985). In the case of visual recognition memory, for example, specific corticocortical pathways in occipitotemporal cortex provide the channels that are necessary for transferring the relevant visual information to the limbic structures involved in storage and retrieval (Mishkin, 1982). On the other hand, the cholinergic pathways that are present in all regions of the brain, could modulate the efficiency (or signal-to-noise ratio) of this information transfer, thereby controlling the state-dependent aspects of learning and memory.

Physiological experiments indicate that the acetylcholine (ACh) released by cortical cholinergic axons has two effects upon pyramidal neurons: a rapid and inhibitory effect probably mediated by GABA-ergic interneurons, and a more prolonged excitatory effect probably reflecting the direct effect of ACh upon postsynaptic cholinergic receptors located on pyramidal neurons (McCormick and Prince, 1985). The latter effect is caused by a reduction of potassium conductance in the membrane of the cholinoceptive neuron (Krnjevic, 1981). The excitatory nature and prolonged time course of this effect have led to the designation of AChE as an excitatory neuromodulator of cortical pyramidal neurons.

The ubiquitous distribution of cholinergic innervation indicates that virtually all cortical regions and therefore all realms of behavior are likely to be influenced by cholinergic transmission. In fact, even the neuronal response of primary somatosensory cortex to simple tactile stimulation can be modulated by the local application of ACh (Metherate et al., 1987). On the other hand, the preferential concentration of cholinergic pathways in limbic and paralimbic areas also explains why NB-Ch4 lesions (or systemically administered cholinoactive drugs) have their greatest impact on learning and memory, behaviors that are closely associated with the limbic system.

The behavioral specializations of the NB-Ch4 complex have been investigated in a number of animal species with the help of single unit recordings and lesion-behavior studies. In the awake and behaving rhesus monkey, these neurons are responsive to the sight and taste of food in a manner that appears to reflect the degree of hunger and even the desirability of the food object (Rolls et al., 1979). This contrasts sharply with the response contingencies of immediately adjacent globus pallidus neurons which fire mostly in conjunction with specific movements (DeLong, 1971). It appears that the NB-Ch4 neurons are important for encoding the existence of rewarding events. This may explain why electrical stimulation in this general area in some patients has elicited reports of a pleasurable sensation (Heath, 1959) and perhaps also why this region sustains self-stimulation in animals (Olds and Milner, 1964).

In the absence of specific cholinotoxins, it is difficult to interpret experiments based on electrolytic or excitotoxin-induced lesions of the NB-Ch4 complex. Such lesions almost certainly include additional damage to adjacent structures. However, some experiments include two types of controls that increase the specificity of the findings. First, they show that the degree of behavioral deficit correlates with the extent of cortical ChAT depletion. Secondly, they show that the behavioral deficits are reversible by the administration of cholinoactive agents. The most closely studied behaviors in these types of experiments have been in the area of memory and learning. In a number of species including rodents and primates, NB-Ch4 lesions have impaired the acquisition and retention of new tasks and also the relearning of discriminations that had been mastered in the presurgical period (Flicker et al., 1983; Ridley et al., 1986; Dunnett et al., 1987).

Additional ablation experiments (in rats and cats) have also suggested that the NB-Ch4 complex is responsible for mediating the low voltage fast activity in the cortical EEG that is induced by cholinergic drugs (Steward et al., 1984) and the cortical vasodilation that is elicited by cerebellar stimulation (Iadecola et al., 1983). Unilateral lesions in Ch4 have also led to a decrease, at least transiently, of glucose utilization in the ipsilateral cerebral cortex (Orzi et al., 1986). Furthermore, the ability of cats to maintain non-REM sleep is impaired by lesions that appear to include the NB-Ch4 complex (Szymusiak and McGinty, 1986). This brief survey indicates the wide spectrum of behaviors associated with the cholinergic corticopetal projection that arises from the NB-Ch4 complex.

Degenerative changes in the NB-Ch4 complex have been reported in a surprisingly large number of neurological diseases. This rapidly growing list now includes Alzheimer's disease, Down's Syndrome, Parkinson's disease, olivopontocerebellar atrophy, Pick's disease, supranuclear ophthalmoplegia and even schizophrenia (for review see Averback, 1981; Casanova et al., 1985; Rogers et al., 1985; Tagliavini and Pilleri, 1985). It remains to be determined to what extent the involvement in NB-Ch4 contributes to the mental state alteration that exists in each of these conditions. The most extensive observations have been reported in conjunction with Alzheimer's and Parkinson's diseases. These conditions involve a substantial loss of NB-Ch4 neurons and an associated loss of cortical ChAT and cholinergic fibers.

The suggestion had been advanced that the cholinergic loss constitutes the central feature of Alzheimer's disease, in a manner that may parallel the relationship between Parkinson's disease and the loss of dopaminergic nigrostriatal projections. The loss of cortical cholinergic innervation in Alzheimer's disease is, indeed, marked, consistent and significantly correlated with the extent of mental state deficit (Perry et al., 1978). It is, therefore, quite likely that the loss of cortical cholinergic pathways contributes to the severity of the memory loss and perhaps to the other features of the dementia. However, in view of the many additional pathological features in Alzheimer's disease (especially the great number of cortical neurofibrillary tangles, neuritic plaques and

8

neuronal loss) it is doubtful that the cholinergic lesion is the pivotal feature or the prime mover in Alzheimer's disease. It is not even known if the loss of cholinergic cortical innervation in Alzheimer's disease represents a primary involvement of the perikarya in the NB-Ch4 complex or if the primary lesion is in the cortical fibers, resulting in a retrograde degeneration within NB-Ch4 (Mesulam, 1986). Recently, it has been shown that the survival of the NB-Ch4 neurons is dependent on the retrograde transport of trophic factors such as nerve growth factor (NGF) from the cerebral cortex. It is conceivable, therefore, that the severe pathological changes characteristic of the cerebral cortex in Alzheimer's disease, may deplete the trophic factors necessary for the upkeep of the NB-Ch4 perikarya and therefore lead to their degeneration (Hefti and Weiner, 1985). In other conditions such as Parkinson's disease where cortical involvement is much less consistent, the involvement of NB-Ch4 may begin at the level of the perikarya and proceed in a corticopetal direction (Bloxam et al., 1984). Thus, the pathophysiological mechanism for the involvement of NB-Ch4 in neurodegenerative conditions, may vary from one disease to another.

A good deal of attention has also been directed to the age-related changes in the NB-Ch4 complex. The most extensive studies have been done in the rodent and show that these neurons display an age-related decrease in volume but probably not in number. A similar situation may exist in the human brain but this is somewhat controversial. It is conceivable that the age-related shrinkage of the NB-Ch4 neurons causes the relatively minor changes of cortical cholinergic innervation and perhaps also to the alterations of memory abilities characteristic of advanced age (see Hornberger et al., 1985; Mesulam et al., 1987b for review).

In summary, the NB-Ch4 complex is highly developed in the primate brain and provides the origin of one of the most important transmitter systems of the forebrain. Anatomical, physiological and behavioral experiments are only beginning to unravel some of the mechanisms that link this nucleus to complex behavior, aging and neurodegenerative diseases.

Acknowledgements. I want to thank Leah Christie for expert secretarial assistance. Supported in part by a Javits Neuroscience Investigator Award of the NINCDS and the Alzheimer's Disease and Related Disorders Association. This review contains paraphrases and quotations from Mesulam and Geula (1988).

Arendt, T., Bigl, V., Tennstedt, A., and Arendt, A. (1985) Neuronal loss in different parts of the nucleus basalis is related to neuritic plaque formation in cortical target areas in Alzheimer's disease. J. Neurosci. 14: 1–14.
Averback, P. (1981) Lesions of the nucleus ansae peduncularis in neuropsychiatric disease. Arch. Neurol. 38: 230–235.
Bloxam, C. A., Perry, E. K., Perry, R. H., and Candy, J. M. (1984) neuropathological and neurochemical correlate of Alzheimer-type and Parkinsonian dementia. In: Wurtman, R. J.,

Corkin, S. H., and Growdon, J. H. (eds), Alzheimer's Disease: Advances in Basic Research and Therapies. Center for Brain Sciences and Metabolism Charitable Trust, Cambridge, MA, pp. 39–52.

Candy, J. M., Perry, R. H., Thompson, J. E., Johnson, M., Oakley, A. E., and Edwardson, J. A. (1985) The current status of the cortical cholinergic system in Alzheimer's disease and Parkinson's disease. J. Anat. 140: 309–327.

Casanova, M. F., Walker, L. C., Whitehouse, P. J. and Price D. L. (1985) Abnormalities of the nucleus basalis in Down's Syndrome. Ann. Neurol. 18: 310–313.

DeLong, M. R. (1971) Activity of pallidal neurons during movement. J. Neurophysiol. 34: 414–427.

Dunnett, S. B., Whishaw, I. Q., Jones, G. H., and Bunch, S. T. (1987) Behavioral, biochemical and histochemical effects of different neurotoxic amino acids injected into nucleus basalis magnocellularis of rats. J. Neurosci. 20: 653–669.

Ellison, D. W., Kowall, N. W., and Martin, J. B. (1987) Subset of neurons characterized by the presence of NADPH-diophorase in human substantia innominata. J. comp. Neurol. 260: 233–245.

Etienne, P., Robitaille, Y., Wood, P., Gauthier, S., Nair, N. P. V., and Quirion, R. (1986) Nucleus basalis neuronal loss, neuritic plaques and choline acetyltransferase activity in advanced Alzheimer's disease. J. Neurosci. 19: 1279–1291.

Flicker, C., Dean, R. L., Watkins, D. L., Fisher, S. K., and Bartus, R. T. (1983) Behavioral and neurochemical effects following neurotoxic lesions of a major cholinergic input to the cerebral cortex in the rat. Pharmac. Biochem. Behav. 18: 973–981.

Gaspar, P., Berger, B., Alvarex, C., Vigny, A., and Henry, J. P., (1985) Catecholaminergic innervation of the septal area in man: Immunocytochemical study using TH and DBH antibodies. J. comp. Neurol. 214: 12–33.

German, D. C., Bruce, G., and Hersh, L. B. (1985) Immunohistochemical staining of cholinergic neurons in the human brain using a polyclonal antibody to human choline acetyltransferase. Neurosci. Lett. 61: 1–5.

Gorry, J. D. (1963) Studies on the comparative anatomy of the ganglion basale of Meynert. Acta anat. 55: 51–104.

Heath, R. G. (1959) Studies in Schizophrenia. Harvard University Press, Cambridge, MA.

Hefti, F., and Weiner, W. J. (1986) Nerve growth factor and Alzheimer's disease. Ann. Neurol. 20: 275–281.

Iadecola, C., Mraovitch, S., Meeley, M. P., and Reis, D. J. (1983) Lesions of the basal forebrain in rat selectively impair vasodilation elicited from cerebellar fastigial nucleus. Brain Res. 279: 41–52.

Johnston, M. V., McKinney, M., and Coyle, J. T. (1979) Evidence for a cholinergic projection to neocortex from neurons in basal forbrain. Proc. natl Acad. Sci. 76: 5392–5396.

Karczmar, A. G. (1975) Cholinergic influences on behavior. In: Waser, P. G. (ed.), Cholinergic Mechanisms. Raven Press, New York, pp. 501–529.

Krnjević, K. (1981) Acetylcholine as a modulator of amino-acid-mediated synaptic transmission. In: The Role of Peptides and Amino Acids as Neurotransmitters. A. Liss, New York, pp. 124–141.

McCormick, D. A., and Prince, D. A. (1985) Two types of muscarinic responses to acetylcholine in mammalian cortical neurons. Proc. natl Acad. Sci. USA 82: 6344–6348.

Mesulam, M.-M. (1985) Patterns in behavioral neuroanatomy: association areas, the limbic system and hemispheric specialization. In: Mesulam, M.-M. (ed.), Principles of Behavioral Neurology. Contemporary Neurology Series. F. A. Davis Co., Philadelphia, pp. 1–70.

Mesulam, M.-M. (1986) Alzheimer's plaques and cortical cholinergic innervation. Neurosci. 17: 275–276.

Mesulam, M.-M. (1987) Asymmetry of neural feedback in the organization of behavioral states. Science 237: 537–538.

Mesulam, M.-M., and Geula, C. (1988) Nucleus basalis (Ch4) and cortical cholinergic innervation in the human brain. J. comp. Neurol., in press.

Mesulam, M.-M., and Mufson, E. J. (1984) Neural inputs into the nucleus basalis of the substantia innominata (Ch4) in the rhesus monkey. Brain 107: 253–274.

Mesulam, M.-M., Mufson, E. J., Levey, A. I., and Wainer, B. H. (1983a) Cholinergic innervation of cortex by the basal forebrain: Cytochemistry and cortical connections of

10

the septal area, diagonal band nuclei, nucleus basalis (substantia innominata) and hypothalamus in the rhesus monkey. J. comp. Neurol. 214: 170–197.

Mesulam, M.-M., Mufson, E. J., and Rogers, J. (1987) Age-related shrinkage of cortically projecting cholinergic neurons: A selective effect. Ann. Neurol. 22: 31–36.

Mesulam, M.-M., Mufson, E. J., and Wainer, B. H. (1986a) Three-dimensional representation and cortical projection topography of the nucleus basalis (Ch4) in the macaque: Concurrent demonstration of choline acetyltransferase and retrograde transport with a stabilized tetramethylbenzidine method for HRP. Brain Res. 367: 301–308.

Mesulam, M.-M., Mufson, E. J., Wainer, B. H., and Levey, A. I. (1983b) Central cholinergic pathways in the rat: An overview based on an alternative nomenclature (Ch1–Ch6). Neuroscience 10: 1185–1201.

Mesulam, M.-M., Rosen, A. D., and Mufson, E. J. (1984) Regional variations in cortical cholinergic innervation: chemoarchitectonics of acetylcholinesterase-containing fibers in the macaque brain. Brain Res. 311: 245–258.

Mesulam, M.-M., Volicer, L., Marquis, J. K. Mufson, E. J., and Green, R. C. (1986b) systematic regional differences in the cholinergic innervation of the primate cerebral cortex: Distribution of enzyme activities and some behavioral implications. Ann. Neurol. 19: 144–151.

Metherate, R., Tremblay, N., and Dykes, R. W. (1987) Acetylcholine permits long-term enhancement of neuronal responsiveness in cat primary somatosensory cortex. Neuroscience 22: 75–81.

Mishkin, M. (1982) A memory system in man and monkey. Phil. Trans. R. Soc. Lond. B 298: 85–92.

Mufson, E. J., Kehr, A. D., Wainer, B. H., and Mesulam, M.-M. (1987) Cortical effects of neurotoxic damage to the nucleus basalis in rats: persistent loss of extrinsic cholinergic input and lack of transsynaptic effect upon the number of somatostatin-containing, cholinesterase-positive and cholinergic cortical neurons. Brain Res. 417: 385–388.

Mufson, E. J., Martin, T. L., Mash, D. C., Wainer, B. H., and Mesulam, M.-M. (1986) Cholinergic projections from the parabigemenal nucleus (Ch8) to the superior colliculus in the mouse: A combined analysis of HRP transport and choline acetyltransferase immunohistochemistry. Brain Res. 370: 144–148.

Nagai, T., McGeer, P. L., Peng, J. H., McGeer, E. G., and Dolman, C. E. (1983) Choline acetyltransferase immunohistochemistry in brains of Alzheimer's disease patients and controls. Neurosci. Lett. 36: 195–199.

Olds, J., and Milner, P. (1954) Positive reinforcement produced by electrical stimulation of septal area and other regions of rat brain. J. comp. Physiol. Psychol. 47: 419–427.

Orzi, F., Diana, G., Palombo, E., Lenzi, G. L., Bracco, L., and Fieschi, C. (1986) Effects of unilateral lesion of the nucleus basalis on local cerebral glucose utilization in the rat. In: Vezzadini, P., Facchini, A., and Labo, G. (eds), Neuroendocrine System and Aging. Eurage, pp. 259–264.

Pearson, R. C. A., Sofroniew, M. V., Cuello, A. C., Powell, T. P. S., Eckenstein, F., Esiri, M. M., and Wilcock G. K. (1984) Persistence of cholinergic neurons in the basal nucleus in a brain with senile demntia of the Alzheimer's type demonstrated by immunohistochemical staining for choline acetyltransferase. Brain Res. 289: 375–379.

Perry, E. K., Tomlinson, B. E., Blessed, G., Bergmann, K., Gibson, P. H., and Perry, R. H. (1978) Correlation of cholinergic abnormalities with senile plaques and mental test scores in senile dementia. Br. Med. J. 2: 1457–1459.

Ramon-Moliner, E., and Nauta, W. J. H. (1966) The isodendritic core of the brain stem. J. comp. Neurol. 126: 311.

Ridley, R. M., Murray, T. K., Johnson, J. A., and Baker, H. F. (1986) Learning impairment following lesion of the basal nucleus of Meynert in the marmoset: Modification by cholinergic drugs. Brain Res. 376: 108–116.

Rogers, J. D., Brogan, D., and Mirra, S. S. (1985) The nucleus basalis of Meynert in neurological disease: A quantitative morphological study. Ann. Neurol. 17: 163–170.

Rolls, E. T., Sanghera, M. K., and Rober-Hall, A. (1979) The latency of activation of neurons in the lateral hypothalamus and substantia innominata during feeding in the monkey. Brain Res. 164: 121–135.

Russschen, F. T., Amaral, D. G., and Price, J. L. (1985) The afferent connections of the substantia innominata in the monkey, Macaca fascicularis. J. comp. Neurol. 24: 1–27.

Saper, C. B., and Chelimskyh, T. C. (1984) A cytoarchitectonic and histochemical study of nucleus basalis and associated cell groups in the normal human brain. Neuroscience 13: 1023–1037.

Steward, D. J., MacFabe, D. F., and Vanderwolf, C. N. (1984) Cholinergic activation of the electrocorticogram: Role of the substantia innominata and effects of atropine and quinuclidinyl benzylate. Brain Res. 322: 219–232.

Szymusiak, R., and McGinty, D. (1986) Sleep suppression following kainic acid-induced lesions of the basal forebrain. Exp. Neurol. 94: 598–614.

Tagliavini, F., and Pilleri, G. (1985) Neuronal loss in the basal nucleus of Meynert in a patient with olivopontocerebellar atrophy. Acta neuropath. 66: 127–133.

Tago, H., Kimura, H., and Maeda, T. (1986) Visualization of detailed acetylcholinesterase fiber and neuron stainig in rat brain by a sensitive histochemical procedure. J. Histochem. Cytochem. 34: 1431–1438.

Vijayashankar, N., and Brody, H. (1979) A quantitative study of the pigmented neurons in the nuclei locus coeruleus and sub coeruleus in man as related to aging. J. Neuropath. exp. Neurol. 38: 490–497.

Wainer, B. H., Levey, A. I., Mufson, E. J., and Mesulam, M.-M. (1984) Cholinergic systems in mammalian brain identified with antibodies against choline acetyltransferase. Neurochem. int. 6: 163–182.

Woolfe, N. J., and Butcher, L. L. (1981) Cholinergic neurons in the caudate-putamen complex proper are intrinsically organized: A combined Evans Blue and acetylcholinesterase analysis. Brain Res. 7: 487–507.

Afferent connections of the forebrain cholinergic projection neurons, with special reference to monoaminergic and peptidergic fibers

Laszlo Zaborszky

Department of Otolaryngology, Box 430, University of Virginia Medical Center, Charlottesville, VA 22908, USA

Summary. Earlier light microscopic data on afferent connections to the cholinergic forebrain neurons are reconsidered in the light of EM cross-identification of neurons and synapses by combinations of tracer and immunocytochemical techniques. Such studies suggest that brain-stem monoaminergic afferents terminate on cholinergic forebrain neurons, and may modulate the activity of choline acetyltransferase levels in the postsynaptic neurons. A monosynaptic relationship between cholinergic forebrain neurons and neuropeptide Y and somatostatin containing axons is also supported by studies using double immunolabeling techniques at the EM level. These peptidergic afferents originate in part from locally arborizing neurons. Based upon the new data a circuit model for basal forebrain cholinergic neurons is proposed.

Introduction

The basal forebrain cholinergic projection (BFC) system has received considerable attention recently as a result of evidence from several disciplines suggesting that cholinergic mechanisms are important in arousal, memory, and learning, and that disruption of this system may be related, at least in part, to the cognitive decline in patients with Alzheimer's disease (for ref. see Buzsaki et al., 1988; Price et al., 1986; van Hoesen and Damasio, 1987). The BFC system in the rat originates in a widely dispersed, more or less continuous collection of aggregated and non-aggregated cells in the basal forebrain (Fig. 1). The medial septum and nucleus of vertical limb of the diagonal band (VDB) provide the cholinergic innervation of the hippocampus, the nucleus of the horizontal limb of the diagonal band (HDB) projects to the olfactory bulb, piriform and entorhinal corticis, while cholinergic neurons located in the ventral pallidum, sublenticular substantia innominata (SI), globus pallidus, nucleus ansa lenticularis—collectively termed the nucleus basalis—innervate neocortical areas and the basolateral amygdala (Sofroniew et al., 1982; Armstrong et al., 1983; Mesulam et al., 1983a, b; Irle and Markowitsch, 1984; Woolf et al., 1984; Rye et al., 1984; Amaral and Kurz, 1985; Carlsen et al., 1985; Wainer et al., 1985; Zaborszky et al., 1986a; Rao et al., 1987; Fisher et al., 1988).

The widespread distribution of cholinergic terminals throughout the cerebral cortex, and the fact that these terminals contact different

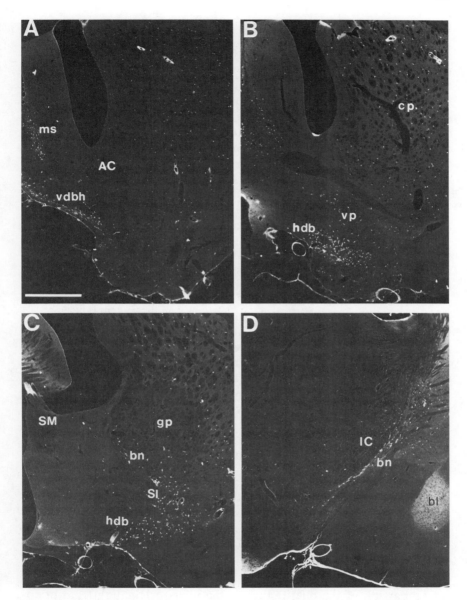

Figure 1. Series of low magnification darkfield photomicrographs of frontal sections of the basal forebrain in rat immunostained for ChAT. *A* rostral; *D* caudal. White dots represent cholinergic cell bodies. AC, anterior commissure; bl, basolateral amygdaloid nucleus; bn, nucleus basalis; cp, caudate putamen; gp, globus pallidus; hdb, horizontal limb of the diagonal band; IC, internal capsule; ms, medial septal nucleus; SI, substantia innominata; SM, stria medullaris; vdbh, vertical limb of the diagonal band pars horizontalis; vp, ventral pallidum. Bar scale: 1 mm.

postsynaptic profiles non-selectively (Houser et al., 1985) suggest that the cortical cholinergic system is a rather diffuse, non-specific system with more general function (Richardson and DeLong, 1988). The extent to which intrinsic cholinergic neurons in the cortex contribute to this 'diffuse' innervation remains unclear. On the other hand, corticopetal cholinergic neurons project topographically to cortical areas (e.g. Kristt et al., 1985; Mesulam et al., 1986; Luiten et al., 1987), and the cortical cholinergic innervation does show some regional variations (Mesulam et al., 1984). Furthermore, it has been shown that individual cholinergic axons have relatively restricted zones of termination and lack divergent collaterals to remote cortical areas (Price and Stern, 1983; Walker et al., 1985; Eckenstein et al., 1988). These facts suggest that cholinergic neurons may be able to selectively influence restricted cortical areas, in addition to having more global effect on cortical functions.

A number of *light microscopic* studies in several species have described sources of input to regions containing cholinergic projection neurons, including cortical and rostral forebrain areas, the amygdala, hypothalamus, thalamus and various brainstem nuclei. These data are summarized in Tables 1 and 2, and give an indication of the magnitude of potential afferent sources to cholinergic forebrain neurons. However, the cholinergic projection neurons are not only related to several prominent ascending and descending fiber tracts, they also interdigitate with other neuronal populations, including GABAergic and peptidergic neurons (Zaborszky et al., 1986b; Mellander et al., 1985). Consequently, in order to determine whether afferent fibers to a region with cholinergic neurons actually establish synaptic contact with the cholinergic neurons, appropriate combinations of tracer and immunocytochemical methods are required at the ultrastructural level. The validation of the different tracer techniques is beyond the scope of this review and the reader is referred to a recent paper on tract-tracing techniques (Zaborszky and Heimer, 1989). The present review will focus on two different transmitter-containing systems: the monoaminergic and peptidergic projections to cholinergic forebrain neurons.

Monoaminergic projections

Comparison of the distribution of the fibers originating from brainstem monoaminergic cell groups with the areas containing the forebrain cholinergic projection neurons suggests the possibility of a direct monoaminergic-cholinergic interaction in the basal forebrain (Swanson and Hartman, 1975; Steinbusch, 1981; Armstrong et al., 1983; tables 1 and 2). Studies using double immunolabeling techniques at the light microscope level showed that tyrosine hydroxylase (TH)-containing terminal varicosities are in close apposition to cholinergic neurons in the dorsal part of the HDB, ventral pallidum, sublenticular SI, and the globus pallidus (Fig. 2). Using EM double immunolabeling technique, we have confirmed that TH-containing terminals do indeed establish symmetric synaptic contact with cholinergic dendrites in the ventral part of the globus pallidus (Zaborszky et al., 1989). As TH is the first enzyme in the catecholamine biosynthetic pathway, the TH immunoreactivity observed in terminals could represent noradrenaline, dopamine or adrenaline. At present there is no direct morphological data to indicate which of these transmitters may be present in terminals that contact the cholinergic neurons. Our studies combining choline acetyltransferase (ChAT) immunocytochemistry and phaseolus vulgaris leucoagglutinin (PHA-L) tracing from brainstem areas known to contain catecholaminergic cell groups (locus coeruleus, A1, A2, A8–A10 cell groups), indicate that both noradrenergic and dopaminergic axons project to the vicinity of cholinergic cells (Fig. 3). Furthermore, light microscopic double labeling experiments suggest that serotoninergic fibers may contact cholinergic neurons as well. However, the exact nature of such contacts can only be determined by the use of appropriate methods at the EM level.

In a subsequent biochemical experiment, ChAT activity was significantly reduced in areas containing cholinergic projection neurons two weeks after neurotoxin (6-OHDA, 5,7-DHT) injections into the ascending monoaminergic bundles (Zaborszky et al., 1989). These findings, together with the data obtained in the tracing studies referred to above, represent converging lines of evidence to suggest that monoaminergic axons transsynaptically affect ChAT levels in postsynaptic neurons. The decline of ChAT activity in the basal forebrain appears to reflect the interruption of a net facilitatory influence of monoamines on the cholinergic neurons. However, other studies suggested an inhibitory influence of the monoaminergic fibers on different neurons in the CNS (Moroni et al., 1982; Foote et al., 1983). One cannot exclude the possibility that the biochemical changes are mediated in part by GABAergic interneurons, which are known to innervate the cholinergic neurons (Robinson et al., 1979; Zaborszky et al., 1986b) and might similarly receive monoaminergic afferents. These effects might be similar to those in the lateral geniculate body, where locus coeruleus-induced potentiation of synaptic transmission is thought to be mediated through local inhibitory interneurons (Nakai et al., 1974).

The functional significance of the monoaminergic input to the basal forebrain cholinergic neurons is unknown. Considering the suggestion that basal forebrain neurons are involved in EEG activation (Buzsaki et al., 1988; Steriade and Llinas, 1988) and the observations that locus coeruleus neurons have characteristic discharge rates in relation to vigilance levels (for ref. see Foote et al., 1983; Hobson and Steriade, 1987), one could speculate that locus coeruleus afferents inform the

Table 1. Afferents to forebrain areas containing cholinergic projection neurons from telencephalic regions

	Terminal area	Method	Species	Authors
Cortex				
prefrontal	SI	F-H,A,HRP	M,	Leichnetz and Astruc, 1976; Mesulam and Mufson, 1984
			R,C	Haring and Wang, 1986; Irle and Markowitsch, 1986
orbitofrontal	SI,VDB	N,HRP,	M	Whitlock and Nauta, 1956; Aggleton et al., 1980
		A	M	Mesulam and Mufson, 1984; Ruschen et al., 1985
temporal	SI	N,A	M	Whitlock and Nauta, 1956; Mesulam and Mufson, 1984
		WGA-HRP,A	M,C	Russchen et al. 1985; Irle and Markowitsch, 1986
anterior cingulate	SI,VDB	WGA-HRP	R	Saper, 1984; Leman and Saper, 1985
	SI,HDB	A,HRP	M	Russchen et al., 1985; Irle and Markowitsch, 1986
retrosplenial	SI,HDB,VDB	WGA-HRP	R	Saper, 1984
infralimbic	SI	HRP,A	M	Russchen et al. 1985
	HDB,VDB	WGA-HRP	R	McLean et al., 1986; Saper, 1984
prelimbic	SI	WGA-HRP,A	M	Russchen et al., 1985
	VDB	A,WGA-HRP	R	Beckstead, 1979; Saper, 1984; Russchen et al., 1985
insular	SI	HRP,A	M,C	Aggleton et al., 1980; Irle and Markowitsch, 1986
				Mesulam and Mufson, 1984
perirhinal	SI,HDB	WGA-HRP	R	Saper, 1984; Leman and Saper, 1984; McLean et al., 1986
piriform, entorhinal	SI	HRP	M	Aggleton et al., 1980
	SI	N,WGA-HRP,A	M	Whitlock and Nauta, 1956; Russchen et al., 1985;
		HRP	C	Irle and Markowitsch, 1986
	HDB	WGA-HRP,As	R	McLean et al., 1986; Fuller et al., 1986
	VDB	WGA-HRP	R	Saper, 1984
hippocampus	HDB,VDB	WGA-HRP	R	Saper, 1984

Rostral forebrain

vertical limb of DB	SI,HDB	A	R	Conrad and Pfaff, 1976; Swanson and Cowan, 1979
horizontal limb of DB	HDB(contra)	HRP	C,M	Irle and Markowitsch, 1986
		N,WGA-HRP PHA-L,FT	R	Price and Powell, 1971; Fuller et al., 1986; McLean et al., 1986; Semba et al., 1988a,
lateral septum	SI,VDB	A	R	Swanson and Cowan, 1979
nucleus accumbens	SI	F-H,A	R,C	Williams et al., 1977; Nauta et al., 1978; Troiano and Siegel, 1978; Groenewegen and Russchen, 1984; Mesulam and
		A	C,M	Mufson, 1984
bed n. st. termin.	SI	PHA-L	R	Grove et al., 1986
		A	R	Swanson, 1976; Swanson and Cowan, 1979; Conrad and Pfaff, 1976
septohippocampal n.	HDB	WGA-HRP	R	McLean et al., 1986
preoptic area	SI,HDB,VDB	A	R	Conrad and Pfaff, 1976; Swanson, 1976; Saper et al., 1979

Amygdala

n. lat. olf. tract	HDB	WGA-HRP	R	McLean et al., 1986
anterior cortical n.	HDB	WGA-HRP	R	McLean et al., 1986
medial n.	SI,HDB	A	R,C	Krettek and Price, 1978
posterior cortical n.	HDB	A	R	Krettek and Price, 1978
basal (mg, pc) n.	SI,HDB	A,PHA-L, WGA-HRP	R,C,M	Krettek and Price, 1978; Kelley et al., 1982; Russchen and Price, 1984; Russchen et al., 1985; McLean et al., 1986; Irle and Markowitsch, 1986; Price et al., 1987
accessory basal n.	VDB	A	M	Ruschen et al., 1985
central n.	HDB	WGA-HRP	R	McLean et al., 1986
	SI,HDB	A,PHA-L	R,C,M	Price and Amaral, 1981; Krettek and Price, 1978; Kelley et al., 1982; Russchen et al., 1985

Table 2. Afferents to forebrain areas containing cholinergic projection neurons from diencephalic and brainstem regions

	Terminal area	Method	Species	Author
Hypothalamus				
ant. hyp. area	SI	A	R	Conrad and Pfaff, 1976; Saper et al., 1976
paraventricular n.	SI	A	R	Conrad and Pfaff, 1976
ventromedial n.	SI	A	M	Jones et al., 1976; Amaral et al., 1982
	SI	A	R	Saper et al., 1976; Krieger et al., 1979
dorsal hypothalamus	SI	HRP	C,M	Irle and Markowitsch, 1986
lateral hyp. area	SI,HDB,VDB	N,WGA-HRP, A	R,M	Price and Powell, 1970; Saper et al., 1979; Berk and Finkelstein, 1982; Amaral and Cowan, 1982; Saper, 1985
caudal magnocell. n.	HDB	WGA-HRP	R	Fuller et al., 1986
mammillary body	HDB	A	G	Shen, 1983
	SI,VDB	A	M	Amaral et al., 1982
supramammillary n.	VDB,HDB	WGA-HRP	M,R	Russchen et al., 1985; Vertes, 1988
Thalamus				
midline, intralam. nn.	HDB,SI	HRP-WGA,As	C,M,R	McLean et al., 1986; Fuller et al., 1986;
mediodorsla n.	HDB,SI	HRP	C	Irle and Markowitsch, 1986
habenular nn.	SI,HDB	A	R	Herkenham and Nauta, 1979
Subthalamus				
fields of Forel	SI	HRP	C,M	Irle and Markowitsch, 1986
zona incerta	SI	HRP,A	C,M,R	Ricardo, 1981; Irle and Markowitsch, 1986
Brainstem				
substantia nigra	SI	A,HRP,FT	M,R	Fallon and Moore, 1978; Aggleton et al., 1980; Takagi et al., 1980; Jones and Yang 1985; Martinez-Murillo et al., 1988; Semba et al., 1988b

ventral tegmental a.	SI,HDB,VDB	A,WGA-HRP	R	Fallon and Moore, 1978; Beckstead et al., 1979; Simon et al., 1979; Vertes 1984b; Jones and Yang, 1985; Fuller et al., 1986; Vertes, 1988
peripeduncular n.	SI	A	M,R	Jones et al., 1976; Jones and Yang, 1985; Russchen et al., 1985 Vertes, 1988
periaqueductal gray	VDB,HDB	WGA-HRP	R	Vertes, 1988
n. of Darkschewitsch	VDB,HDB	WGA-HRP	R	
midbrain, pontine ret. formation	HDB,SI	WGA-HRP,A	R	Edwards and deOlmos, 1976; Eberhart et al., 1985; Vertes, 1984b; McLean et al., 1986; Semba et al., 1988b; Vertes, 1988
parabrachial nn.	SI,HDB,VDB	A,HRP	R,M	Norgren, 1976; Saper and Loewy, 1980; Takagi et al., 1980; Irle and Markowitsch, 1986; Vertes, 1988; Semba et al., 1988b
pedunculopontine n.	HDB	WGA-HRP,Ch	R	McLean et al., 1986; Fuller et al., 1986; Jones and Beaudet, 1987; Semba et al., 1988b; Vertes, 1988
laterodorsal tegm. n.	HDB,VDB	WGA-HRP PHA-L,Ch	R	Takagi et al., 1980; McLean et al., 1986; Fuller et al., 1986; Satoh and Fibiger, 1986; Jones and Beaudet, 1987; Semba et al. 1988b; Vertes, 1988
locus coeruleus	SI,HDB,VDB,	A WGA-HRP	R R,M	Jones and Moore, 1977; Takagi et al., 1980; Vertes, 1984a; 1988; Hering and Wang, 1986; McLean et al, 1986, Fuller et al., 1986; Irle and Markowitsch, 1986; Martinez-Murillo et al., 1988; Semba et al., 1988b
raphe nuclei	SI,HDB,VDB	A,WGA-HRP, Ch	R	Conrad et al., 1974; Azmitia and Segal, 1978; Takagi et al., 1980; Jones and Yang, 1985; McLean et al, 1986; Jones and Beaudet, 1987; Semba et al., 1988b; Vertes and Martin, 1988
solitary region	SI,HDB	A,WGA-HRP	R	Ricardo and Koh, 1978; Semba et al., 1988b; Vertes, 1988
ventrolat. medulla	SI,HDB,VDB	A,WGA-HRP	R	Loewy et al., 1980; McKellar and Loewy, 1982; Semba et al., 1988b; Vertes, 1988

A, autoradiography; As, 3H-D-aspartate; C, cat; Ch, 3H-choline; F-H, Fink-Heimer; FT, fluorescent tracer; WGA-HRP, wheatgerm agglutinin-horseradish peroxidase; M, monkey; N, Nauta; PHA-L, phaseolus vulgaris leucoagglutinin; R, rat; ms, magnocellular; pc, parvocellular.

Figure 2. Tyrosine hydroxylase (TH) immunoreactive puncta (arrows) around cholinergic cells in the ventral pallidum. Double immunolabeling using DAB for ChAT and BDHC for TH according to the method of Levey et al. (1986). Bar scale: 100 μm.

cholinergic system about the onset of behaviorally significant stimuli originating from the external environment. Another explanation is that the effect of norepinephrine on cholinergic neuronal functional activity is similar to that exerted on other regions in the CNS, i.e. enhancing reliability and efficacy of synaptic transmission for other input converging on the same neuron during the period of simultaneous coactivation (Foote et al., 1983).

Peptidergic afferents

The various peptide-containing fiber systems show characteristic distributions in the basal forebrain; different populations of cholinergic neurons are likely to be contacted by different afferent peptidergic fibers (for the distribution of different peptides see Palkovits, 1984). For example, cholinergic neurons in the ventral pallidum and globus pallidus may have rich neurotensin and enkephalin innervation, but seem to be contacted only occasionally by other peptidergic afferents. On the other hand, cholinergic neurons in the substantia innominata may receive a substantial input from several different peptidergic systems. So far only enkephalin (Chang et al., 1987), substance P (Bolam et al., 1986), NPY and somatostatin (Zaborszky and Braun, in preparation; Figs 4–6) have been localized in boutons contacting cholinergic forebrain neurons. In addition, our light microscopic double labeling studies have revealed a hitherto unknown local NPY and somatostatin

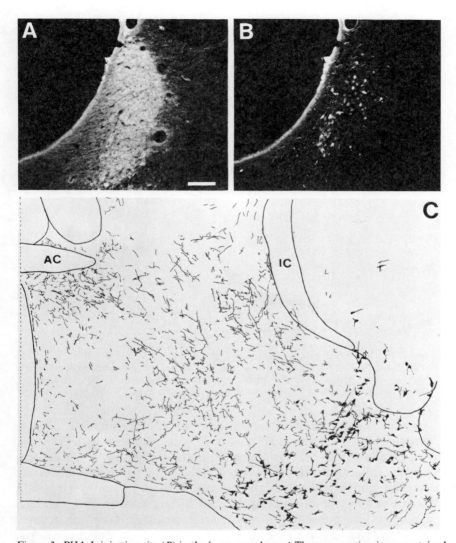

Figure 3. PHA-L injection site (*B*) in the locus coeruleus. *A* The same section immunostained with an antibody against dopamine-beta-hydroxylase (DBH). Double immunofluorescence (FITC/RITC) technique. Note that all PHA-L labeled cell bodies are confined to the heavy DBH-positive area. Bar scale: 100 μm. *C* Line drawing approximately between the levels of figures 1B and 1C to show the distribution of PHA-L labeled fibers and terminal varicosities around cholinergic neurons following injection of the tracer into the locus coeruleus. Antibody to DBH was kindly provided by Dr R. Grzanna. From the material of Zaborszky and Cullinan.

neuronal system (Fig. 5). A single peptidergic neuron with its locally arborizing axoncollaterals apparently innervate a number of cholinergic neurons. Conversely, the very same cholinergic neuron may receive axonterminals from several NPY or somatostatin neurons.

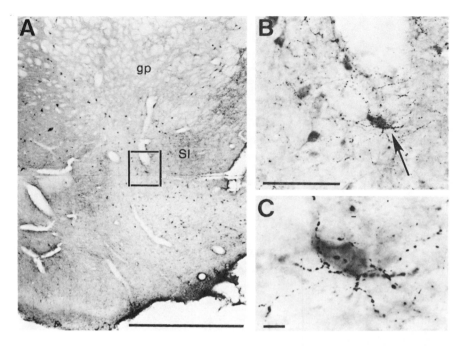

Figure 4. Frontal section immunostained for ChAT and neuropeptide Y (NPY) using DAB for ChAT and NiDAB for NPY. Boxed area in *A* is shown with higher magnification in *B*. *B* The cholinergic cell body labeled by arrow is heavily innervated by NPY terminals. *C* Higher magnification view of the same neuron. Bar scale: *A*: 1 mm, *B*: 100 μm, *C*: 10 μm.

Considering the graded transmitter release in peptidergic neurons (Swanson, 1983; Iversen, 1983), as well as the different metabolic and membrane potential changes which are induced in postsynaptic neurons by peptidergic afferents (Kow and Pfaff, 1988), the above-mentioned divergence-convergence relationships could result in spatio-temporally different patterns of information transfer through synapses on cholinergic neurons. Further complexity is introduced if we take into account that peptidergic neurons often contain more than one neuroactive substance. For example, galanin has been shown to be colocalized with acetylcholine (ACh) in hippocampopetal cholinergic neurons (Melander et al., 1985). Since NPY and somatostatin (Chronwall et al., 1984) and somatostatin and GABA (Somogyi et al., 1984) already have been shown to be colocalized in several places in the forebrain, it is possible that all three substances could be colocalized, at least in a population of interneurons.

The functional significance of peptidergic afferents to cholinergic neurons is supported by a number of pharmacological experiments in which the delivery of peptides into the basal forebrain resulted in changes in ACh turnover in the cortex or hippocampus (Wenk, 1984;

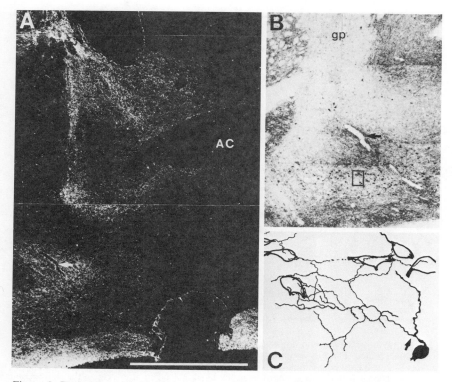

Figure 5. Dark (*A*) and brightfield (*B*) photograph of the same section immunostained for somatostatin (SS) and ChAT using nickel enhanced DAB (NiDAB) for SS and DAB for ChAT. Arrows point to the same vessel. *C* Camera lucida drawing of boxed area from *B* to show the axonal arborization of an SS-positive neuron (black) around cholinergic neurons (open symbols). Arrow points to the bifurcation of the axon. Bar scale: 1 mm.

Wood and McQuade, 1986; Constantinidis et al., 1988). Furthermore, changes in the level of somatostatin, neurotensin, NPY, alfa-MSH, and galanin in basal forebrain areas rich in cholinergic neurons have been reported in brains of patients with Alzheimer's disease (Constantinidis et al., 1988).

Circuit model

The hypothetical circuit diagram in Figure 7 illustrates the possible flow of information through the cholinergic forebrain neurons to the cerebral cortex. Some key features of the diagram will be discussed under separate headings.

A) Parallel processing of information through forebrain cholinergic neurons. Through a series of combined light-electron microscopic studies, in

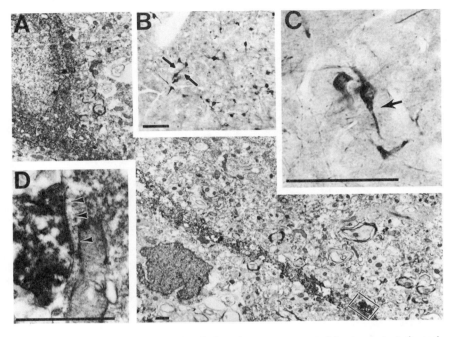

Figure 6. Electronmicrograph (*A*) of a cholinergic neuron contacted by a somatostatin-positive axon terminal. Double immunostaining using DAB for ChAT and NiDAB for somatostatin. *B* The identified neuron is located in the substantia innominata indicated by two arrows. *C* Higher magnification photo from the blockface taken after cutting the thin section shown in *A*. *D* Enlarged view of the boxed area in *A*. Arrowheads point to the postsynaptic side. Bar scale: *A*, *D*: 1 μm; *B*, *C*: 100 μm.

which different tracer and immunocytochemical procedures were used to identify the transmitter of both the afferent fiber system and the postsynaptic target, we have established that cholinergic forebrain neurons receive afferent synaptic connections from the basolateral amygdala (Zaborszky et al., 1984), the brainstem monoaminergic cell groups (Zaborszky et al., 1989) and the lateral hypothalamus (Zaborszky and Cullinan, 1989), all of which were shown to provide a significant 'diffuse' cortical innervation (for ref. see Saper, 1987). Whether such parallel processing through the forebrain cholinergic projection system reflects a general organizational principle of diffuse corticopetal systems remains to be determined. It is also unknown whether, within a given population of corticopetal ascending neurons, the same cells that project to the cortex also innervate forebrain cholinergic neurons through their collaterals. In addition, it is unclear whether these neurons influence the same cortical fields through their direct and indirect projections (i.e. through the cholinergic system). In view of recent evidence indicating that some of the cholinergic neurons in the SI project to the reticular nucleus of the thalamus (Hallanger et al., 1987; Steriade et al., 1987a),

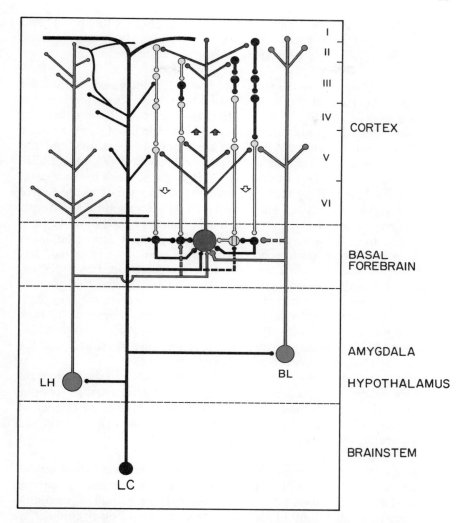

Figure 7. Circuit model to show the flow of information through the basal forebrain cholinergic neurons (red-colored neuron in the center). From the afferents, only those are included here which have been identified on the basis of combined EM studies. Interrupted lines in the basal forebrain indicate connections which have been inferred only from pharmacological or indirect morphological data. The intracortical circuitry is a simplified version of the model of Oshima (1983). Inhibitory cells are drawn in full black, excitatory neurons are in yellow. The neuron symbolized with hatching corresponds to a peptidergic (NPY or somatostatin) neuron. The model does not take into account the possible overlap of the cortical arborization of the different types of nonspecific axons. BL, basolateral amygdala; LH, lateral hypothalamus; LC, locus coeruleus.

known to influence the cortex through its connections with specific and intralaminar thalamic nuclei (Steriade et al., 1987b), the conduction line through the cholinergic neurons may be even more complex.

B) Local connectivity in the basal forebrain. We have shown that the cholinergic neurons in the ventral pallidum receive GABAergic input (Zaborszky et al., 1986b). Other studies (Ingham et al., 1988) confirmed this observation, and indicated that the GABAergic input could be quite substantial, since about 50% of the synapses on the soma and the proximal dendrite of putative cholinergic neurons are GABA-containing. Considering the large number of GABAergic neurons in the ventral pallidum itself, it is likely that they constitute at least part of the source of the GABAergic input. Since many of the forebrain GABAergic neurons receive catecholaminergic input (Leranth et al., 1988), we have postulated that GABAergic neurons receiving input from the locus coeruleus project to corticopetal cholinergic neurons. As noted, our biochemical data are compatible with this interpretation. Pharmacological and electrophysiological studies (Okamoto and Wada, 1984; Levine et al., 1986) also suggest that the excitatory influence from the amygdala is mediated, at least partly, to the cholinergic forebrain neurons through GABAergic interneurons. Peptidergic interneurons (somatostatin, neuropeptide Y, etc.) may further modulate synaptic transmission at the level of the cholinergic neurons as discussed above.

There is also evidence to suggest that collaterals of basal forebrain corticopetal neurons contribute to local integrative synaptic actions (Kristt et al., 1985; Semba et al., 1987). However, since in neither of these experiments was the chemical identity of the cells determined unequivocally, this issue requires further study.

C) Cortical processing and feedback to cholinergic neurons. According to Houser et al. (1985), cholinergic synapses are present in all cortical layers. The most common postsynaptic elements were small- to medium-sized dendritic shafts of unidentified origin. In addition, cholinergic terminals form synaptic contacts with apical and basilar dendrites of pyramidal neurons in layer V, as well as with the somata of nonpyramidal, possibly GABAergic interneurons of layer III. Using the model of Oshima and co-workers (Oshima, 1983) it could be envisaged that cholinergic afferents, like other nonspecific afferents in the cortex, elicit a cascade of excitatory and inhibitory processes in the supragranular layers, resulting in complex combinations of excitatory, inhibitory, disfacilitatory and disinhibitory responses in layers V and VI.

To date, only one EM study (Lemann and Saper, 1985) has addressed the identification of cortical afferent synapses on basal forebrain neurons, indicating that the cholinergic neurons might receive direct feedback from the cortex. Since this study did not determine the chemical nature of the postsynaptic neuron, it remains unclear whether this is indeed the case. The situation, however, may be analogous to the septohippocampal system, where hippocampofugal axons terminate on GABAergic interneurons (Leranth, this volume).

Concluding remarks

The overall picture that emerges from these considerations may still be rather vague, but it nevertheless gives an impression of how BFC neurons, together with GABAergic and other local interneurons, are involved in information transfer through the basal forebrain to cortical areas. This basal forebrain cholinergic-GABAergic-peptidergic integrative system receives multichannel input, and presents a topographically organized output to the cortex.

Future studies are likely to modify this simplistic scheme, particularly those defining precise target of the cholinergic neurons, along with the identification of afferent connections both in qualitative and quantitative terms. Such studies will determine the level of specificity in the organization of the cholinergic system, crucial in assessing its importance in modulating cortical functions. These data will also be essential in our understanding of the deficiencies in the information processing occurring in disease states.

Acknowledgements. I would like to thank Prof. L. Heimer and Prof. J. R. Wolff for helpful discussions. Special thanks are due to W. E. Cullinan who collaborated on several of our studies. My appreciation is due also to Prof. T. Tömböl who helped in the evaluation of the somatostatin material. Lee F. Snavely is gratefully acknowledged for skilful technical assistance. The original research summarized in this review is supported by USPHS Grants NS 23945 and 17743.

Aggleton, J. P., Burton, M. J., and Passingham, R. E. (1980) Cortical and subcortical afferents to the amygdala of the rhesus monkey (Macaca Mulatta). Brain Res. 190: 347–368.

Amaral, D. G., and Kurz, J. (1985) An analysis of the origins of the cholinergic and noncholinergic septal projections to the hippocampal formation. J. comp. Neurol. 240: 37–59.

Amaral, D. G., Veazey, R. B., and Cowan, W. M. (1982) Some observations on hypothalamo-amygdaloid connections in the monkey. Brain Res. 252: 13–27.

Armstrong, D. M., Saper, C. B., Levey, A. I., Wainer, B. H., and Terry, R. D. (1983) Distribution of cholinergic neurons in rat brain: Demonstrated by the immunocytochemical localization of choline acetyltransferase. J. comp. Neurol. 216: 53–68.

Azmita, E. C., and Segal, M. (1978) An autoradiographic analysis of the differential ascending projections of the dorsal and median raphe nuclei in the rat. J. comp. Neurol. 179: 641–688.

Beckstead, R. M. (1979) An autoradiographic examination of corticocortical and subcortical projections of the mediodorsal-projection (prefrontal) cortex in the rat. J. comp. Neurol. 184: 43–62.

Beckstead, R. M., Domesick, V. B., and Nauta, W. J. H. (1979) Efferent connections of the substantia nigra and ventral tegmental area in the rat. Brain Res. 175: 191–217.

Berk, M. L., and Finkelstein, J. A. (1982) Efferent connections of the lateral hypothalamic area of the rat: An autoradiographic investigation. Brain Res. Bull. 8: 511–526.

Bolam, J. P., Ingham, C. A., Izzo, P. N., Levey, A. I., Rye, D. B., Smith, A. D., and Wainer, B. H. (1986) Substance P-containing terminalis in synaptic contact with cholinergic neurons in the neostriatum and basal forebrain: A double immunocytochemical study in the rat. Brain Res. 397: 279–289.

Buzsáki, G., Bickford, R. G., Ponomareff, G., Thal, L. J., Mandel, R., and Gage, F. H. (1988) Nucleus basalis and thalamic control of neocortical activity in the freely moving rat. J. Neurosci. 8: 4007–4026.

Carlsen, J., Zaborszky, L., and Heimer, L. (1985) Cholinergic projections from the basal forebrain to the basolateral amygdaloid complex: A combined retrograde fluorescent and immunohistochemical study. J. comp. Neurol. 234: 155–167.

Chan, H. T., Penny, G. R., and Kitai, S. T. (1987) Enkephalinergic-cholinergic interaction in the rat globus pallidus: A pre-embedding double-labeling immunocytochemistry study. Brain Res. 426: 197–203.

Chronwall, B. M., Chase, T. N., and O'Donohue, T. L. (1984) Coexistence of neuropeptide Y and somatostatin in rat and human cortical and rat hypothalamic neurons. Neurosci. Lett. 52: 213–217.

Conrad, L. C. A., and Pfaff, D. W. (1976) Efferents from medial basal forebrain and hypothalamus in the rat. II. An autoradiographic study of the anterior hypothalamus. J. comp. Neurol. 169: 221–262.

Conrad, L. C. A., Leonard, C. M., and Pfaff, D. W. (1974) Connections of the median and dorsal raphe nuclei in the rat: An autoradiographic and degeneration study. J. comp. Neurol. 156: 179–206.

Constantinidis, J., Bouras, C., and Vellet, P. G. (1988) Neuropeptides in Alzheimer's and in Parkinson's Disease. Mt Sinai J. Med. 55: 102–115.

Eberhart, J. A., Morell, J. I., Krieger, M. S., and Pfaff D. W. (1985) An autoradiographic study of projections ascending from the midbrain central gray, and from the region lateral to it in the rat. J. comp. Neurol. 241: 285–310.

Eckenstein, F. P., Baughman, R. W., and Quinn, J. (1988) An anatomical study of cholinergic innervation in rat cerebral cortex. Neuroscience 25: 457–474.

Edwards, S. B., and de Olmos, J. S. (1976) Autoradiographic studies of the projections of the midbrain reticular formation: Ascending projections of nucleus cuneiformis. J. comp. Neurol. 165: 417–432.

Fallon, J. H., and Moore, R. Y. (1978) Catecholamine innervation of the basal forebrain. IV. Topography of the dopamine projection to the basal forebrain and neostriatum. J. comp. Neurol. 180: 545–580.

Fisher, R. S., Buchwald, N. A., Hull, C. D., and Levine, M. S. (1988) GABAergic basal forebrain neurons project to the neocortex: The localization of glutamic acid decarboxylase and choline acetyltransferase in feline corticopetal neurons. J. comp. Neurol. 272: 489–502.

Foote, S. L., Bloom, F. E., and Aston-Jones, G. (1983) Nucleus locus ceruleus: New evidence of anatomical and physiological specificity. Physiol. Rev. 63: 844–914.

Fuller, T. A., Carnes, K. M., and Price J. L. (1986) Afferents to the horizontal diagonal band of rat. Soc. Neurosci. Abstr. 12: 351.

Groenewegen, H. J., and Russchen, F. T. (1984) Organization of the efferent projections of the nucleus accumbens to pallidal, hypothalamic and mesencephalic structures. A tracing and immunohistochemical study in the cat. J. comp. Neurol. 223: 347–368.

Grove, E. A., Domesick, V. B., Nauta, W. J. H. (1986) Light microscopic evidence of striatal input to intrapallidal neurons of cholinergic cell group Ch4 in the rat: A study employing the anterograde tracer Phaseolus vulgaris leucoagglutinin (PHAL). Brain Res. 367: 379–384.

Hallanger, A. E., Levey, A. L., Lee, H. J., Rye, D. B., and Wainer, B. H. (1987) The origins of cholinergic and other subcortical afferents to the thalamus in the rat. J. comp. Neurol. 262: 105–124.

Haring, J. H., and Wang, R. Y. (1986) The identification of some sources of afferent input to the rat nucleus basalis magnocellularis by retrograde transport of horseradish peroxidase. Brain Res. 366: 152–158.

Herkenham, M., and Nauta, W. J. H. (1979) Efferent connections of the habenular nuclei in the rat. J. comp. Neurol. 187: 19–48.

Hobson, J. A., and Steriade, M. (1986) Neuronal basis of behavioral state control. In: Mountcastle, V. B., Bloom, F. E., and Geiger, S. R. (eds), Handbook of Physiology. The nervous system. Intrinsic regulatory systems of the brain, Sect. 1, Vol. 4. Am. physiol. Soc., Bethesda, pp. 701–823.

Houser, C. R., Crawford, G. D., Salvaterra, P. M., and Vaughn, J. E. (1985) Immunocytochemical localization of choline acetyltransferase in rat cerebral cortex: A study of cholinergic neurons and synapses. J. comp. Neurol. 234: 17–34.

Ingham, C. A., Bolam, J. P., and Smith, A. D. (1988) GABA-immunoreactive synaptic boutons in the rat basal forebrain: Comparison of neurons that project to the neocortex with pallidosubthalamic neurons. J. comp. Neurol. 273: 263–282.

Irle, E., and Markowitsch, H. J. (1984) Basal forebrain efferents reach the whole cerebral cortex of the cat. Brain Res. Bull. 12: 493–512.

Irle, E., and Markowitsch, H. J. (1986) Afferent connections of the substantia innominata/basal nucleus of Meynert in carnivores and primates. J. Hirnforsch. 27: 343–367.

Iversen, L. L. (1983) Amino acids and peptides: Fast and slow chemical signals in the nervous system? Proc. R. Soc. Lond. B 221: 245–260.

Jones, B. E., and Beaudet, A. (1987) Retrograde labeling of neurons in the brain stem following injections of [³H] choline into the forebrain of the rat. Exp. Brain Res. 65: 437–448.

Jones, B. E., and Yang, T.-Z. (1985) The efferent projections from the reticular formation and the locus coeruleus studied by anterograde and retrograde axonal transport in the rat. J. comp. Neurol. 242: 56–92.

Jones, B. E., and Moore, R. Y. (1977) Ascending projections of the locus coeruleus in the rat. II. Autoradigraphic study. Brain Res. 127: 23–53.

Kelley, A. E., Domesick, V. B., and Nauta, W. J. H. (1982) The amygdalostriatal projection in the rat—An anatomical study by anterograde and retrograde tracing methods. Neuroscience 7: 615–630.

Kow, L.-M., and Pfaff, D. W. (1988) Neuromodulatory actions of peptides. Ann Rev. Pharmac. Toxic. 28: 163–188.

Krettek, J. E., and Price, J. L. (1978) Amygdaloid projections to subcortical structures within the basal forebrain and brainstem in the rat and cat. J. comp. Neurol. 178: 225–254.

Krieger, M. S., Conrad, L. C. A., and Pfaff, D. W. (1979) An autoradiographic study of the efferent connections of the ventromedial nucleus of the hypothalamus. J. comp. Neurol. 183: 785–816.

Kristt, D. A., McGowan, R. A., Martin-MacKinnon, N., and Solomon, J. (1985) Basal forebrain innervation of rodent neocortex: Studies using acetylcholinesterase histochemistry, Golgi and lesion strategies. Brain Res. 337: 19–39.

Leichnetz, G. R., and Astruc, J. (1977) The course of some prefrontal corticofugals to the pallidum, substantia innominata, and amygdaloid complex in monkeys. Exp. Neurol. 54: 104–109.

Leman, W., and Saper, C. B. (1985) Evidence for a cortical projection to the magnocellular basal nucleus in the rat: An electron microscopic axonal transport study. Brain Res. 334: 339–343.

Leranth, C., MacLusky, N. J., Shanabrough, M., and Naftolin, F. (1988) Catecholaminergic innervation of LHRH and GAD immunopositive neurons in the rat medial preoptic area: An electron microscopic double immunostaining and degeneration study. Neuroendocrinology 48: 591–602.

Levey, A. I., Bolam, J. P., Rye, D. B., Hallanger, A. E., Demuth, R. M., Mesulam, M.-M., and Wainer, B. H. (1986) A light and electron microscopic procedure for sequential double antigen localization using diaminobenzidine and benzidine dihydrochloride. J. Histochem. Cytochem. 34: 1449–1457.

Levine, M. S., Cepeda, C., and Buchwald, N. A. (1968) Electrophysiological properties of basal forebrain neurons: Responses to microphoretic application of putative neurotransmitters. Soc. Neurosci. Abstr. 12: 1468.

Loewy, A. D., Wallach, J. H., and McKellar, S. (1981) Efferent connections of the ventral medulla oblongata in the rat. Brain Res. Rev. 3: 63–80.

Luiten, P. G. M., Gaykema, R. P. A., Traber, J., and Spencer, D. G. (1987) Cortical projection patterns of magnocellular basal nucleus subdivisions as revealed by anterogradely transported Phaseolus vulgaris leucoagglutinin. Brain Res. 413: 229–250.

Martinez-Murillo, R., Semenenko, F., and Cuello, A. C. (1988) The origin of tyrosine hydroxylase immunoreactive fibers in the regions of the nucleus basalis magnocellularis of the rat. Brain Res. 451: 227–236.

McKellar, S., and Loewy, A. D. (1982) Efferent projections of the A1 catecholamine cell group in the rat: An autoradiographic study. Brain Res. 241: 11–29.

McLean, J. H., Nickell, W. T., and Shipley, M. T. (1986) Afferent connections to the horizontal limb of diagonal band. Soc. Neurosci. Abstr. 12: 351.

Melander, T., Staines, W. A., Hökfelt, T., Rökaeus, Å., Eckenstein, F., Salvaterra, P. M., and Wainer, B. H. (1985) Galanin-like immunoreactivity in cholinergic neurons of the septum-basal forebrain complex projecting to the hippocampus of the rat. Brain Res. 360: 130–138.

Mesulam, M.-M., Mufson, E. J., Wainer, B. H., and Levey, A. I. (1983a) Central cholinergic pathways in the rat: An overview based on an alternative nomenclature (Ch1-Ch6). Neuroscience 10: 1185–1201.

Mesulam, M.-M., Mufson, E. J., Levey, A. I., and Wainer, B. H. (1983b) Cholinergic innervation of cortex by the basal forebrain: Cytochemistry and cortical connections of the septal area, diagonal band nuclei, nucleus basalis (substantia innominata), and hypothalamus in the rhesus monkey. J. comp. Neurol. 214: 170–197.

Mesulam, M.-M., and Mufson, E. J. (1984) Neural inputs into the nucleus basalis of the substantia innominata (Ch4) in the rhesus monkey. Brain 107: 253–274.

Mesulam, M.-M., Rosen, A. D., and Mufson, E. J. (1984) Regional variations in cortical cholinergic innervation: Chemoarchitectonics of acetylcholinesterase-containing fibers in the macaque brain. Brain Res. 311: 245–258.

Moroni, F., Bianchi, C., Moneti, G., Tanganelli, S., Spidelieri, G., Guandalini, P., and Beani, L. (1982) Release of GABA from the guinea-pig neocortex induced by electrical stimulation of the locus coeruleus or by norepinephrine. Brain Res. 232: 216–221.

Nakai, Y., and Takaori, S. (1974) Influence of norepinephrine containing neurons derived from the locus coeruleus on lateral geniculate neuronal activities of cats. Brain Res. 71: 47–60.

Nauta, W. J. H., Smith, G. P., Faull, R. L. M., and Domesick, V. B. (1978) Efferent connections and nigral afferents of the nucleus accumbens septi in the rat. Neuroscience 3: 385–401.

Norgren, R. (1976) Taste pathways to hypothalamus and amygdala. J. comp. Neurol. 166: 17–30.

Okamoto, M., and Wada, J. A. (1984) Reversible suppression of amygdaloid kindled convulsion following unilateral gabaculine injection in to the substantia innominata. Brain Res. 305: 389–392.

Oshima, T. (1983) Intracortical organization of arousal as a model of dynamic neuronal processes that may involve a set for movements. In: Desmedt, J. E. (ed.), Motor Control Mechanisms in Health and Disease. Raven Press, New York, pp. 287–300.

Palkovits, M. (1984) Distribution of neuropeptides in the central nervous system: A review of biochemical mapping studies. Prog. Neurobiol. 23: 151–189.

Price, D. L., Whitehouse, P. J., and Struble, R. G. (1986) Cellular pathology in Alzheimer's and Parkinson's Diseases. TINS 9: 29–33.

Price, J. L., and Stern, R. (1983) Individual cells in the nucleus basalis-diagonal band complex have restricted axonal projections to the cerebral cortex in the rat. Brain Res. 269: 352–356.

Price, J. L., and Amaral, D. G. (1981) An autoradiographic study of the projections of the central nucleus of the monkey amygdala. J. Neurosci. 1: 1242–1259.

Rao, Z. R., Shiosaka, S., and Tohyama, M. (1987) Origin of cholinergic fibers in the basolateral nucleus of the amygdaloid complex by using sensitive double-labeling technique of retrograde biotinized tracer and immunocytochemistry. J. Hirnforsch. 28: 553–560.

Ricardo, J. A. (1981) Efferent connections of the subthalamic region in the rat. II. The zona incerta. Brain Res. 214: 43–60.

Ricardo, J. A., and Koh, E. T. (1978) Anatomical evidence of direct projections from the nucleus of the solitary tract to the hypothalamus, amygdala and other forebrain structures in the rat. Brain Res. 153: 1–26.

Richardson, R. T., and DeLong, M. R. (1988) A reappraisal of the functions of the nucleus basalis of Meynert. TINS 11: 264–267.

Robinson, S. E., Malthe-Sorenssen, D., Wood, P. L., and Commissiong, J. (1979) Dopaminergic control of the septal-hippocampal cholinergic pathway. J. Pharm. exp. Ther. 476–479.

Rye, D. B., Wainer, B. H., Mesulam, M. M., Mufson, E. J., and Saper, C. B. (1984) Cortical projections arising from the basal forebrain: A study of cholinergic and non-cholinergic components employing combined retrograde tracing and immunohistochemical localization of choline acetyltransferase. Neuroscience 13: 627–643.

Russchen, F. T., Amaral, D. G., and Price, J. L. (1985) The afferent connections of the substantia innominata in the monkey, Macaca Fascicularis. J. comp. Neurol. 242: 1–27.

Russchen, F. T., and Price, J. L. (1984) Amygdalostriatal projections in the rat. Topographical organization and fiber morphology shown using the lectin PHA-L as an anterograde tracer. Neurosci. Lett. 47: 15–22.

Saper, C. B. (1987) Diffuse cortical projection systems: Anatomical organization and role in cortical function. In: Mountcastle, V. B., Plum, F. , and Geiger, S. (eds), Handbook of Physiology, Section 1. The nervous system, Vol. 5, part 1. Am. physiol. Soc., Bethesda, pp. 169–210.

Saper, C. B. (1984) Organization of cerebral cortical afferent systems in the rat. I. Magnocellular basal nucleus. J. comp. Neurol. 222: 1–30.

Saper, C. B. (1985) Organization of cerebral cortical afferent systems in the rat. II. Hypothalamocortical projections. J. comp. Neurol. 237: 21–46.

Saper, C. B., and Loewy, A. D. (1980) Efferent connections of the parabrachial nucleus in the rat. Brain Res. 197: 291–317.

Saper, C. B., Swanson, I. W., and Cowan, W. M. (1979) An autoradiographic study of the efferent connections of the lateral hypothalamic area in the rat. J. comp. Neurol. 183: 689–706.

Saper, C. B., Swanson, L. H., and Cowan, W. M. (1976) The efferent connections of the ventromedial nucleus of the hypothalamus of the rat. J. comp. Neurol. 169: 408–442.

Satoh, K., and Fibiger, H. C. (1986) Cholinergic neurons of the laterodorsal tegmental nucleus: Efferent and afferent connections. J. comp. Neurol. 253: 277–302.

Semba, K, Reiner, P. B., McGeer, E. G., and Fibiger, H. C. (1987) Morphology of cortically projecting basal forebrain neurons in the rat as revealed by intracellular iontophoresis of horseradish peroxidase. Neuroscience 20: 637–651.

Semba, K., Reiner, P. B., McGeer, E. G., and Fibiger, H. C. (1988a) Non-cholinergic basal forebrain neurons project to the contralateral basal forebrain in the rat. Neurosci. Lett. 84: 23–28.

Semba, K., Reiner, P. B., McGeer, E. G., and Fibiger, H. C. (1988b) Brainstem afferents to the magnocellular basal forebrain studied by axonal transport, immunohistochemistry, and electrophysiology in the rat. J. comp. Neurol. 267: 433–453.

Shen, C. I. (1983) Efferent projections from the mammillary complex of the guinea pig: An autoradiographic study. Brain Res. Bull. 11: 43–59.

Simon, H., Le Moal, M., and Calas A. (1979) Efferents and afferents of the ventral tegmental-A10 region studied after local injection of [^{3}H] leucin and horseradish peroxidase. Brain Res. 178: 17–40.

Sofroniew, M. V., Eckenstein, F., Thoenen, H., and Cuello, A. C. (1982) Topography of choline acetyltransferase-containing neurons in the forebrain of the rat. Neurosci. Lett. 33: 7–12.

Somogyi, P., Hodgson, A. J., Smith, A. D., Nunzi, M. G., Gorio, A., and Wu, J.-Y. (1984) Different populations of GABAergic neurons in the visual cortex and hippocampus of cat contain somatostatin- or cholecystokinin-immunoreactive material. J. Neurosci. 4: 2590–2603.

Steinbusch, H. W. M. (1981) Distribution of serotonin-immunoreactivity in the central nervous system of the rat—cell bodies and terminals. Neuroscience 6: 557–618.

Steriade, M., and Llinás, R. R. (1988) The functional states of the thalamus and the associated neuronal interplay. Physiol. Rev. 68: 649–742.

Steriade, M., Domich, L., Oakson, G., and Deschenes, M. (1987) The deafferented reticular thalamic nucleus generates spindle rhythmicity. J. Neurophysiol. 57: 260–273.

Steriade, M., Parent, A., Paré, D., and Smith, Y. (1987) Cholinergic and noncholinergic neurons of cat basal forebrain project to reticular and mediodorsal thalamic nuclei. Brain Res. 408: 372–376.

Swanson, L. W. (1983) Neuropeptides—new vistas on synaptic transmission. TINS 6: 294–295.

Swanson, L. W., and Cowan, W. M. (1979) The connections of the septal region in the rat. J. comp. Neurol. 186: 621–656.

Swanson, L. W., and Hartman, B. K. (1975) The central adrenergic system. An immunofluorescence study of the location of cell bodies and their efferent connections in the rat utilizing dopamine-β-hydroxylase as a marker. J. comp. Neurol. 163: 467–506.

32

Swanson, L. W. (1976) An autoradiographic study of the efferent connections of the preoptic region in the rat. J. comp. Neurol. 167: 227–256.

Takagi, H., Shiosaka, S., Tohyama, M., Senba, E., and Sakanaka, M. (1980) Ascending components of the medial forebrain bundle from the lower brain stem in the rat, with special reference to raphe and catecholamine cell groups. A study by the HRP method. Brain Res. 193: 315–337.

Troiano, R., and Siegel, A. (1978) Efferent connections of the basal forebrain in the cat: The nucleus accumbens. Exp. Neurol. 61: 185–197.

Van Hoesen, G. W., and Damasio, A. R. (1987) Neuroal correlates of cognitive impairment in Alzheimer's disease. In: Mountcastle, V. B., Plum, F., and Geiger, S. R. (eds), Handbook of Physiology, Sect. 1, vol. 5, Part 2. Am. physiol. Soc., Bethesda, pp. 871–898.

Vertes, R. P. (1984a) A lectin horseradish peroxidase study of the origin of ascending fibers in the medial forebrain bundle of the rat. The lower brainstem. Neuroscience 11: 651–668.

Vertes, R. P. (1984b) A lectin horseradish peroxidase study of the origin of ascending fibers in the medial forebrain bundle of the rat. The upper brainstem. Neuroscience 11: 669–690.

Vertes, R. P. (1988) Brainstem afferents to the basal forebrain in the rat. Neuroscience 24: 907–935.

Vertes, R. P., and Martin, G. F. (1988) Autoradiographic analysis of ascending projections from the pontine and mesencephalic reticular formation and the median raphe nucleus in the rat. J. comp. Neurol. 275: 511–541.

Wainer, B. H., Levey, A. I., Rye, D. B., Mesulam, M.-M., and Mufson, E. J. (1985) Cholinergic and non-cholinergic septohippocampal pathways. Neurosci. Lett. 54: 45–52.

Walker, L. C., Kitt, C. A., DeLong, M. R., and Price, D. L. (1985) Noncollateral projections of basal forebrain neurons to frontal and parietal neocortex in primates. Brain Res. Bull. 15: 307–314.

Whitlock, D. C., and Nauta, W. J. H. (1956) Subcortical projections from the temporal neocortex in Macaca mulatta. J. comp. Neurol. 106: 183–212.

Williams, D. I., Crossman, A. R., and Slater, P. (1977) The efferent projections of the nucleus accumbens in the rat. Brain Res. 130: 217–227.

Wood, P. L., and McQuade, P. (1986) Substantia innominata—cortical cholinergic pathway: Regulatory afferents. Adv. Behav. Biol. 30: 999–1006.

Woolf, N. J., Eckenstein, F., and Butcher, L. L. (1984) cholinergic systems in the rat brain. I. projections to the limbic telencephalon. Brain Res. Bull. 13: 751–784.

Zaborszky, L., Leranth, Cs., and Heimer, L. (1984) Ultrastructural evidence of amygdalofugal axons terminating on cholinergic cells of the rostral forebrain. Neurosci. Lett. 52: 215–225.

Zaborszky, L., Carlsen, J., Brashear, H. R., and Heimer, L. (1986a) Cholinergic and GABAergic afferents to the olfactory bulb in the rat with special emphasis on the projection neurons in the nucleus of the horizontal limb of the diagonal band. J. comp. Neurol. 243: 488–509.

Zaborszky, L., Heimer, L., Eckenstein, F., and Leranth, Cs. (1986b) GABAergic input to cholinergic forebrain neurons: An ultrastructural study using retrograde tracing of HRP and double immunolabeling. J. comp. Neurol. 250: 282–295.

Zaborszky, L., and Cullinan, W. E. (1989) Hypothalamic axons terminate on forebrain cholinergic neurons: An ultrastructural double-labeling study using PHA-L tracing and ChAT immunocytochemistry. Brain Res., 479: 177–184.

Zaborszky, L., and Heimer, L. (1989) Combinations of tracer techniques, especially HRP and PHA-L, with transmitter identification for correlated light and electron microscopic studies. In: Heimer, L., and Zaborszky, L. (eds), Neuroanatomical Tract-Tracing Methods II: Recent Progress. Plenum Press, in press.

Zaborszky, L., Luine, V. N., Cullinan, W. E., Allen, D. L., and Heimer, L. (1989) Catecholaminergic-cholinergic interactions in the basal forebrain: Morphological and biochemical studies. Neuroscience, submitted.

Central cholinergic synapses: The septohippocampal system as a model

Michael Frotscher

Institute of Anatomy, Johann Wolfgang Goethe University, Theodor-Stern-Kai 7, D-6000 Frankfurt am Main, Federal Republic of Germany

Summary. This chapter deals with the septohippocampal cholinergic projection in the rodent brain. A monoclonal antibody against choline acetyltransferase (ChAT) was used to label the cells of origin in the medial septum/diagonal band complex (MSDB) and cholinergic fibers and synapses in the hippocampal formation. The target neurons of septohippocampal cholinergic afferents were identified by combining immunocytochemistry with Golgi impregnation or by double immunolabeling. Types of cholinergic synapses very similar to those found in the hippocampal formation were also observed in other brain regions containing cholinergic neurons and terminals. This lends support to the view detailed in this chapter that the septohippocampal projection is a useful model for studies of cholinergic synaptic mechanisms in the CNS.

Introduction

There are several reasons why the septohippocampal cholinergic pathway is a useful model system of central cholinergic synaptic transmission:

1) The *cells of origin* are well defined by studies combining retrograde horseradish peroxidase (HRP) tracing and immunocytochemistry for choline acetyltransferase (ChAT), the acetylcholine-synthesizing enzyme (Amaral and Kurz, 1985; cf. Fig. 1). Also, the fine structure of the cholinergic septal neurons and their synaptic organization has been described in some detail (Bialowas and Frotscher, 1987). It has been shown that the septal cholinergic neurons are large, multipolar cells comparable with the cholinergic neurons in the basal nucleus of Meynert and in the neostriatum. The perikaryal cytoplasm is densely filled with cytoplasmic organelles and, as a rule, the nucleus exhibits several deep indentations of its membrane. Amaral and Kurz (1985) have differentiated three groups of septohippocampal cholinergic neurons, i.e., a dorsal ChAT group which corresponds to the lateral portion of the medial septal nucleus (Ch 1 of Mesulam et al., 1983), a ventral ChAT group which forms a major component of the vertical limb of the nucleus of the diagonal band (Broca) (Ch 2 of Mesulam et al., 1983), and an intermediate group of ChAT-positive cells. The three groups of neurons were found to project to different parts of the hippocampal formation (Amaral and Kurz, 1985). Several studies have shown that

34

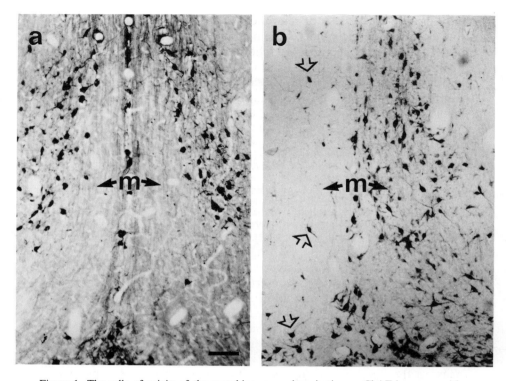

Figure 1. The cells of origin of the septohippocampal projection. *a* ChAT-immunopositive neurons in the medial septum. Note that the medial part (m) of the medial septal nucleus is almost free of cholinergic cells. Scale: 100 μm. *b* Retrograde labeling of medial septal neurons after unilateral injection of HRP into the hippocampus. This time the medial part of the medial septal nucleus contains labeled cells suggesting a noncholinergic septohippocampal projection. Open arrows pointing to retrogradely labeled cells in the contralateral medial septum. This indicates a small crossed septohippocampal projection. Scale in *a* applies also to *b*.

only part of the septohippocampal projection is cholinergic (Lynch et al., 1978; Baisden et al., 1984; Amaral and Kurz, 1985). As demonstrated in Figure 1b, the medial portion of the medial septal nucleus is labeled after HRP injection into the ipsilateral hippocampus, whereas it is unstained following immunocytochemistry for ChAT (Fig. 1a). There is evidence that GABAergic septal neurons project to the hippocampal formation (Köhler et al., 1984). A few cells in the contralateral medial septum are also retrogradely labeled indicating a small crossed septohippocampal projection (Fig. 1b).

2) Immunohistochemical experiments with anti-ChAT have revealed the *course and distribution of septohippocampal cholinergic fibers*. Fine varicose fibers are stained in all layers of the hippocampus and fascia dentata but are most numerous in the vicinity of the cell layers where they form a three-dimensional network around the cell bodies of pyramidal neurons and granule cells (Houser et al., 1983; Clarke, 1985; Frotscher

and Leranth, 1985). It is of relevance for the septohippocampal projection as a cholinergic model system that, at least as far as the dorsal hippocampus is concerned, all cholinergic fibers seem to reach the hippocampal formation via the fimbria-fornix. Transection of the fimbria-fornix, which is relatively easy to perform, results in an almost complete loss of cholinergic fibers in the hippocampus proper and fascia dentata. This strongly suggests that the few ChAT-positive neurons intrinsic to the hippocampal formation (Houser, 1983; Frotscher et al., 1986) contribute little, if at all, to the cholinergic innervation of hippocampus and fascia dentata. Moreover, the hippocampal ChAT-positive neurons do not seem to sprout after removal of the septo-hippocampal cholinergic input (Blaker et al., 1988; Frotscher, 1988). Thus, the dorsal hippocampus can be deprived of its cholinergic input by a fimbria-fornix lesion. The situation is less clear in the case of the temporal hippocampus which receives an additional septal projection via ventral routes (Milner and Amaral, 1984).

3) The *types of synaptic contacts* formed by cholinergic terminals in the hippocampus have been described (Clarke, 1985; Frotscher and Leranth, 1985, 1986; Leranth and Frotscher, 1987), and much is known about the *physiological effects of acetylcholine* on hippocampal neurons (e.g., Dodd et al., 1981; Benardo and Prince, 1982; Cole and Nicoll, 1984). I will not deal with the latter studies here as they are treated in several other chapters of this book. As a morphologist I will rather focus on the fine structure of cholinergic synapses in the hippocampus proper and fascia dentata and the various types of target neurons involved. Identification of the target cells of cholinergic terminals was done by combining electron microscopic immunocytochemistry for ChAT with the Golgi/EM technique (Fairén et al., 1977) or by electron microscopic double immunolabeling procedures.

The brief description given above has already demonstrated that some details are known as to the cells of origin of the cholinergic innervation of hippocampus and fascia dentata and the course and distribution of cholinergic fibers. A detailed knowledge about a system is one of the prerequisites for its use as a model. One may ask, on the other hand, to what extent contacts of cholinergic terminals in the hippocampal formation represent general characteristics of central cholinergic synapses. Therefore, cholinergic synapses in the hippocampus and in other CNS regions are compared in this chapter, and an attempt is made to elaborate some general fine structural characteristics of central cholinergic synapses.

Materials and methods

Young adult male Sprague Dawley rats were used for the present experiments. The animals were fixed in ether anesthesia (30 mg/kg) by transcardial perfusion of isotonic saline followed by a fixative containing 4% paraformaldehyde, 0.08% glutaraldehyde and 15% saturated picric acid in phosphate buffer, pH 7.4 (Somogyi and Takagi, 1982). The brains were removed from the skull and Vibratome sections containing the following regions were cut: medial septum/diagonal band complex (MSDB), lateral septum, neostriatum, amygdaloid complex, and hippocampus. The sections were carefully washed in phosphate buffer and immunostained for ChAT as described in more detail elsewhere (Frotscher and Leranth, 1985). Briefly, the sections were freeze-thawed, washed in phosphate buffer, and immunostained using a monoclonal anti-ChAT antibody (Boehringer type I, see Eckenstein and Thoenen, 1982; dilution 1:9) and the peroxidase-antiperoxidase technique. Following postfixation in osmium tetroxide and dehydration the sections were embedded in Araldite and processed for correlated light and electron microscopy (Frotscher and Leranth, 1985).

Electron microscopic double labeling procedures

Double labeling was performed to allow for the identification of the target neurons of the cholinergic terminals in the hippocampus and fascia dentata. Two approaches were employed: 1) immunostaining for ChAT combined with Golgi impregnation and gold-toning (Fairén et al., 1977) using a modification of the section Golgi impregnation described in detail elsewhere (Frotscher and Leranth, 1986; Frotscher and Zimmer, 1986). This procedure was used for the identification of pyramidal neurons and granule cells as targets of cholinergic terminals. 2) double immunostaining for ChAT and glutamate decarboxylase (GAD), the GABA-synthesizing enzyme. With this procedure we wanted to identify GABAergic inhibitory neurons as target cells of cholinergic fibers. The ChAT-positive terminals were immunostained with the PAP technique as described above and the GAD-immunoreactive neurons were labeled with a biotinylated secondary antibody and avidinated ferritin as a contrasting electron-dense marker (Leranth and Frotscher, 1987). Osmication, dehydration, and embedding were performed as mentioned above for the single immunolabeling experiments.

Additional material used in the present study included tracer experiments with HRP (Fig. 1b) and lesion studies (fimbria-fornix transections) to disrupt the septohippocampal pathway (cf. Frotscher, 1988).

Cholinergic synapses in the hippocampus and fascia dentata

In the electron microscope, the thin varicose ChAT-positive fibers were identified as unmyelinated axons and terminals which formed synaptic contacts with characteristic membrane specializations. This is noteworthy because acetylcholine at central sites has been regarded as a *modulator* rather than a neurotransmitter (Krnjević, 1984). ChAT-positive boutons established both symmetric (Gray type II) and asymmetric (Gray type I) synaptic contacts. This again is interesting because the two different types of contacts were traditionally held to be functionally different. Synapses with asymmetric contacts were generally regarded as excitatory whereas synapses forming symmetric contacts were associated with inhibition. Asymmetric contacts of ChAT-positive boutons were mainly observed on the *heads* of small spines but occasionally also on dendritic shafts. Symmetric contacts were found on dendritic shafts (Fig. 2a), cell bodies, and the *neck* of large complex spines, especially those spines in the dentate molecular layer (Fig. 2b). Asymmetric contacts on the heads of small spines and symmetric contacts on dendritic shafts were the most frequently observed types of cholinergic synapses in the hippocampus and fascia dentata.

Identification of the target neuron was relatively easy for axosomatic contacts on the cell bodies of pyramidal neurons and granule cells because the perikarya of both types of neurons exhibit some characteristic fine structural features. As a rule, the type of target cell could not be identified for contacts on peripheral dendrites. In this case we had to rely on double labeling. ChAT immunocytochemistry combined with Golgi impregnation revealed that cholinergic terminals establish synaptic contacts on cell bodies, dendritic shafts and spines of identified pyramidal neurons and granule cells (Fig. 2b). In the case of spine synapses, both the asymmetric contacts on spine heads and the symmetric contacts on spine necks were observed on the same identified neuron. Double labeling with anti-GAD revealed symmetric synaptic contacts of cholinergic terminals on cell bodies and dendritic shafts of GABAergic nonpyramidal neurons. Having established that all major cell groups in the hippocampal formation receive a cholinergic innervation, it remains to be elucidated whether or not the effects of acetylcholine are different on the different types of target cells or at the different types of synaptic contacts.

Very similar types of cholinergic synapses were observed in the other analyzed CNS regions, i.e. in the septal region (Bialowas and Frotscher, 1987), the amygdaloid complex (see also the chapter by Nitecka and Frotscher), and the neostriatum. Again, symmetric contacts on dendritic shafts and asymmetric contacts on the head of small spines were most frequently found. Figure 2c shows symmetric synaptic contacts of cholinergic terminals on a ferritin-labeled GAD-immunoreactive dendrite in the neostriatum. Such a cholinergic-GABAergic interconnection was observed in all CNS regions studied so far and seems to be a characteristic feature of central cholinergic innervation. Also, the symmetric contacts on the neck of large spines were found in all analyzed

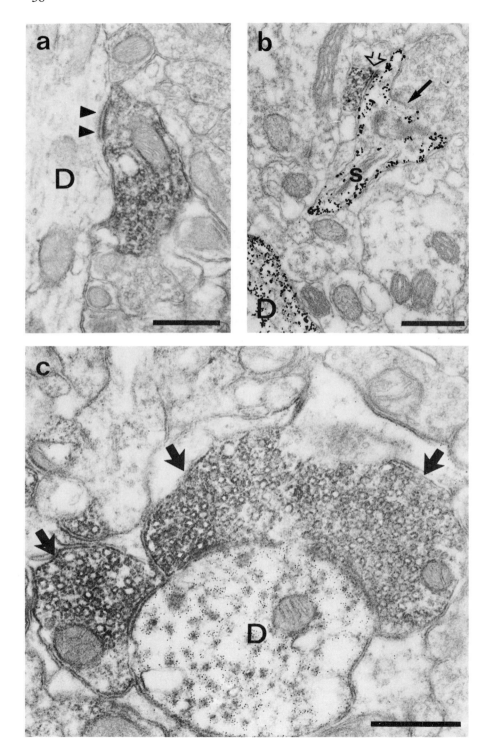

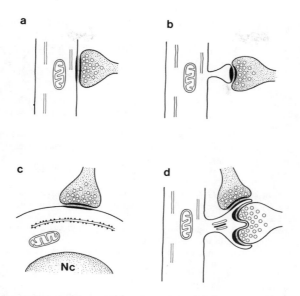

Figure 3. Types of cholinergic synapses in the CNS. *a* Symmetric synaptic contact on dendritic shaft. *b* Asymmetric contact on the spine head of a small spine. *c* Symmetric contact on cell body. *d* Symmetric contact on the back side of a large complex spine. The characteristic cup-shaped portion of the spine is contacted by a nonimmunoractive terminal which forms an asymmetric synaptic contact. Nc, nucleus.

brain regions. As in the hippocampal formation (Fig. 2b), the head of these spines is contacted by an immunonegative bouton forming an asymmetric synaptic contact. The close vicinity of the two contacts strongly suggests mutual interaction of the two different terminals on the same postsynaptic element.

As summarized in Figure 3, the present study has shown that cholinergic terminals in the CNS form a variety of synaptic contacts. However, the types of synapses observed in the various CNS regions studied were very similar to those in the hippocampus and fascia dentata, thereby making the septohippocampal pathway an appropriate model system for studies of cholinergic synaptic mechanisms. As shown in the chapter by B. H. Gähwiler, the septohippocampal cholinergic projection is already an *in vitro* model for the study of central cholinergic synaptic transmission.

Figure 2. Examples of cholinergic synapses in hippocampus and neostriatum. *a* Symmetric synaptic contact (arrow heads) of ChAT-positive terminal on dendritic shaft (D) in stratum radiatum of hippocampal field CA3. Scale: 0.5 μm. *b* ChAT-immunoreactive bouton forming symmetric synaptic contact (open arrow) on the back side of a large complex spine arising from an identified (Golgi-impregnated and gold-toned) dentate granule cell. The characteristic excavated side of the spine is contacted by a large, nonimmunoreactive terminal. Arrow pointing to a spinule which invades this immunonegative bouton. D, parent dendrite; s, spine apparatus. Scale: 0.5 μm (from Frotscher and Leranth, 1986). *c* ChAT-positive boutons (arrows) establishing symmetric synaptic contacts on ferritin-labeled, GAD-immunoreactive dendrite (D) in the neostriatum. Scale: 0.5 μm.

40

Acknowledgements. Parts of this study were carried out in collaboration with Drs K. Lübbers and C. Leranth. The author wishes to thank E. Thielen, E. Schreiber and B. Krebs for excellent technical assistance. This study was supported by the Deutsche Forschungsgemeinschaft (SFB 45, Fr 620/1–3).

Amaral, D. G., and Kurz, J. (1985) An analysis of the origins of the cholinergic and noncholinergic projections to the hippocampal formation of the rat. J. comp. Neurol. 240: 37–59.

Baisden, R. H., Woodruff, M. L., and Hoover, D. B. (1984) Cholinergic and non-cholinergic septo-hippocampal projections: A double-label horseradish peroxidase-acetylcholinesterase study in the rabbit. Brain Res. 290: 146–151.

Benardo, L. S., and Prince, D. A. (1982) Cholinergic excitation of mammalian hippocampal pyramidal cells. Brain Res. 249: 315–331.

Bialowas, J., and Frotscher, M. (1987) Choline acetyltransferase-immunoreactive neurons and terminals in the rat septal complex: A combined light and electron microscopic study. J. comp. Neurol. 259: 298–307.

Blaker, S. N., Armstrong, D. M., and Gage, F. H. (1988) Cholinergic neurons within the rat hippocampus: Response to fimbria-fornix transection. J. comp. Neurol. 272: 127–138.

Clarke, D. J. (1985) Cholinergic innervation of the rat dentate gyrus: An immunocytochemical and electron microscopical study. Brain Res. 360: 349–354.

Cole, A. E., and Nicoll, R. A. (1984) The pharmacology of cholinergic excitatory responses in hippocampal pyramidal cells. Brain Res. 305: 283–290.

Dodd, J., Dingeldine, R., and Kelly, J. S. (1981) The excitatory action of acetylcholine on hippocampal neurons of the guinea-pig and rat maintained in vitro. Brain Res. 207: 109–127.

Eckenstein, F., and Thoenen, H. (1982) Production of specific antisera and monoclonal antibodies to choline-acetyltransferase. Characterization and use for identification of cholinergic neurons. Embo J. 1: 363–368.

Fairén, A., Peters, A., and Saldanha, J. (1977) A new procedure for examining Golgi impregnated neurons by light and electron microscopy. J. Neurocytol. 6: 311–337.

Frotscher, M. (1988) Cholinergic neurons in the rat hippocampus do not compensate for the loss of septohippocampal cholinergic fibers. Neurosci. Lett. 87: 18–22.

Frotscher, M., and Leranth, C. (1985) Cholinergic innervation of the rat hippocampus as revealed by choline acetyltransferase immunocytochemistry: A combined light and electron microscopic study. J. comp. Neurol. 239: 237–246.

Frotscher, M., and Leranth, C. (1986) The cholinergic innervation of the rat fascia dentata: Identification of target structures on granule cells by combining choline acetyltransferase immunocytochemistry and Golgi impregnation. J. comp. Neurol. 243: 58–70.

Frotscher, M., Schlander, M., and Leranth, C. (1986) Cholinergic neurons in the hippocampus: A combined light and electron microscopic immunocytochemical study in the rat. Cell Tissue Res. 246: 293–301.

Frotscher, M., and Zimmer, J. (1986) Intracerebral transplants of the rat fascia dentata: A Golgi/electron microscope study of dentate granule cells. J. comp. Neurol. 246: 181–190.

Houser, C. R., Crawford, G. D., Barber, R. P., Salvaterra, P. M., and Vaughn, J. E. (1983) Organization and morphological characteristics of cholinergic neurons: An immunocytochemical study with a monoclonal antibody to choline acetyltransferase. Brain Res. 266: 97–119.

Krnjević, K. (1984) Neurotransmitters in cerebral cortex. A general account. In: Jones, E. G., and Peters, A. (eds), Functional Properties of Cortical Cells. Plenum, New York, pp. 39–61 (Cerebral cortex, vol. 2).

Köhler, C., Chan-Palay, V., and Wu, J. Y. (1984) Septal neurons containing glutamic acid decarboxylase immunoreactivity project to the hippocampal region in the rat brain. Anat. Embryol. 169: 41–44.

Leranth, C., and Frotscher, M. (1987) Cholinergic innervation of hippocampal GAD- and somatostatin-immunoreactive commissural neurons. J. comp. Neurol. 261: 33–47.

Lynch, G., Rose, G., and Gall, C. (1978) Anatomical and functional aspects of the septo-hippocampal projections. In: Functions of the Septo-hippocampal System. Ciba Foundation Symp 58. Elsevier, New York, pp. 5–24.

Mesulam, M.-M., Mufson, E.J., Wainer, B. H., and Levey, A. I. (1983) Central cholinergic pathways in the rat: An overview based on an alternative nomenclature (Ch1–Ch6). Neuroscience 10: 1185–1201.

Milner, T. A., and Amaral, D. G. (1984) Evidence for a ventral septal projection to the hippocampal formation of the rat. Exp. Brain Res. 55: 579–585.

Somogyi, P., and Takagi, H. (1982) A note on the use of picric acid-paraformaldehyde-glutaraldehyde fixative for correlated light and electron microscopic immunocytochemistry. Neuroscience 7: 1779–1784.

Cholinergic-GABAergic synaptic interconnections in the rat amygdaloid complex: An electron microscopic double immunostaining study

L. Nitecka[a] and M. Frotscher

Institute of Anatomy, Johann Wolfgang Goethe University, Theodor-Stern-Kai 7, D-6000 Frankfurt am Main, Federal Republic of Germany

Summary. A correlated light and electron microscopic immunocytochemical study was performed to analyze 1) the distribution of cholinergic and GABAergic perikarya and terminals in the rat amygdala, and 2) the cholinergic innervation of GABAergic neurons in some amygdaloid nuclei. We will demonstrate here that cholinergic terminals establish synaptic contacts with GABAergic neurons in the basolateral amygdaloid region. These GABAergic neurons in turn are supposed to exert an inhibitory influence on the centromedial amygdaloid region. Our data suggest that the amygdaloid nuclei provide a useful model for studies of cholinergic-GABAergic synaptic interconnections in the CNS.

Introduction

Previous biochemical, histochemical and immunohistochemical studies have shown that there is a specific distribution of cholinergic and GABAergic markers in the various nuclei of the amygdaloid body (Hall and Geneser-Jensen, 1971; Nitecka et al., 1971; Palkovits et al., 1974; Nitecka, 1975; Ben-Ari et al., 1976, 1977; Rotter et al., 1979; Arimatsu et al., 1981; Carlsen and Heimer, 1986; Hellendall et al., 1986; Ottersen et al., 1986; Nitecka and Ben-Ari, 1988; Nitecka and Frotscher, 1989). These data also suggest interaction between cholinergic and GABAergic neurons since interconnections between the various amygdaloid nuclei are known. It is relatively well established that cholinergic afferents to the amygdala arise from the magnocellular basal nucleus and terminate mainly in the basal dorsal nucleus and in the lateral olfactory tract nucleus. These nuclei, on the other hand, exhibit numerous GABAergic neurons (Emson et al., 1979; Woolf and Butcher, 1982; Carlsen et al., 1985; Ottersen et al., 1986; Nitecka and Ben-Ari, 1988). This characteristic distribution of cholinergic and GABAergic elements in the amygdala suggests that this brain region is a useful model to study cholinergic-GABAergic interactions.

In the present study we have used antibodies against the transmitter-synthesizing enzymes choline acetyltransferase (ChAT) and glutamate

---

[a] L. Nitecka is on leave of absence from the Department of Anatomy, Medical Academy, Debinki 1, 80211 Gdansk, Poland.

decarboxylase (GAD) and double immunolabeling procedures to analyze cholinergic-GABAergic interconnections in some nuclei of the
amygdaloid complex.

Materials and methods

Young adult male Sprague-Dawley rats (200–250 g b.wt) were used
for the present correlated light and electron microscopic studies. Single
and double immunostaining for ChAT and GAD were performed as
described in more detail elsewhere (Frotscher et al., 1984; Frotscher and
Leranth, 1985, 1986; Leranth and Frotscher, 1987). We have used a
monoclonal antibody against ChAT (Boehringer Mannheim type I,
dilution 1:9) and a polyclonal antibody against GAD (Oertel et al.,
1982) and the peroxidase-antiperoxidase technique or ABC method
(Hsu et al., 1981). In the case of double immunostaining the immunoreactivity for ChAT was visualized with the PAP technique and GAD-
immunolabeling with a biotinylated secondary antibody and avidinated
ferritin (Leranth and Frotscher, 1987).

Results

The terminology of the amygdaloid nuclei established by Johnston
(1923) has been adopted with modifications resulting from recent
neuroanatomical and histochemical studies localizing acetylcholinesterase activity (Nitecka, 1975; Price, 1981). Traditionally, the
amygdaloid complex has been subdivided into two main parts, i.e. the
basolateral 'inhibitory' region and the centromedial region known to
have an excitatory effect on various behavioral reactions and on endocrine and vegetative functions (e.g. Kaada, 1972).
The schematic diagram (Fig. 1) shows the subdivision of the amygdaloid complex as well as topographical relationships between its nuclei.

ChAT-immunoreactive elements in the amygdaloid nuclei

ChAT-immunopositive perikarya. Two types of ChAT-immunopositive
neurons were observed: 1) small ovoid perikarya which were randomly
distributed throughout the whole amygdaloid region. Their long dendrites could be recognized best in those nuclei which did not exhibit an
intense neuropil labeling. 2) Large, strongly immunoreactive neurons
which resembled the cholinergic neurons in the nucleus basalis
(Meynert) and in the medial septum/diagonal band complex. These
large cholinergic neurons were restricted to nuclei of the centromedial

44

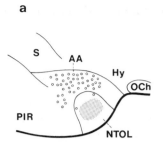

a

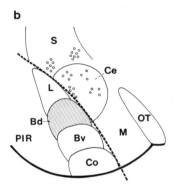

b

Figure 1. Schematic diagram demonstrating the topography of the amygdaloid nuclei in frontal sections through the rat brain. *a* Anterior pole of amygdala. *b* Midrostrocaudal part of the amygdaloid complex. Dotted areas indicate regions of intense terminal staining for ChAT. Circles represent large cholinergic neurons. Broken line separating the basolateral and centromedial regions of amygdala. AA, anterior amygdaloid area; Bd, basal dorsal nucleus of amygdala; Bv, basal ventral nucleus of amygdala; Ce, central nucleus of amygdala; Co, cortical nucleus of amygdala; Hy, hypothalamus; L, lateral nucleus of amygdala; M, medial nucleus of amygdala; NTOL, lateral olfactory tract nucleus of amygdala; OCh, optic chiasm; S, striatum; OT, optic tract; PIR, piriform cortex.

region of amygdala. They were particularly numerous in the anterior amygdaloid area, the lateral and medial zones of the central nucleus as well as in the peripheral zone of the lateral olfactory tract nucleus. Their long dendrites invaded adjacent nuclei, for instance the intercalated nuclei, the central zone of the central nucleus, the anterior pole of the basal dorsal nucleus and the central portion of the lateral olfactory tract nucleus.

ChAT-immunoreactive terminals. Immunolabeling of terminals varied in the different amygdaloid nuclei. Intense fiber staining was observed in the basal dorsal nucleus and in the central zone of the lateral olfactory tract nucleus (Fig. 1). In general, the nuclei of the centromedial region exhibited less immunostaining of terminals than the nuclei of the basolateral region.

GAD-immunoreactive elements in the amygdaloid nuclei

Our findings on GAD-immunoreactivity in the amygdala were in good agreement with previous studies employing antibodies against GABA (Ottersen et al., 1986; Nitecka and Ben-Ari, 1988). Thus, we will not provide a detailed description here. In principle, there are two differently labeled areas: 1) the basolateral region mainly exhibiting labeled cell bodies, and 2) the centromedial region which shows intense labeling of fibers and terminals. It appears that the centromedial region is the main target of GABAergic amygdaloid connections.

Synaptic contacts between cholinergic terminals and GABAergic neurons

In our double labeling experiments ChAT-immunoreactive terminals were identified by the PAP technique whereas GAD-positive postsynaptic elements were labeled with ferritin. Two amygdaloid nuclei were chosen to study cholinergic-GABAergic synaptic interconnections, i.e. the basal dorsal nucleus and the central zone of the lateral olfactory tract nucleus. As mentioned above, both nuclei displayed intense immunostaining of cholinergic terminals and numerous GABAergic cell bodies. Our observations in both nuclei will be described together.

In general, ChAT-positive peroxidase-labeled boutons were numerous in both nuclei. They formed symmetric synaptic contacts on spines, cell bodies and dendritic shafts (Fig. 2). In the case of cholinergic synapses on spines both symmetric and asymmetric contacts were noted. Asymmetric contacts were established on the heads of relatively small spines. Symmetric contacts, which were difficult to recognize, occurred on the necks of large complex spines. The heads of these large spines were occupied by nonimmunoreactive boutons. Symmetric and asymmetric spine contacts were similarly observed in other brain regions such as the hippocampus and neostriatum (Frotscher and Leranth, 1986; Olbrich and Frotscher, 1987; cf. Frotscher, this volume).

Occasionally the postsynaptic element (cell body, dendritic shaft or spine) contained fine ferritin grains and was thus identified as GAD-immunoreactive (Fig. 2). It is needless to mention that cholinergic synapses on GABAergic neurons were only rarely observed. Most of these contacts were found on cell bodies and dendritic shafts which is not surprising since the present GAD antibody mainly stains cell bodies and proximal dendrites in addition to terminal boutons.

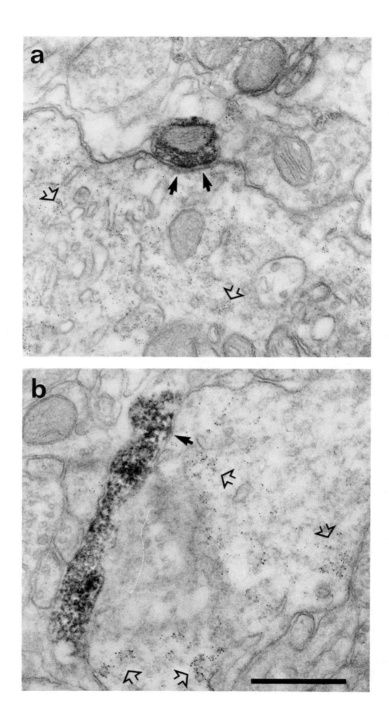

Functional implications

The results of the present study provide morphological evidence for cholinergic-GABAergic interactions in the amygdaloid complex. Two nuclei, i.e. the basal dorsal nucleus and the lateral olfactory tract nucleus were chosen for analysis because both nuclei were found to contain a dense network of cholinergic terminals but also numerous GABAergic perikarya in single immunolabeling experiments. Double immunostaining for ChAT and GAD in fact revealed cholinergic synapses on GABAergic neurons in both nuclei. A contrasting pattern of ChAT and GAD immunoreactivity, i.e. intense staining of GABAergic terminals and very few cholinergic fibers, was observed in the centromedial amygdaloid region. These findings together with data from the literature may lead to some functional conclusions:

1) It is well established that the basal dorsal nucleus receives a massive direct projection from magnocellular basal forebrain neurons located in the substantia innominata (Emson et al., 1979; Woolf and Butcher, 1982; Carlsen et al., 1985). These cholinergic afferents may activate GABAergic inhibitory neurons in the basal dorsal nucleus (Femano et al., 1979). The cholinergic synapses on GABAergic neurons described in the present report are the morphological substrate of such a functional connection.

2) Evidence has been provided that the basal dorsal nucleus gives rise to intraamygdaloid projections terminating in the nuclei of the centromedial region (Kamal and Tömböl, 1975; Nitecka et al., 1981; McDonald, 1984; Ottersen, 1982; Millhouse and De Olmos, 1983; Price and Amaral, 1984). Many of these fibers are likely to be GABAergic since the basal dorsal nucleus contains numerous GABAergic perikarya and the centromedial region shows intense labeling of GAD-immunoreactive fibers and terminal-like puncta.

3) Cholinergic activation of GABAergic neurons in the basal dorsal nucleus and subsequent inhibition of centromedial neurons via intraamygdaloid GABAergic connections may play an important role in vegetative and hormonal processes. The centromedial region is known to have an 'excitatory' effect on hypothalamic centers and thus on vegetative and endocrine functions. Inhibition of centromedial neurons via intraamygdaloid GABAergic projections may reduce this excitatory effect in addition to the well-known direct inhibitory effect of the basolateral region on hypothalamic activity.

Figure 2. Cholinergic-GABAergic synaptic interconnection in the rat amygdala. *a* ChAT-immunoreactive terminal in the basal dorsal nucleus establishing symmetric synaptic contact (arrows) on the cell body of a GABAergic neuron identified as such by the presence of fine ferritin grains. *b* ChAT-positive bouton which forms what appears to be a symmetric contact (arrow) on a ferritin labeled, GAD-immunoreactive dendrite in the lateral olfactory tract nucleus. Open arrows pointing to accumulations of ferritin grains. Scale: 0.5 μm.

Of course, some of the above considerations are speculative. However, they may serve as a working hypothesis until further details are known about intraamygdaloid connections. Our results at least suggest that cholinergic-GABAergic interaction is an important element of central cholinergic synaptic transmission.

Acknowledgements. The authors wish to thank B. Krebs, E. Thielen and E. Schreiber for technical assistance. This work was supported by the Deutsche Forschungsgemeinschaft (SFB 45).

Arimatsu, Y., Seto, A., and Amano, T. (1981) An atlas of α-bungarotoxin binding sites and structures containing acetylcholinesterase in the mouse central nervous system. J. comp. Neurol. 198: 603–631.

Ben-Ari, Y., Kanazawa, I., and Zigmond, R. E. (1976) Regional distribution of glutamate decarboxylase and GABA within the amygdaloid complex and stria terminalis system of the rat. J. Neurochem. 26: 1276–1283.

Ben-Ari, Y., Zigmond, R. E., Shute, C. C., and Lewis, P. R. (1977) Regional distribution of choline acetyltransferase and acetylcholinesterase within the amygdaloid complex and stria terminalis system. Brain Res. 120: 435–445.

Carlsen, J., and Heimer, L. (1986) A correlated light and electron microscopic immunocyto-chemical study of cholinergic terminals and neurons in the rat amygdaloid body with special emphasis on the basolateral amygdaloid nucleus. J. comp. Neurol. 244: 121–136.

Carlsen, J., Zaborszky, L., and Heimer, L. (1985) Cholinergic projections from the basal forebrain to the basolateral amygdaloid complex: A combined retrograde fluorescent and immunohistochemical study. J. comp. Neurol. 234: 155–167.

Emson, P. C., Paxinos, G., Le Gal La Salle, G., Ben-Ari, Y., and Silver, A. (1979) Choline acetyltransferase and acetylcholinesterase containing projection from the basal forebrain to the amygdaloid complex of the rat. Brain Res. 165: 271–282.

Femano, P. A., Edinger, H. M., and Siegel, A. (1979) Evidence of a potent excitatory influence from substantia innominata on basolateral amygdaloid units: A comparison with the insular-temporal cortex and lateral olfactory tract stimulation. Brain Res. 177: 361–366.

Frotscher, M., Leranth, C., Lübbers, K., and Oertel, W. H. (1984) Commissural afferents innervate glutamate decarboxylase immunoreactive non-pyramidal neurons in the guinea pig hippocampus. Neurosci. Lett. 46: 137–143.

Frotscher, M., and Leranth, C. (1985) Cholinergic innervation of the rat hippocampus as revealed by choline acetyltransferase immunocytochemistry: A combined light and electron microscopic study. J. comp. Neurol. 239: 237–246.

Frotscher, M., and Leranth, C. (1986) The cholinergic innervation of the rat fascia dentata: Identification of target structures on granule cells by combining choline acetyltransferase immuncytochemistry and Golgi impregnation. J. comp. Neurol. 243: 58–70.

Hall, E., and Geneser-Jensen, F. A. (1971) Distribution of acetylcholinesterase and monamine oxidase in the amygdala of the guinea pig. Z. Zellforsch. Mikrosk. Anat. 120: 204–221.

Hellendall, R. P., Godfrey, D. A., Ross, C. D., Armstrong, D. M., and Price, J. L. (1986) The distribution of choline acetyltransferase in the rat amygdaloid complex and adjacent cortical areas, as determined by quantitative micro-assay and immunohistochemistry. J. comp. Neurol. 249: 486–49.

Hsu, S. M., Raine, L., and Fanger, H. (1981) The use of avidin-biotin-peroxidase complex (ABC) in immunoperoxidase techniques: A comparison between ABC and unlabeled antibody (peroxidase) procedures. J. Histochem. Cytochem. 29: 577–590.

Johnston, J. B. (1923) Further contribution to the study of evolution in the forebrain. J. comp. Neurol. 35: 337–481.

Kaada, B. R. (1972) Stimulation and regional ablation of the amygdaloid complex with reference to functional representations. In: Eleftheriou, B. E. (ed.), The Neurobiology of the Amygdala. Plenum Press, New York, pp. 205–281.

Kamal, A. M., and Tömböl, T. (1975) Golgi studies on the amygdaloid nuclei of the cat. J. Hirnforsch. 16: 175–209.

Leranth, C., and Frotscher, M. (1987) Cholinergic innervation of hippocampal GAD- and somatostatin-immunoreactive commissural neurons. J. comp. Neurol. 261: 33–47.

McDonald, A. J. (1984) Neuronal organization of the lateral and basolateral amygdaloid nuclei in the rat. J. comp. Neurol. 222: 589–606.

Millhouse, D. E., and De Olmos, J. (1983) Neuronal configuration in lateral and basolateral amygdala. Neuroscience 10: 1269–1300.

Nitecka, L. (1975) Comparative anatomic aspects of localization of acetylcholinesterase activity in the amygdaloid body. Folia morphol. (Warsz.) 34: 167–185.

Nitecka, L., and Ben-Ari, Y. (1988) Distribution of GABA-like immunoreactivity in the rat amygdaloid complex. J. comp. Neurol. 266: 45–55.

Nitecka, L., Amerski, L., and Narkiewicz, O. (1981) The organization of intraamygdaloid connections: An HRP study. J. Hirnforsch. 22: 3–7.

Nitecka, L., and Frotscher, M. (1989) Organization and synaptic interconnections of GABAergic and cholinergic elements in the rat amygdaloid nuclei: Single and double immunolabeling studies. J. comp. Neurol. 279: 470–488.

Nitecka, L., Narkiewicz, O., and Zawistowska, H. (1971) Acetylcholinesterase activity in the nuclei of the amygdaloid complex in thre rat. Acta neurobiol. exp. 31: 383–388.

Oertel, W. H., Schmechel, D. E., Mugnaini, E., Tappaz, M. L., and Kopin, I. J. (1982) Immunocytochemical localization of glutamate decarboxylase in the rat cerebellum with a new antiserum. Neuroscience 6: 2715–2735.

Olbrich, H. G., and Frotscher, M. (1987) Cholinerge Synapsen in Neostriatum und Hippocampus—ein Vergleich. Verh. Anat. Ges. 81: 893–894.

Ottersen, O. P. (1982) Connections of the amygdala of the rat. IV. Corticoamygdaloid and intraamygdaloid connections as studied with axonal transport of horseradish peroxidase. J. comp. Neurol. 205: 30–48.

Ottersen, O. P., Fischer, B. O., Rinvik, E., and Storm-Mathisen, J. (1986) Putative amino acid transmitters in the amygdala. In: Schwarcz, R., and Ben-Ari, Y. (eds), Excitatory Amino Acids and Seizure Disorders. Plenum Press, New York, pp. 53–66.

Palkovits, M., Saavedra, J. M., Kobayashi, R. M., and Brownstein, M. (1974) Choline acetyltransferase content of limbic nuclei of the rat. Brain Res. 79: 443–450.

Price, J. L. (1981) Toward a consistent terminology for the amygdaloid complex. In: Ben-Ari, Y. (ed.), Amygdaloid Complex. Elsevier North Holland, New York, pp. 13–18.

Price, J. L., and Amaral, D. G. (1984) An autoradiographic study of the projections of the central nucleus of the monkey amygdala. J. Neurosci. 1: 1242–1259.

Rotter, A., Birdsall, N. J. M., Burden, A. S. Y., Field, P. M., Hulme, E. C., and Raisman, G. (1979) Muscarinic receptors in the central nervous system of the rat. I. Technique for autoradiographic localization of the binding of (^{3}H) propylbenzilylcholine mustard and its distribution in the forebrain. Brain Res. Rev. 1: 141–165.

Woolf, N. J., and Butcher, L. L. (1982) Cholinergic projections to the basolateral amygdala: A combined Evans Blue and acetylcholinesterase analysis. Brain Res. Bull. 8: 751–763.

Topography of βNGF receptor-positive and AChE-reactive neurons in the central nervous system

G. Raivich and G. W. Kreutzberg

*Department of Neuromorphology, Max-Planck-Institut für Psychiatrie,
Am Klopferspitz 18A, D-8033 Martinsried, Federal Republic of Germany*

Summary. Recent reports have led to widespread interest in the role of β-nerve growth factor (βNGF) in the central nervous system. To learn more about the action of βNGF in the central nervous system we have mapped the distribution of βNGF receptors and compared it with that of acetylcholinesterase (AChE), a sensitive enzyme marker for cholinergic neurons.

In situ autoradiography revealed strong and saturable βNGF binding to several groups of neurons in basal forebrain and brainstem. They also contain significant levels of mRNA coding for βNGF receptors. βNGF receptors and AChE are codistributed on the medial septal nuclei and in the basal forebrain, including the striatum. In the brainstem, βNGF receptors are present on the neurons in the lower part of the reticular formation and in cochlear nuclei but do not correspond to the distribution of AChE reactivity.

Introduction

The β-nerve growth factor is a very basic, 27 kD protein, which is known primarily for its neurotrophic action in the peripheral nervous system on sympathetic and sensory neurons (Levi-Montalcini and Angeletti, 1968; Thoenen and Barde, 1980). More recently, three separate lines of evidence have suggested that βNGF possesses a similar regulatory role in the central nervous system. βNGF has been shown to induce choline acetyltransferase (ChAT), the key enzyme of acetylcholine synthesis, in the basal forebrain *in vivo* (Gnahn et al., 1983; Mobley et al., 1985) as well as *in vitro* (Honegger and Lenoir, 1982; Martinez et al., 1985). Secondly, βNGF-like neurotrophic activity (Björklund and Stenevi, 1979; Collins and Crutcher, 1985), protein (Korsching et al., 1985) and mRNA (Korsching et al., 1985; Shelton and Reichard, 1986) are present in the innervation fields of these central cholinergic neurons. Finally, radioactively labeled βNGF is retrogradely transported to the cholinergic nuclei of the basal forebrain (Schwab et al., 1979; Seiler and Schwab, 1984), indicating a specific, receptor-mediated (Taniuchi et al., 1986) uptake of βNGF by the cholinergic neurites in their innervation targets. A number of recent reports have also documented the presence of specific βNGF binding sites and βNGF receptor protein on neuronal perikarya in the developing (Raivich et al., 1985; Raivich et al., 1987) and the adult central nervous system

(Richardson et al., 1986; Taniuchi et al., 1986; Raivich and Kreutzberg, 1987; Springer et al., 1987).

To learn more about the significance of βNGF-mediated action in the central nervous system, it is essential to know about the potential target sites (i.e. receptors) on which this protein may act. In this article we describe the localization of these βNGF receptors in the central nervous system, the distribution of their sites of synthesis and their relationship to the distribution of acetylcholinesterase-reactive (AChE-reactive) central cholinergic neurons (Eckenstein and Sofroniew, 1983; Levey et al., 1983).

Methods

The *in situ* (^{125}I)βNGF binding to rat brain tissue sections was performed as described in Raivich and Kreutzberg (1987). The level of nonspecific (^{125}I)βNGF binding, performed in the presence of a 100-fold surplus of unlabeled βNGF, is shown in Figures 1b, 3b and 4b. Equilibrium-binding displacement experiments showed a dissociation constant of approximately 0.2 nM to the slowly dissociating, *in situ* (^{125}I)βNGF binding sites (Raivich and Kreutzberg, 1986). Northern blot quantitation of βNGF receptor mRNA levels were performed as described by Heumann et al. (1984) for βNGF mRNA. To compare the distribution of AChE-reactive neuronal perikarya and (^{125}I)βNGF binding, adjacent sections from diisopropylfluorophosphate-treated animals were stained for AChE (Butcher and Woolf, 1984) or labeled with (^{125}I)βNGF.

Results

βNGF receptor distribution in the forebrain

In situ receptor autoradiography performed on tissue sections of the rat brain reveals heavy and specific (^{125}I)βNGF binding on the neuronal perikarya in the forebrain cholinergic nuclei of the medial septal nucleus (Fig. 1a), the ventral (Fig. 1a) and horizontal (Fig. 2) limb of the diagonal band of Broca, and on the basal nucleus of Meynert (Fig. 2). Some widely dispersed, specifically labeled neuronal perikarya are also present in the caudatoputamen (striatum). As shown in Table 1, there is a very close correlation between the density of (^{125}I)βNGF-labeled neuronal perikarya and the level of βNGF receptor mRNA. No specific labeling was found in other forebrain regions, the olfactory lobe, cerebral cortex, thalamus and hypothalamus, or on the neural retina, which contains, however, a surprisingly high level of βNGF receptor

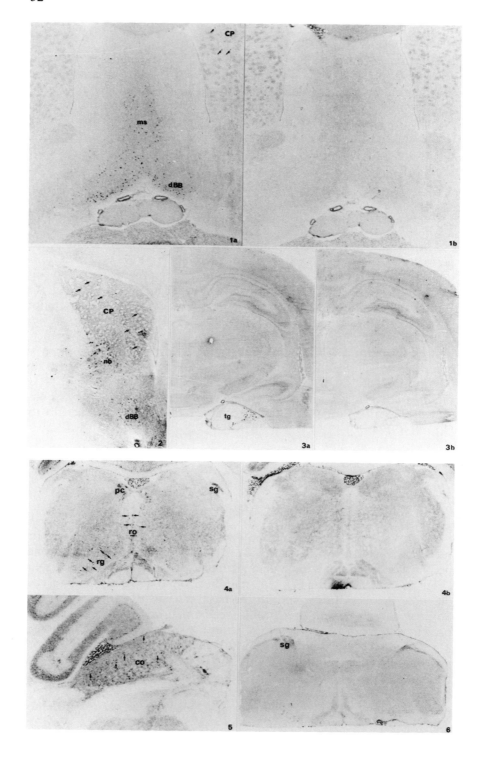

Table 1. NGF receptor mRNA levels in the nervous system (Northern Blot Analysis)

	NGFR-mRNA/lane	Tissue/lane	Recovery	NGFR-mRNA/tissue
CNS	(in pg)	(in mg)	(in %)	(in ng/g wet weight)
Septum	1.75	28	17	0.357
DBB	8.0	20	50	0.800
Ncl. Meynert	3.6	13	40	0.690
Striatum	0.4	27	40	0.037
Hippocampus	0.18	34	40	0.013
Cortex	0.19	28	47	0.015
Retina	1.8	10	13	1.38
Periventricular				
Ncll.	2.15	25	52	0.167
Cochlear Ncl.	2.10	6	53	0.667
Brainstem	1.45	24	37	0.162
Cerebellum	5.2	23	33	0.696
Spinal Cord	0.46	26	27	0.056
PNS				
Sciatic Nerve	3.1	36	60	0.144
Trigeminal Ggl. A	25.5	22	31	3.72
Trigeminal Ggl. B	26.7	25	27	3.98
Mean ± SD	26.1 ± 0.8	23.5 ± 2.0	29 ± 3	3.85 ± 0.20

Quantitation of βNGF receptor (NGFR) mRNA levels in different regions of the central and peripheral nervous system. The NGFR-mRNA (3.9 kb) levels were measured by calibration against different concentrations of synthetic 3.7 kb NGFR-mRNA fragment run on adjacent, spare lanes of the same agarose gel. RNA recovery was quantitated by adding 20 pg of a 2.4 kb NGFR-mRNA fragment to the tissue before RNA extraction and measuring its level on the Northern blot filter. With the exception of retina and cerebellum, there is a very close correlation between high levels of βNGF receptor mRNA and the density of neurons bearing in situ (^{125}I)βNGF binding sites.

mRNA (Table 1) and which is ontogenetically derived from dien-cephalon. No specific (^{125}I)βNGF binding was observed on the cholin-ergic fiber tracts (e.g. septohippocampal fibers), nor could a diffuse specific labeling, similar to that on substantia gelatinosa (see below) be shown on the cholinergic innervation targets such as the hippocampus.

Figures 1–6. *In situ* autoradiography of (^{125}I)βNGF binding to frontal sections of the rat central nervous system and trigeminal ganglia. Figures 1, 2 and 3—anterior, middle and posterior forebrain; 4—upper myelencephalon, 5—metencephalon and 6—lower myel-encephalon. The level of nonspecific (^{125}I)βNGF binding is shown in Figures 1b, 3b and 4b.

Abbreviations used in figure legends: ms; medial septal nucleus; dBB, diagonal Band of Broca; CP, caudatoputamen (striatum); nb, basal nucleus of Meynert; tg, trigeminal ganglia; pv, periventricular nuclei; rg, gigantocellular reticular nucleus; ro, obscure raphe nucleus; rl, lateral reticular nucleus; tld, dorsolateral tegmental nucleus; co, cochlear nuclei; n.V., trigeminal motor nucleus; n.V., facial motor nucleus; lc, locus coeruleus; sg, substantia gelatinosa; CNS, central nervous system; ncl, nucleus; PNS, peripheral nervous system; ggl, ganglion.

54

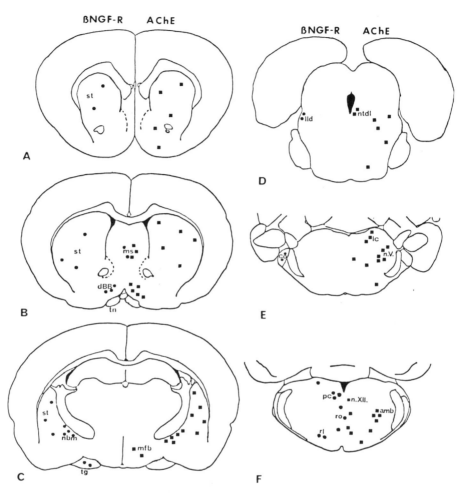

Figure 7. Schematic illustration of the distribution patterns of the (^{125}I)βNGF-labeled and intensely AChE-reactive neurons on representative transverse sections of the rat central nervous system at the level of the frontal cortex (A), frontoparietal cortex (B), parietooccipital cortex (C), midbrain (D), metencephalon (E) and myelencephalon (F). Strong spatial correlation between AChE and the β-NGF receptor is found only in the forebrain. Reproduced from Raivich and Kreutzberg (1987) Neuroscience. 20: 23–36 with permission from Pergamon Press PLC.

βNGF receptor distribution in the brainstem

With the exception of a few neuronal perikarya in the dorsolateral lemniscus (see Fig. 7d), the specific neuronal βNGF labeling is restricted to the cochlear nuclei (Fig. 5) and the lower part of the reticular formation. Heavy (^{125}I)βNGF binding is observed on the densely packed neuronal perikarya in the periventricular column slightly above the hypoglossal nucleus, the gigantocellular and paragigantocellular reticular nuclei and the lower part of brainstem raphe (Fig. 4a). In the lower part of the brainstem and in the spinal cord, a moderate and specific, somewhat diffuse (^{125}I)βNGF binding is observed to the substantia gelatinosa (Fig. 6), which contains very densely packed neurite terminals of βNGF receptor-bearing sensory ganglionic neurons (e.g. trigeminal ganglia in Fig. 3a and Table 1). As in the rat forebrain, a close correlation is observed between the density of (^{125}I)βNGF-labeled neuronal perikarya and the level of βNGF receptor mRNA. A second exception to this generally good correlation is the high level of βNGF receptor mRNA in the rat cerebellum, where we did not observe specific perikaryal (^{125}I)βNGF binding.

Codistribution of neuronal βNGF receptors and AChE reactivity

In the forebrain, there is a good correlation between the distribution of (^{125}I)βNGF-labeled and AChE-reactive neuronal perikarya (see Fig. 7a–c). Almost all strongly AChE-reactive neurons are also located in the above-mentioned cholinergic nuclei and striatum, although they are slightly more numerous than the (^{125}I)βNGF-labeled neuronal profiles. The ratios of AChE-reactive: (^{125}I)βNGF-labeled neurons are approximately 1.1–1.2:1 in the medial septum, 1.2:1 in the diagonal band of Broca and basal nucleus of Meynert, and 1.2–1.4:1 in the striatum.

This correlation was not observed in the brainstem (see Fig. 7d–f). Strongly AChE-reactive neurons were not found in the periventricular column of heavily (^{125}I)βNGF-labeled neurons except where its lower part overlaps the upper part of the motor hypoglossal nucleus. The heavily (^{125}I)βNGF-labeled cochlear nuclei also did not contain strongly AChE-reactive profiles. No specific (^{125}I)βNGF-labeling was observed on the cholinergic trigeminal, facial, glossopharyngeal or vagal motor nuclei. It was also absent on the strongly AChE-reactive but non-cholinergic substantia nigra and locus coeruleus, pointing to the differential expression of brainstem βNGF receptors and AChE-reactivity.

Discussion

The data described here show strong βNGF receptor expression in a number of different brain regions. In the rat forebrain βNGF receptors are clearly codistributed with cholinergic neurons. This distribution pattern of βNGF receptors is in agreement with what is known at present about the central cholinergic function of βNGF. βNGF strongly increases the levels of choline acetyltransferase, which is the key enzyme in acetylcholine synthesis in the forebrain but not in the brainstem cholinergic neurons (Honegger and Lenoir, 1982; Gnahn et al., 1983). More recently, high concentrations of intraventricularly applied βNGF were also shown to protect the forebrain cholinergic neurons from cell death following axotomy (Hefti, 1986). By comparison, nothing is known about the possible noncholinergic function of βNGF in the central nervous system (Olson et al., 1979; Schwab, 1979). Here, the identification of the neurotransmitter type and the innervation targets of the βNGF receptor-bearing, noncholinergic brainstem neurons would be an important step in understanding its action in this region.

The present report shows a good correlation between the distribution of βNGF receptor mRNA and the presence of high affinity and slowly dissociating in situ (^{125}I)βNGF binding on the neuronal perikarya in the central nervous system. This is not surprising, since βNGF receptor-bearing neurons must be able to synthesize βNGF receptors. Similarly, the CNS regions which contain only very low levels of βNGF receptor mRNA, for example, the cerebral cortex or the hippocampus, lack βNGF receptor-bearing neuronal perikarya. However, according to Taniucchi et al. (1986), cortex and hippocampus do contain high concentrations of the βNGF receptor protein. In view of the very low levels of the βNGF receptor mRNA in these regions, it might be assumed that most of these βNGF receptors are not locally synthesized but rather 'imported' and localized on the cholinergic neurite terminals where they mediate the βNGF uptake and its transport to their perikarya in the basal forebrain.

Somewhat surprising, however, is the situation in rat cerebellum and neural retina. Both regions contain very high levels of βNGF receptor mRNA, but lack specific in situ (^{125}I)βNGF binding. This has been confirmed in in vivo studies (Riopelle et al., 1987), which showed the absence of high affinity (^{125}I)βNGF binding on cerebellar membranes. On the other hand, the cerebellum has been shown to contain a relatively high concentration of the βNGF receptor protein (Taniuchi et al., 1986), and a recent immunohistochemical study (Pioro and Cuello, 1988) has demonstrated the presence of βNGF receptor-like molecules on the Purkinje cells in the adult rat cerebellum. These findings have to be interpreted cautiously. One should be reminded that βNGF binding studies (Sutter et al., 1979; Schechter and Bothwell, 1981; Green et al., 1986) have demonstrated the existence of two different types of neuronal βNGF receptors: a high affinity (Kd = 0.02–0.2 nM), slowly dissociating binding site (Type I) and a lower affinity (Kd = 2 nM), very rapidly dissociating binding site (Type II). The present data suggest that cerebellum and retina may contain only the rapidly dissociating (Type II) βNGF receptors, which would allow (^{125}I)βNGF to be washed from these binding sites before or during the fixation preceding the autoradiography. Interestingly, all βNGF-dependent neurons studied so far have been shown to contain the Type I βNGF receptor species (Sutter et al.,

1979; Schechter and Bothwell, 1981; Green et al., 1986; Riopelle et al., 1987), the presence of Type II receptors alone does not appear to confer sensitivity to physiological levels of βNGF (Green et al., 1986). It is therefore questionable whether the retinal and cerebellar neurons are physiologically sensitive to βNGF as are the neurons in the forebrain cholinergic nuclei and, probably, in the cochlear nuclei and the reticular formation.

In summary, the present report has described and compared the distribution of neuronal high affinity βNGF receptors and their mRNA in the cholinergic as well as in the non-cholinergic nuclei of the central nervous system. At present, very little is known about role of βNGF in the CNS, apart from its well-described action on the cholinergic neurons of the basal forebrain. Information about the chemical nature of the non-cholinergic βNGF receptor-bearing neurons in the brainstem and how they are affected by βNGF will therefore play an important role in gaining a more complete understanding of βNGF function in the central nervous system.

Acknowledgements. We thank Ms Waltraud Komp for her expert technical assistance, Dr Hermann Rohrer for his generous gift of unlabeled βNGF, Dr Eric Shooter and Dr Rolf Heumann for providing us with the NGF receptor cDNA and cRNA probes and Dr Martin Reddington for reading the manuscript.

Björklund, A., and Stenevi, U. (1981) In vivo evidence for a hippocampal neurotrophic factor specifically released after septal deafferentation. Brain Res. 229: 403–428.

Butcher, L. L., and Woolf, N. J. (1984) Histochemical distribution of AChE in the central nervous system: clues to the localization of the cholinergic neurones. In: Björklund, A., Hökfelt, T., and Kuhar, M. J. (eds), Handbook of Chemical Neuroanatomy, Vol. 3. Elsevier, Amsterdam, pp. 1–50.

Collins, F., and Crutcher, K. A. (1985) Neurotrophic activity in the adult rat hyppocampal formation. Regional distribution and increase after septal lesion. J Neurosci. 5: 2809–2814.

Eckenstein, F., and Sofroniew, M. W. (1983) Identification of central cholinergic neurons containing both cholineacetyltransferase and acetylcholinesterase and of central neurons containing only acetylcholinesterase. J. Neurosci. 3: 2286–2291.

Gnahn, H., Hefti, F., Heumann, R., Schwab, M. E., and Thoenen, H. (1983) NGF-mediated increase of choline acetyltransferase (ChAT) in the neonatal forebrain: evidence for a physiological role of bNGF in the brain? Dev. Brain Res. 9: 45–52.

Green, S. H., Rydel, R. E., Connolly, J. L., and Greene, L. A. (1986) PC12 cell mutants that possess low- but not high affinity nerve growth factor receptors neither respond nor internalize nerve growth factor. J. Cell Biol. 102: 830–843.

Hefti, F. (1986) Nerve growth factor promotes survival of septal cholinergic neurons after fimbrial transection. J. Neurosci. 6: 2155–2162.

Heumann, R., Korsching, S., Scott, J., and Thoenen, H. (1984) Relationship between the levels of nerve growth factor (NGF) and its messenger mRNA in sympathetic ganglia and peripheral target tissues. EMBO J. 3: 3183–3189.

Honegger, P., and Lenoir, D. (1982) Nerve growth factor (NGF) stimulation of cholinergic telencephalic neurons in aggregating cell cultures. Dev. Brain Res. 3: 229–238.

Korsching, S., Auburger, G., Heumann, R., Scott, J., and Thoenen, H. (1985) Levels of nerve growth factor and its mRNA in the central nervous system correlate with cholinergic innervation. EMBO J. 4: 1389–1394.

Levey, A. I., Wainer, B. H., Mufson, E. J., and Mesulam, M.-M. (1983) Colocalization of acetylcholinesterase and cholineacetyl-transferase in the rat cerebrum. Neuroscience 9: 9–22.

58

Levi-Montalcini, R., and Angeletti, P. U. (1968) The nerve growth factor. Physiol. Rev. 48: 534–569.

Martinez, H. J., Dreyfus, C. F., Jonakeit, M., and Black, I. B. (1985) Nerve growth factor promotes cholinergic development in brain striatal cultures. Proc. natl Acad. Sci. USA 82: 7777–7781.

Mobley, W. C., Rutkowski, J. L., Tennekoon, G. I., Buchanan, K., and Johnston, M. V. (1985) Choline acetyltransferase activity in the striatum of neonatal rats increased by nerve growth factor. Science 229: 284–287.

Olson, L., Ebendal, T., and Seiger, A. (1979) NGF and anti-NGF: evidence against effects on fiber growth in locus coeruleus from cultures of perinatal CNS cultures. Dev. Neurosci. 2: 160–176.

Pioro, E. P., and Cuello, A. C. (1988) Purkinje cells of adult cerebellum express nerve growth factor receptor immunoreactivity: light microscopical observations. Brain Res. 455: 182–186.

Raivich, G., and Kreutzberg, G. W. (1987) The localization and distribution of high affinity β-nerve growth factor binding sites of the adult rat. A light microscopic autoradiographic study using (^{125}I)β-nerve growth factor. Neuroscience 20: 23–36.

Raivich, G., Zimmermann, A., and Sutter, A. (1985) The spatial and temporal pattern of βNGF receptor expression in the developing chick embryo. EMBO J. 4: 637–644.

Raivich, G., Zimmermann, A., and Sutter, A. (1987) Nerve growth factor receptor expression in chicken cranial development J. comp. Neurol. 256: 229–245.

Richardson, P. M., Verge Issa, V. M. K., and Riopelle, R. J. (1986) Distribution of neuronal receptors for nerve growth factor in the rat. J. Neurosci. 6: 2312–2321.

Riopelle, R. J., Verge Issa, V. M. K., and Richardson, P. M. (1987) Properties of receptors for nerve growth factor in the mature rat nervous system. Molec. Brain Res. 3: 45–53.

Schwab, M. E., Otten, U., Agid, Y., and Thoenen, H. (1979) Nerve growth factor (NGF) in the rat CNS: absence of specific retrograde transport and tyrosine hydroxylase induction in locus coeruleus and substantia nigra. Brain Res. 168: 473–483.

Schechter, A. L., and Bothwell, M. A. (1981) Nerve growth factor receptors on PC12 cells. Evidence for two receptor classes with differing cytoskeletal associated characteristics. Cell 24: 867–874.

Seiler, M., and Schwab, M. E. (1984) Specific retrograde transport of nerve growth factor from the neocortex to nucleus basalis in the rat. Brain Res. 300: 33–39.

Shelton, D. L., and Reichard, L. F. (1986) Studies on the expression of the β-nerve growth factor (NGF) gene in the central nervous system: level and regional distribution of NGF mRNA suggest that NGF functions as a trophic factor for several distinct populations of neurons. Proc. natl Acad. Sci. USA 83: 2714–2718.

Springer, J. E., Koh, S., Tayrien, M. W., and Loy, R. (1987) Basal forebrain magnocellular neurons stain for nerve growth factor receptor: correlation with cholinergic cell bodies and effects of axotomy. J. Neurosci. Res. 17: 111–118.

Sutter, A., Riopelle, R. J., Harris-Warrick, R. M., and Shooter, E. M. (1979) Nerve growth factor receptors: characterization of two distinct classes of binding sites on the chick embryo sensory ganglia cells. J. biol. Chem. 254: 5972–5982.

Taniuchi, M., Schweitzer, J. B., and Johnson, E. M. (1986) Nerve growth factor molecules in rat brain. Proc. natl Acad. Sci. USA 83: 1950–1954.

Thoenen, H., and Barde, Y. A. (1980) Physiology of nerve growth factor. Physiol. Rev. 60: 1284–1335.

Zimmermann, A., and Sutter, A. (1983) Nerve growth factor (NGF) receptors on glial cells. Cell-cell interaction between neurons and Schwann cells in cultures of chick sensory ganglia. EMBO J. 2: 879–885.

Chol-1: A cholinergic-specific ganglioside of possible significance in central nervous system neurochemistry and neuropathology

E. Borroni, E. Derrington and V. P. Whittaker

Arbeitsgruppe Neurochemie, Max-Planck-Institut für biophysikalische Chemie, Postfach 2841, D-3400 Göttingen, Federal Republic of Germany

Summary. By use of an antiserum raised against presynaptic plasma membrane purified from the purely cholinergic electromotor system of *Torpedo marmorata* we have been able to identify a group of antigenically-related minor gangliosides (collectively designated Chol-1) that appear to be exclusively localized on cholinergic neurons. The cholinergic-specificity of these antigens has been shown by the following findings:

a) The anti-Chol-1 antiserum induces a selective complement-mediated lysis of the cholinergic subpopulation of mammalian brain synaptosomes;

b) Section of the fimbria, which causes a massive degeneration of cholinergic terminals in the hippocampus, leads to a concomitant depletion of the level of the Chol-1 gangliosides in the hippocampus;

c) The anti-Chol-1 serum can be used to immunostain cholinergic elements in the central and peripheral nervous systems of the rat.

The discovery of a cell surface cholinergic-specific antigen has provided a new and effective tool with which to study the cholinergic neuron. For instance, we have immuno-isolated cholinergic synaptosomes from rat cortex and used this preparation to study transmitter coexistence. Our results indicate that approximately 75% of the cortical cholinergic neurons also express the neuropeptide VIP. Furthermore, we are investigating the expression of Chol-1 in patients affected by diseases such as ALS which primarily involve central cholinergic neurons.

The existence of cell-type-specific antigens represents the selective localization of (a) molecule(s) on a specific subpopulation of cells. Presumably the functions of these molecules are relevant uniquely to a cell of that type and therefore important because they contribute to the functional identity of that cell type. By use of appropriate immunological techniques antibodies directed against cell-type-specific antigens can be raised even if the antigen itself has not been identified or characterized yet. These antibodies can then be used to identify the cell-type-specific antigen to study its cellular localization, its topographical and temporal expression and possibly its cellular function. Furthermore, when the cell-type-specific antigen happens to be exposed on the cell surface, this antibody can also be used to immunoaffinity-purify the cells expressing the antigen.

By immunizing sheep with synaptosomal membranes derived from the purely cholinergic innervation of the electric organ of *Torpedo marmorata* we have raised an antiserum which recognizes two minor gangliosides that appear to be selectively localized on cholinergic neurons. These antigenic gangliosides have been designated Chol-1.

Here we present evidence for the cholinergic-specificity of Chol-1 and for its gangliosidic nature and review some of the applications of the anti-Chol-1 antiserum in the study of the cholinergic neuron. Furthermore, we discuss preliminary investigations we have made into the possibility that Chol-1 may be involved in the etiology of diseases affecting the cholinergic system.

The cholinergic specificity and gangliosidic nature of the Chol-1 antigen

Incubation of mammalian synaptosomes with anti-TSM serum followed by exposure to complement leads to a selective lysis of the cholinergic subpopulation of synaptosomes. The specificity of this lysis is shown by the fact that the complement-antiserum complex causes an almost complete release in the supernatant of the cholinergic cytoplasmic enzyme choline acetyltransferase (ChAT), none of a range of soluble markers specific for other types of nerve terminals and about 5% of a general cytoplasmic marker, lactate dehydrogenase (LDH; Richardson 1983). This indicates the presence of a cholinergic-specific cell-surface antigen on mammalian nerve terminals which has been conserved from *Torpedo* to mammals (Ferretti and Borroni, 1986). Since no evidence of immunoreactivity against protein antigens could be established either by immunoprecipitation or by Western blotting the possibility that the antigen recognized by anti-TSM is a lipid was investigated. Immunostaining of thin-layer-chromatograms (TLC) showed that the antiserum recognized two immunoreactive bands migrating on TLC just below GQ and GT1b, respectively (Fig. 1a, b; Ferretti and Borroni, 1986; Derrington and Borroni, 1989). The presence of sialic acid in these glycolipid bands and hence their gangliosidic nature was confirmed by demonstrating that partial digestion with neuraminidase—an enzyme that removes sialic acid—caused a change in their migration on TLC (Fig. 2). Furthermore, when the digestion is carried out under conditions that lead to a complete release of sialic acid, the immunoreactivity was lost indicating that a sialic acid residue is probably part of the epitope (Derrington and Borroni, 1989; Derrington et al., 1989).

Evidence that the Chol-1 ganglioside antigens are located selectively on cholinergic nerve endings can be derived from two fundamentally different types of experiments. Firstly it has been demonstrated that ganglioside fractions enriched in the immunoreactive bands (a) inhibit the cholinergic-specific lysis of synaptosomes induced by the anti-TSM antiserum (Fig, 1c; Ferretti and Borroni, 1986; Derrington et al., 1989), (b) when immobilized on glass beads they can be used to affinity-purify

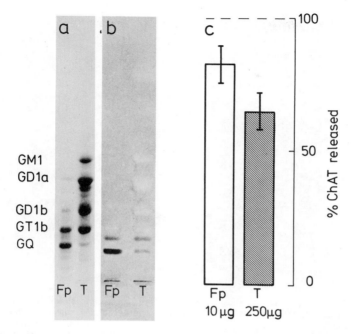

Figure 1. *a* TLC of a polysialoganglioside fraction (F_P) and of total forebrain ganglioside (T) visualized by Ehrlich's reagent. *b* Immunostaining with anti-Chol-1 serum of F_P and T. Two immunoreactive bands migrating near the major gangliosides GT1b and GQ are detected. *c* Preincubation of the anti-Chol-1 serum with T or F_P leads to an inhibition of the specific lysis of the cholinergic synaptosome. Fraction F_P, which is enriched in the antigen as demonstrated by the more intense staining on TLC, has a higher inhibitory activity.

a subpopulation of immunoglobulins from the whole anti-TSM anti-serum with increased anti-cholinergic synaptosome titer as measured by their ability to lyse cholinergic synaptosomes per μg immunoglobulin (Ferretti and Borroni, 1986). Secondly, specific lesions have been used to perturb the cholinergic innervation of the hippocampus. Section of the fimbria disrupts the main cholinergic input to the hippocampus leading to a massive loss of cholinergic markers 1–2 weeks post-lesion in the ipsilateral hippocampus. Using a novel quantitative method of TLC ELISA (Derrington et al., 1989) it has been demonstrated that such lesions cause falls in the concentration of Chol-1 equivalent to those exhibited by the classical cholinergic marker ChAt (Fig. 3; Derrington et al., 1989). The slightly slower rate of disappearance being the result of a difference in metabolism of gangliosides and cytoplasmic proteins in degenerating terminals. By contrast, lesions to the entorhinal cortex cause the degeneration of glutaminergic synapses primarily in the ipsi-lateral dentate gyrus. This provokes the sprouting of cholinergic neurons and leads to a rise in the level of cholinergic markers (Lynch et al., 1972; Ulas et al., 1986) including Chol-1 (Derrington et al., 1989). The

62

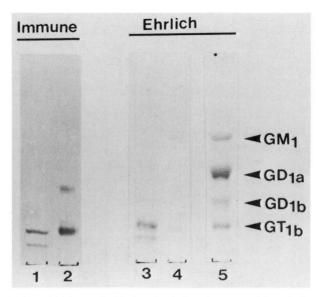

Figure 2. TLC of a polysialoganglioside fraction before (lanes 1 and 3) and after (lanes 2 and 4) mild neuraminidase treatment. The plates were immunostained with anti-Chol-1 serum (lanes 1 and 2) and Ehrlich's reagent (lanes 3 and 4). The change in migration of the immunoreactive bands following neuraminidase treatment indicates the presence of sialic acid in this glycolipid and hence its gangliosidic nature. Lane 5, standard ganglioside.

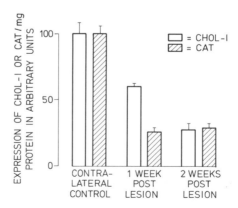

Figure 3. Time-course of ChAT and Chol-1 following lesion of the fimbria. The concentrations of Chol-1 and ChAT present in the ipsilateral (lesioned) hippocampus are expressed as percentage of the contralateral (control). Though the Chol-1 immunoreactivity falls more slowly than ChAT activity it is clear that the deafferentiation of the major cholinergic input to the hippocampus causes a fall in Chol-1 concomitant with that of ChAT. Blocks and bars are means ± SD of 6 experiments.

evidence mentioned above strongly supports the cholinergic specificity of the Chol-1 gangliosides. Though cell-surface carbohydrates have been demonstrated to be specific markers for various neural cell types (Raff et al., 1978) and subclasses of neurons (Dodd and Jessell, 1985) the Chol-1 gangliosides are so far unique in that they specify neurons of one transmitter type. It is intriguing to speculate that other minor gangliosides may eventually be found which extend this kind of specificity to other transmitters.

Applications of the anti-Chol-1 antiserum

The Chol-1 gangliosides, as specific cell-surface antigens of the cholinergic neurons, provide a useful tool for the study of this subset of neurons. Antibodies to the Chol-1 antigen have been successfully used to detect cholinergic neurons in the central and peripheral nervous systems of the rat (Obrocki and Borroni, 1988). Some examples of staining are given in Figures 4 and 5. In Fig. 4, serial sections from rat spinal cord were stained alternatively with monoclonal anti-ChAT antibodies (a, c) and with anti-Chol-1 antiserum (b). Of 24 cells in Figure 4b judged to be Chol-1-positive, 21 cells were ChAT-positive in adjacent sections (Fig. 4a, c); the remaining Chol-1-positive cells in Figure 4b may well be cholinergic cells lying wholly within the section. In Figure 5, adjacent sections of the trochlear nucleus stained (a) for ChAT and (b) for Chol-1, 60% of the Chol-1-positive cells in (b) were identified in (a) and the remaining 40% in the adjacent section (not shown). Thus the anti-Chol-1 antiserum provides a promising new tool for tracing a minor constituent specific for cholinergic neurons.

Though other cell markers exist for cholinergic neurons (e.g. anti-ChAT and acetylcholine antibodies) the location of the Chol-1 antigen on the cell surface of cholinergic nerve endings is particularly exciting in that it has allowed their immuno-isolation from mixed mammalian-brain synaptosome preparations (Richardson et al., 1984). Preparations of intact, metabolically active nerve terminals homogenous with respect to the transmitter they contain represent useful tools for pharmacological and biochemical studies. Such preparations have proved particularly useful in establishing the co-release of acetylcholine and ATP (Richardson and Brown, 1987). We have used immuno-isolated cholinergic synaptosomes to study the coexistence of the neuropeptide VIP and acetylcholine in the some nerve terminals (Agoston et al., 1988). Table 1 shows that the purification of cholinergic synaptosomes also results in the purification of VIP. The ChAT to LDH ratio was $14.8 \pm 1.4\%$ indicating the high degree of purity of the cholinergic nerve terminal preparation. The purification of VIP amounted to 7.6-fold, which suggests that, given the 14.8-fold purification of ChAT, approximately

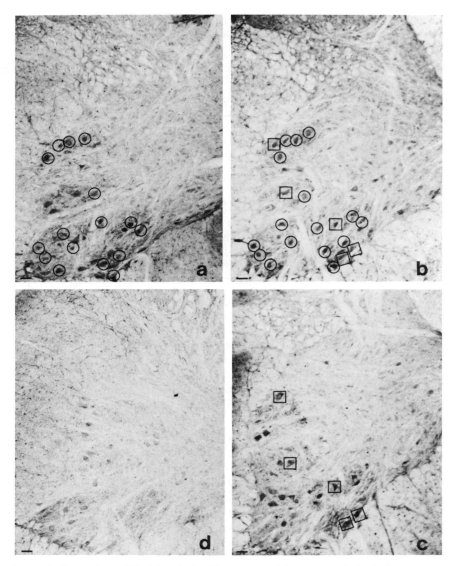

Figure 4. Comparison of Chol-1 and ChAT immunoreactivity in rat cervical spinal cord; (a-c) are 40-μm-thick serial sections stained for ChAT (a, c) and Chol-1 (b). In (b) neurons framed by circles appear in (a) and those framed by squares appear in (c). (d) Section treated with non-immune sheep serum. Bars: 50 μm.

51% of the purified cholinergic nerve terminals contain both VIP and ChAT. The yield of VIP during the immunopurification was 3.3 ± 0.1% compared to 4.0 ± 0.2% for ChAT; this suggests that approximately 75% of the VIP present in cortical nerve terminals is copurifying with ChAT. The application of immunopurification of cholinergic nerve

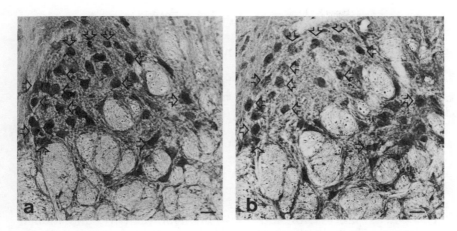

Figure 5. Adjacent section of the trochlear nucleus stained for (a) ChAT and (b) Chol-1. The hollow arrows point to immunopositive cells present in both sections; the remaining Chol-1-positive cells in (b) were found in the opposite section (not shown). Bars: 30 μm.

Table 1. Transmitter and enzyme content of rat cortical immuno-isolated cholinergic synaptosomes

Fraction	ACh (mmol/mg)	VIP (pmol/mg)	ChAT (nmol/h/mg)	LDH (μmol/h/mg)
Parent (P₂¹)	105 ± 007	14 ± 01	258 ± 19	227 ± 16
Immuno-isolated	20 ± 2	11 ± 1	3822 ± 211	220 ± 17
Purification factor	19	8	15	10

Results are the mean ± SEM of three or more experiments; [1]The P_2 fraction was prepared as described by Gray and Whittaker (1962).

terminals followed by specific RIA of peptides offers a new and quantitative tool for studying transmitter coexistence. The possibility that anti-Chol-1 antibodies can be used to immuno-isolate cholinergic cell bodies is under investigation. Such preparations may be used to produce cholinergic-specific cDNA libraries that may lead to a better understanding of the molecular biology of the cholinergic neuron.

Chol-1 and neuropathologies

The significance of cell-surface glycoconjugates in pathologies are manifold. Breakdown in the ganglioside metabolism leads to the central nervous system disorder of Tay Sachs syndrome. Changes in the relative concentration of glycolipids are associated with oncogenesis. Cholera toxin has been shown to bind specifically to the ganglioside GM1 (Holmgren et al., 1975) and tetanus toxin binds primarily to the gangliosides GD1b and GT1b (v. Heyningen, 1974). Furthermore it has been shown that various infective agents (bacteria and viruses) have the capacity to bind to carbohydrate sequences exposed on the external cell surface and in doing so they may act as an adjuvant eliciting the production of antibodies against the glycoconjugate receptor of the infective agent (Feizi, 1987). These antibodies may then lead to auto-immune pathologies as in the case of auto-immune hemolytic anemia (Feizi, 1987).

In view of the role played by glycoconjugates in various pathologies, we investigated the possibility that the Chol-1 gangliosides may be involved in the etiology of diseases preferentially affecting cholinergic neurons. We have examined sera from typical (high anti-acetylcholine receptor titer) and atypical (low anti-acetylcholine receptor titer) myasthenia gravis patients for anti-Chol-1 immunoreactivity, and though several patients showed signs of anti-presynaptic membrane titer (as measured by their ability to lyse central cholinergic synaptosomes) none showed titer against the Chol-1 gangliosides. Amyotrophic Lateral Sclerosis (ALS) is typified by the degeneration of motoneuron cell bodies in the spinal cord. We are presently investigating whether Chol-1 can be suitable marker for pathologically induced degeneration as suggested by the results of our lesion study (Derrington et al., 1989) or whether anti-Chol-1 antibodies may actually be involved in the pathogenesis of this disease. Previous reports have shown that ChAT is depleted in post-mortem spinal cord from ALS patients (Gillberg et al., 1982). Though preliminary experiments show that the level of Chol-1 correlates with that of ChAT in such tissues our studies have not sufficiently advanced for us to draw specific conclusions concerning the possible role of Chol-1 in this pathology. Finally, there is some evidence that the B subunit of botulinum toxin can inhibit the cholinergic-specific lysis induced by the anti-Chol-1 antiserum suggesting that Chol-1 may be associated with the whole or part of the receptor for this toxin subunit (Evans et al., 1988). However, no direct evidence showing toxin binding to the Chol-1 gangliosides was obtained and furthermore the selectivity of the anti-Chol-1 serum used was not demonstrated. Though these results are very interesting since botulinum toxin is known to irreversibly inhibit the release of acetylcholine, further investigations are required to demonstrate the association of the Chol-1 gangliosides with the acceptor for botulinum toxin.

Acknowledgments. EB was supported by the Deutsche Forschungsgemeinschaft (grant No. Wh 1/5-1) and ED by a stipend from Fidia Pharmaforschung GmbH.

Agoston, D. V., Borroni, E., and Richardson, P. J. (1988) Cholinergic surface antigen Chol-1 is present in a subclass of VIP-containing rat cervical synaptosomes. J. Neurochem. 50: 1659–1662.

Derrington, E. A., and Borroni, E. (1989) The developmental expression of the cholinergic-specific gangliosidic antigen Chol-1 in the central and peripheral nervous system of the rat. Submitted.

Derrington, E. A., Masco, M., and Whittaker, V. P. (1989) Confirmation of the cholinergic specificity of the Chol-1 gangliosides in mammalian brain using affinity-purified antisera and lesions affecting the cholinergic input to the hippocampus. Submitted.

Dodd, J., and Jessell, T. (1985) Lactoseries carbohydrates specify subsets of dorsal root ganglion neurons projecting to the superficial dorsal horn of rat spinal cord. J. Neurosci. 5: 3278–3294.

Evans, D. M., Richardson, P. J., Fine, A., Mason, W. G., and Dolly, J. O. (1988) Relationship of acceptors for botulinum neurotoxins (types A and B) in rat cns with the cholinergic marker, Chol-1. Neurochem. Int. 13: 25–36.

Feizi, T. (1987) Significance of carbohydrate components of cell surface. In: Evered, D., and Whelan, J. (eds), Autoimmunity and Autoimmune Disease, Ciba Foundation Symposium 129. Wiley and Sons, Chichester, pp. 43–53.

Ferretti, P., and Borroni, E. (1986) Putative cholinergic-specific gangliosides in guinea pig forebrain. J. Neurochem. 46: 1888–1894.

Gillberg, P. G., Aquilonius, S.-M., Eckernäs, S.-A., Lundqvist, G., and Winblad, B. (1982) Choline acetyltransferase and substance P-like immunoreactivity in the human spinal cord: changes in amyotrophic lateral sclerosis. Brain Res. 250: 394–397.

Gray, E. G., and Whittaker, V. P. (1962) The isolation of nerve endings derived by homogenization and centrifugation. J. Anat. 96: 79–88.

Heyningen, W. E. van (1974) Gangliosides as membrane receptors for tetanus toxin, cholera toxin and serotonin. Nature 249: 415–417.

Holmgren, J., Lönnroth, I., Mansson, J. E., and Svennerholm, L. (1975) Interaction of cholera toxin and membrane GM1 ganglioside of small intestine. Proc. natl Acad. Sci. USA 72: 2520–2524.

Lynch, G., Matthews, D. A., Mosko, W., Parks, T., and Cotman, C. (1972) Induced acetylcholinesterase-rich layer in rat dentate gyrus following entorhinal lesions. Brain Res. 42: 311–318.

Obrocki, J., and Borroni, E. (1988) Immunochemical evaluation of a cholinergic-specific ganglioside antigen (Chol-1) in the central nervous system of the rat. Exp. Brain Res. 72: 71–82.

Raff, M. C., Fields, K. L., Hakomori, S. I., Mirsk, R., and Pruss, R. M. (1978) Cell-type specific markers for distinguishing and studying neurons and major classes of glial cells in culture. Brain Res. 174: 283–308.

Richardson, P. J. (1983) Presynaptic distribution of the cholinergic-specific antigen Chol-1 and 5′-nucleotidase in rat brain, as determined by complement-mediated release of neurotransmitters. J. Neurochem. 41: 640–648.

Richardson, P. J., and Brown, S. J. (1987) ATP release from affinity-purified rat cholinergic nerve terminals. J. Neurochem, 48: 622–630.

Richardson, P. J., Siddle, K., and Luzio, J. P. (1984) Immunoaffinity purification of intact metabolically active, cholinergic nerve terminals from mammalian brain. Biochem. J. 219: 647–657.

Ulas, J., Gradkowska, M., Jezierska, M., Skup, M., Skangiel-Kramska, J., and Oderfeld-Nowak, B. (1986) Bilateral changes in glutamate uptake, muscarinic receptor binding and acetylcholinesterase level in the rat hippocampus after unilateral entorhinal cortex lesions. Neurochem. Int. 9: 255–263.

Pharmacological characterization of muscarinic responses in rat hippocampal pyramidal cells

P. Dutar* and R. A. Nicoll**

Departments of Pharmacology and Physiology, University of California, San Francisco, CA 94143-0450, USA

Summary. Intracellular recording from hippocampal CA1 pyramidal cells was used to characterize the pharmacological properties of muscarinic responses. Results obtained with the M1 antagonist pirenzepine and the M2 antagonist gallamine suggest that an M1 muscarinic receptor is involved in the muscarinic-induced membrane depolarization and blockade of the afterhyperpolarization (AHP). On the other hand, an M2 receptor may be involved in the cholinergic depression of the EPSP and the blockade of the potassium current termed the M-current. Pretreatment of hippocampi with pertussis toxin did not prevent any of the muscarinic responses suggesting that a pertussis toxin-sensitive G-protein is not involved. The M-current, in contrast to the other muscarinic actions, was unaffected by muscarinic agonists which are weak at increasing phosphoinositide (PI) turnover and actually blocked the action of full agonists. This finding suggests that stimulation of PI turnover may be involved in the blockade of the M-current. Although activation of protein kinase C with phorbol esters has little effect on the M-current, intracellular application of inositol trisphosphate did reduce the M-current. We were unable to establish any clear relationship between biochemical effector systems and the muscarinic receptor subtypes.

Introduction

Cholinergic agonists act in hippocampal pyramidal cells through muscarinic receptors to induce a variety of different responses including: 1) a long duration depolarization associated with action-potential discharge and a decrease in a potassium conductance, 2) a blockade of the afterhyperpolarization (AHP) which follows a series of action potentials, 3) a blockade of a voltage- and time-dependent potassium current called the M-current, 4) a presynaptic depression of the excitatory and inhibitory post-synaptic potential (EPSPs/IPSPs) triggered by electrical stimulation of afferents to CA1 pyramidal cells (Nicoll, 1988). However, little is known concerning the nature of the receptor subtypes involved in these actions and on the biochemical effectors linked to these effects. Recent results using molecular cloning have identified at least four functional muscarinic receptor clones (Bonner et al., 1987) and pharmacological studies have described two muscarinic receptor subtypes termed M1 and M2 based on their relative sensitivity to the antagonist

*Present address: Laboratoire de Neurophysiologie Pharmacologique, INSERM U 161, 2 rue d'Alésia, F-75014 Paris, France.
**To whom all correspondence should be addressed.

pirenzepine (PZ) (Hammer et al., 1980). Muscarinic receptors have also been characterized in terms of the coupling of the receptors to second messenger systems. In some cases, pertussis toxin-sensitive GTP-binding proteins (G-proteins) have been identified as coupling muscarinic receptors to a variety of effectors (see Stryer and Bourne, 1986); in addition, muscarinic agonists have been divided into two groups in the brain according to their ability to activate the phosphoinositide (PI) turnover (Fisher et al., 1983, 1984). With these data in mind, we attempted to classify the various muscarinic responses recorded from CA1 pyramidal neurons in terms of muscarinic receptor subtypes, membrane effector and second messenger systems involved.

These electrophysiological data have been obtained from CA1 hippocampal neurons in slices using conventional intracellular recording or single-electrode voltage clamp techniques and have been detailed in recent publications (Dutar and Nicoll, 1988a, b).

Classification in terms of M1 and M2 muscarinic receptors

Electrophysiological studies on classifying muscarinic receptor subtypes in the CNS are still controversial: M1 receptors are involved in the depolarization of neocortical (McCormick and Prince, 1985) or hippocampal (Müller and Misgeld, 1986; Benson et al., 1988) pyramidal cells. M2 receptors appear to be involved in the activation of neocortical interneurons (McCormick and Prince, 1985) and locus coeruleus neurons (Egan and North, 1985), and also in hyperpolarizing responses induced by cholinergic agonists (McCormick and Prince, 1986; Egan and North, 1986). The muscarinic blockade of the afterhyperpolarization (AHP) which follows a series of action potentials may involve an M2 receptor (Constanti and Sim, 1987; Müller and Misgeld, 1986) although some evidence implicates an M1 receptor (Cole and Nicoll, 1984). Finally, the blockade of the M-current may be mediated by M2 receptors in neurons of the pyriform cortex (Constanti and Sim, 1987). In our study we elicited the cholinergic responses by applying carbachol (carb) in the superfusion medium (see Fig. 1). Carb at a low concentration (1 μM) depolarized the membrane, blocked the AHP and depressed the EPSP in every neuron tested. At higher concentrations (e.g. 20 μM), it blocked the M-current. We first examined the action of pirenzepine on these responses. At a low concentration (0.3 μM), PZ antagonized preferentially the carbachol-induced depolarization and blockade of the AHP. At a higher concentration (1 μM) PZ had no evident selectivity. However in some cells, the depression of the EPSP and the blockade of the M-current were less sensitive to PZ. Thus we were unable to clearly differentiate the various responses with PZ. However, gallamine, which binds with higher affinity to the M2 receptor in rat brain (Burke, 1986)

70

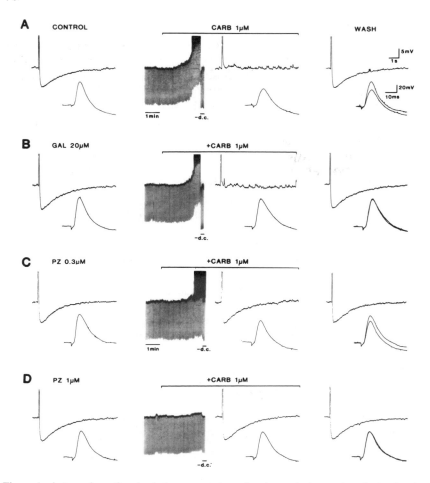

Figure 1. Antagonism of carbachol responses by gallamine and pirenzepine. In **A**, the chart recorder trace on the left shows the AHP in response to a brief (60 ms) depolarizing current pulse at a fast chart speed and below this trace, the EPSP induced by stimulation of stratum radiatum. The chart speed is then reduced and constant current hyperpolarizing pulses are applied at 0.3 Hz. Carbachol (Carb 1 μM) applied by superfusion in the bath during the time indicated by the bar induces a strong depolarization associated with an increase in input resistance (as shown by repolarizing the membrane potential to the control value (-d.c.)), a blockade of the AHP and a depression of the EPSP. These effects recover after washing carbachol from the bath. The EPSP recorded in carbachol is superimposed on the trace recorded after washing. In *B*, gallamine is applied for 10 min before starting carbachol application. It failed to antagonize the depolarization and the blockade of the AHP. In contrast, the depression of the EPSP is antagonized. In *C*, after washing out the gallamine for 20 min, pirenzepine (PZ) is applied at 0.3 μM for 10 min before the carbachol application. PZ strongly antagonizes the depolarization and the blockade of the AHP but fails to antagonize the depression of the EPSP. In *D*, at 1 μM, PZ antagonized all the cholinergic effects. Membrane potential = -66 mV, the voltage and time calibrations in *A* apply to all records.

did permit differentiation of the effects. Gallamine ($20 \, \mu M$) could produce an antagonism of the depression of the EPSP and the blockade of the M-current but had no effect on the depolarization and the blockade of the AHP. Taken together, the results using PZ and gallamine suggest the involvement of an M2 receptor subtype for the cholinergic blockade of the M-current and for the depression of the EPSP. In contrast, the membrane depolarization and the blockade of the AHP may be mediated by M1 receptors.

Classification in terms of second messenger systems

Implication of a G-protein

Various muscarinic effects are thought to involve the activation of a G-protein (see Dunlap et al., 1987, for refs). To test the possibility that a G-protein may mediate the muscarinic responses in the hippocampus, we performed experiments in rats pretreated with pertussis toxin (PTX). PTX is able to ADP-ribosylate the alpha subunit of a G-protein known to interact with some muscarinic receptors (Kurose et al., 1983) and, as a consequence, blocks the muscarinic action. Hippocampal slices from these rats were recorded 3 days after the injection. The effect of PTX was checked by testing the effect of serotonin or baclofen on pyramidal neurons, the hyperpolarizing responses obtained for these compounds in normal rats being blocked in rats pretreated with PTX (Andrade et al., 1986). The action of carbachol was clearly preserved in PTX-pretreated slices. Thus carb was able to depolarize the membrane, block the AHP, and block the M-current. The depression of the EPSP was less constant but occurred in a majority of cases.

Thus, it is unlikely that a PTX-sensitive G-protein (Gi or Go) is required for any of the muscarinic responses described.

Intracellular messengers

Blockade of the slow AHP. Cholinergic agonists block the AHP and accommodation of the spike discharge. These events are primarily due to the activation of a calcium-activated potassium conductance (gK(Ca)). Norepinephrine is known to block the slow AHP by reducing the gK(Ca). This effect is mediated by stimulation of adenylate cyclase and the subsequent increase in intracellular cyclic AMP (Madison and Nicoll, 1986). However, such an intracellular mechanism is unlikely to apply to muscarinic actions, since muscarinic receptors are negatively coupled to adenylate cyclase.

Another possible intracellular mechanism might be the activation of phosphoinositide (PI) turnover. Indeed cholinergic agonists are able to stimulate the PI turnover leading to the formation of two main compounds: diacylglycerol (DG) which activates protein kinase C (PKC) (see Nishizuka, 1986) and inositol trisphosphate (IP3) known to have a role in the intracellular release of calcium (Berridge, 1984). Phorbol esters, which directly activate the PKC, mimic some cholinergic actions in blocking the AHP and accommodation of spike discharge (Baraban et al., 1985; Malenka et al. 1986). In contrast, phorbol analogues which do not activate PKC do not mimic these cholinergic actions. These results suggest that stimulation of PI turnover and the subsequent production of DG and activation of PKC might be involved in the cholinergic blockade of the AHP.

Blockade of the M-current. To examine the possible role of PI turnover in mediating the suppression of the M-current, we have relied on the finding that cholinergic agonists can be classified according to their ability to stimulate PI turnover in the brain (Fisher et al., 1983, 1984). We tested these agonists (20 μM) using the 'voltage-clamp' technique (see Fig. 2). Oxotremorine-M and carbachol classified as full agonists at stimulating PI turnover caused an inward shift in the holding current and abolished the M-current. In contrast, oxotremorine, pilocarpine and arecoline, which are weak agonists for stimulating PI tunover had little or no effect on the M-current. However, weak and strong agonists, at a low concentration (1 μM) had similar effects in depolarizing the membrane and blocking the AHP. Evidence was presented above for the involvement of PI turnover in blocking the AHP. Thus, the fact that the weak agonists for stimulating PI turnover block the AHP to the same extent as the full agonists is surprising. This observation does not entirely rule out the involvement of PI turnover in the blockade of the AHP because it is possible that considerable receptor reserve exists for this action or that diacylglycerol is produced by other metabolic pathways not linked to IP3 generation.

Based on the differential blockade of the M-current by the weak and strong agonists, our results suggest that PI turnover may be involved in this cholinergic effect. As described above, stimulation of PI turnover leads to the formation of DG and IP3. Activation of PKC by phorbol esters does not block the M-current in hippocampus (Malenka et al., 1986; Dutar and Nicoll, 1988b) (Fig. 3A) while it does reduce the M-current in other preparations (Brown and Adams, 1987; Brown and Higashida, 1988). This suggests that an increase in DG is not responsible for the depression of the M-current in this structure. Therefore, we considered the possibility that the IP3 might be responsible for the blockade of the M-current. We found that intracellular application of IP3 (applied by diffusion from the recording electrode) although technically difficult, caused a reproducible depression of the M-current while other characteristics of the cell as the Q-current appeared unchanged

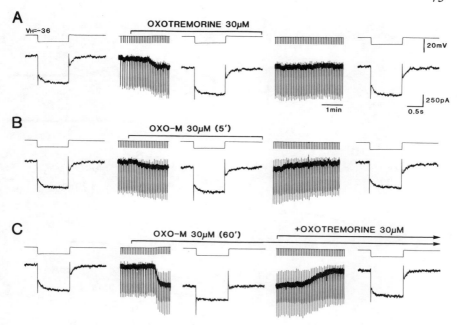

Figure 2. Oxotremorine antagonizes the effect of oxotremorine-M (OXO-M) on the M-current. In all records, the cell was held in voltage clamp at a potential of -36 mV and stepped to -48 mV to elicit the M-current. In *A*, oxotremorine (30 μM), a weak agonist for stimulating the PI turnover, applied by superfusion to the bath induced an inward current but failed to block the M-current. In *B*, OXO-M (30 μM), a full agonist for stimulating the PI turnover, applied 5 min after stopping oxotremorine induced a small inward current, but failed to block the M-current. *C* Only after 60 min of washing out the oxotremorine did OXO-M strongly block the M-current. Oxotremorine added to the bath during the OXO-M application quickly reversed the depressing effect of OXO-M. Horizontal bars indicate the duration of application of the drugs. *A*, *B*, and *C* are continuous recordings from the same cell. Resting membrane potential $= -58$ mV; calibration in top trace applies to others. TTX was added to the bath. (from Dutar and Nicoll, 1988b).

(Fig. 3B). Thus IP3 or one of its metabolites may play a role in the cholinergic blockade of the M-current in hippocampal neurons. It is important to note that in sympathetic ganglion cells (Brown and Adams, 1987; Pfaffinger et al., 1988) and neuroblastoma × glioma hybrid cells (Brown and Higashida, 1988) injection of IP3 has no effect on the M-current suggesting that a different mechanism may be involved (see Brown, 1988).

The effect of IP3 as an inducer of calcium release has been demonstrated in various non-neuronal cell types (see Abdel-Latif, 1986). More recently, an increase in cytosolic Ca^{++} concentration has been demonstrated in hippocampal neuron in response to muscarinic receptor stimulation (Kudo et al., 1988), one component being due to mobilization of Ca^{++} from intracellular stores. We tested whether the effect of muscarinic agonists on the suppression of the M-current could be due to

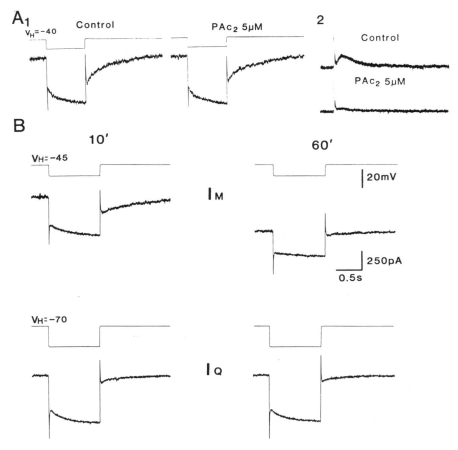

Figure 3. Effect of phorbol ester and inositol trisphosphate (IP3) in CA1 pyramidal cells. Application of phorbol 12, 13 diacetate (PAc2) (dissolved in dimethyl sulfoxide, DMSO) for 6 min had no effect on the M-current (*A*1) but blocked the AHP current (*A*2). The M-current was evoked by a 1-s, 13-mV hyperpolarizing pulse from a holding potential of −40 mV. The AHP current was evoked by a brief depolarizing step. Resting membrane potential = −58 mV. *B* shows the effect of intracellular application of IP3. The cell was recorded in voltage clamp 10 min (left column) and 60 min (right column) after inpalement with an electrode filled with 3M KCl plus 60 mM IP$_3$. The M-current was elicited by applying a 1-s hyperpolarizing step of 12 mV from a holding potential of −45 mV. The M-current was strongly depressed 60 min after the beginning of the diffusion of IP$_3$ into the cell (*B*, top trace). The Q-current evoked from the same cell by a 22-mV hyperpolarizing pulse from a holding potential of −70 mV was unchanged (bottom trace). Resting membrane potential = −57 mV.

an indirect effect through mobilization of intracellular calcium by IP3. We found that following the injection of the calcium chelators, EGTA or BAPTA, it was still possible to induce the cholinergic blockade of the M-current suggesting that this effect does not require a rise in intracellular calcium. However, due to technical limitations, it has been impossible to study directly the effect of IP3 in the presence or absence of calcium in the cell. Additional experiments are necessary to answer this question.

In conclusion, the cholinergic responses elicited in CA1 pyramidal neurons can be differentiated in terms of muscarinic receptor subtypes and in terms of biochemical effector systems. However, the present results in conjunction with other data in the literature do not permit a correlation between receptor subtypes and second messenger systems.

Abdel-Latif, A. A. (1986) Calcium mobilizing receptors, polyphosphoinositides, and the generation of second messengers. Pharmac. Rev. 38: 227–272.

Andrade, R., Malenka, R. C., and Nicoll, R. A. (1986) A G-protein couples serotonin and GABA$_B$ receptors to the same channel in hippocampus. Science 234: 1261–1265.

Barbaran, J. M., Snyder, S. H., and Alger, B. E. (1985) Protein kinase C regulates ionic conductance in hippocampal pyramidal neurons: electrophysiological effects of phorbol esters. Proc. natl. Acad. Sci. USA 82: 2538–2542.

Benson, D. M., Blitzer, R. D., and Landau, E. M. (1988) An analysis of the depolarization produced in guinea-pig hippocampus by cholinergic receptor stimulation. J. Physiol. 404: 479–496.

Berridge, M.J. (1984) Inositol trisphosphate and diacylglycerol as second messengers. Biochem. J. 220: 345–360.

Bonner, T. I., Buckley, N. J., Young, A. L., and Braun, M. R. (1987) Identification of a family of muscarinic acetylcholine receptor genes. Science 237: 527–532.

Brown, D. A. (1988) M-currents: an update. Trends Neurosci. 11: 294–299.

Brown, D. A., and Adams, P. R. (1987) Effects of phorbol dibutyrate on M-currents and M-current inhibition in bullfrog sympathetic neurons. Cell. molec. Neurobiol. 7: 255–269.

Brown, D. A., and Higashida, H. (1988) Inositol 1,4,5-trisphosphate and diacylglycerol mimic bradykinin effects on mouse neuroblastoma × glioma hybrid cells. J. Physiol. 397: 185–207.

Burke, R. E. (1986) Gallamine binding to muscarinic M1 and M2 receptors, studied by inhibition of (^{3}H) pirenzepine and (^{3}H) quinuclidinylbenzilate binding to rat brain membranes. Molec. Pharmac. 30: 58–68.

Cole, A. E., and Nicoll, R. A. (1984) The pharmacology of cholinergic excitatory responses in hippocampal pyramidal cells. Brain Res. 305: 283–290.

Constanti, A., and Sim, J. A. (1987) Calcium-dependent potassium conductance in guinea-pig olfactory cortex neurons in vitro. J. Physiol. 387: 173–194.

Dunlap, K., Holz, G. G., and Rane, S. G. (1987) G-proteins as regulators of ion channel function. Trends Neurosci. 10: 241–244.

Dutar, P., and Nicoll, R. A. (1988a) Stimulation of phosphatidylinositol (PI) turnover may mediate the muscarinic suppression of the M-current in hippocampal pyramidal cells. Neurosc. Lett. 85: 89–94.

Dutar, P., and Nicoll, R. A. (1988b) Classification of muscarinic responses in hippocampus in terms of receptor sub-types and second messenger systems: electrophysiological studies in vitro. J. Neurosci. 8: 4214–4224.

Egan, T. M., and North, R. A. (1985) Acetylcholine acts on a M2 muscarinic receptor to excite rat locus coeruleus neurons. Br. J. Pharmac. 85: 733–735.

Egan, T. M., and North, R. A. (1986) Acetylcholine hyperpolarizes central neurons by acting on an M2 muscarinic receptor. Nature 319: 405–407.

Fisher, S. K., Klinger, P. D., and Agranoff, B. W. (1983) Muscarinic agonists binding and phospholipid turnover in brain. J. Biol. Chem. 258: 7358–7363.

Fisher, S. K., Figueiredo, J. C., and Bartus, R. T. (1984) Differential stimulation of inositol phospholipid turnover in brain by analogs of oxotremorine. J. Neurochem. 43: 1171–1179.

Hammer, R., Berrie, C. P., Birdsall, N. J. M., Burgen, A. S. V., and Hulme, E. C. (1980) Pirenzepine distinguishes between different subclasses of muscarinic receptors. Nature 283: 90–92.

Kudo, Y., Ogura, A., and Iijima, T. (1988) Stimulation of muscarinic receptor in hippocampal neuron induces characteristic increase in cytosolic free Ca^{2+} concentration. Neurosci. Lett. 85: 345–350.

76

Kurose, H., Katada, T., Amano, T., and Ui, M. (1983) Specific uncoupling by islet-activating protein, pertussis toxin, of negative signal transduction via β-adrenergic, cholinergic, and opiate receptors in neuroblastoma × glioma hybrid cells. J. biol. Chem. 258: 4870–4875.

Madison, D. V., and Nicoll, R. A. (1986) Cyclic adenosine 3′,5′-monophosphate mediates beta-receptor actions of noradrenaline in rat hippocampal pyramidal cells. J. Physiol. 372: 245–259.

Malenka, R. C., Madison, D. V., Andrade, R., and Nicoll, R. A. (1986) Phorbol esters mimic some cholinergic actions in hippocampal pyramidal neurons. J. Neurosci. 6: 475–480.

McCormick, D. A., and Prince, D. A. (1985) Two types of muscarinic response to acetylcholine in mammalian cortical neurons. Proc. natl Acad. Sci. USA 82: 6344–6348.

McCormick, D. A., and Prince, D. A. (1986) Acetylcholine induces burst firing in thalamic reticular neurons by activating a potassium conductance. Nature 319: 402–405.

Müller, W., and Misgeld, U. (1986) Slow cholinergic excitation of guinea pig hippocampal neurons is mediated by two muscarinic receptor subtypes. Neurosci. Lett. 67: 107–112.

Nicoll, R. A. (1988). The coupling of neurotransmitter receptors to ion channels in the CNS. Science 241: 545–551.

Nishizuka, Y. (1986) Studies and perspective of protein kinase C. Science 233: 305–312.

Pfaffinger, P. J., Leibowitz, M. D., Bosma, M. M., Almers, W., and Hille, B. (1988). M-current suppression by agonists: the role of the phospholipase C pathway. Biophys. J. 53: 637a.

Stryer, L., and Bourne, H. R. (1986) G-protein: a family of signal transducers. A. Rev. cell. Biol. 2: 391–419.

Mediation of acetylcholine's excitatory actions in central neurons

N. Agopyan, K. Krnjević* and J. Leblond

Anaesthesia Research Department, McGill University, 3655 Drummond Street, Montréal, Québec H3G 1Y6, Canada

Summary. In experiments on the hippocampus *in situ* (in rats under urethane), neither cyclic GMP nor H-8 (an antagonist of cyclic nucleotide-dependent kinases) had much effect on CA1/CA3 population spikes or on the excitatory action of ACh. This is further evidence against the idea that cyclic nucleotides play a major role as cholinergic second messengers. On the other hand, the results of tests with a PKC antagonist sphinganine are in keeping with some involvement of PKC in cholinergic actions. (Another PKC antagonist, H-7, proved to be a very powerful excitant, probably via *disinhibition*). Preliminary experiments on CA1 neurons in hippocampal slices (by single electrode voltage clamp), confirmed previous reports that carbachol depresses A- and C-type K currents, as well as inward Ca^{2+} currents; though the latter effect was sometimes mainly due to frequency-dependent inactivation of Ca currents. It is suggested that a single, primary muscarinic action, the acceleration of phosphinositide turnover, may account for a variety of secondary effects: on the one hand, via activation of PKC, a number of possible PKC-mediated actions, such as block of the slow AHP; on the other, via IP_3 formation, a block of I_M and a rise in cycloplasmic free Ca^{2+} that may cause inactivation of both Ca^{2+}-inward currents, and Ca^{2+}-dependent G_Ks.

Introduction

Apart from a few sites where transmission is likely to be nicotinic (Eccles et al., 1954; Masland et al., 1984; Phillis, 1971; McCormick and Prince, 1988; De La Garza et al., 1987), the most widespread central excitatory effects of ACh are mediated by muscarinic receptors (Krnjević and Phillis, 1963 a, b; Krnjević, 1975; Brown, 1984; McCormick and Prince, 1988). The unusual characteristics of this excitatory action, including its slow and prolonged time course, voltage dependence, and variability, could be explained by an unprecedented mechanism of action, a selective blockage of K^+ channels (Krnjević et al., 1971). In more recent experiments, a variety of subsequently identified K currents were indeed found to be blocked by ACh (including I_M (Halliwell and Adams, 1982)); I_A (Nakajima et al., 1986); I_{AHP} (slow after hyperpolarization) (Benardo and Prince, 1982; Madison et al., 1987); a voltage-independent leak current (Madison et al., 1987) and probably I_c (Belluzzi et al., 1985).

*To whom correspondence should be addressed.

However, various studies indicate some other possible actions of ACh. In the hippocampus, a presynaptic muscarinic action causes disfacilitation at dendritic (excitatory) synapses in the hippocampus (Yamamoto and Kawai, 1967; Hounsgaard, 1978; Rovira et al., 1983), and an even more striking disinhibition at the somatic level (Ben-Ari et al., 1981; Krnjević et al., 1981; Haas, 1982; Segal, 1983). These observations could be explained by a block of Ca^{2+} inward current, in neurons first observed in sympathetic ganglia (Akasu and Koketsu, 1982; Belluzzi et al., 1985; Wanke et al., 1987) and very recently in isolated hippocampal cells (Gähwiler and Brown, 1987).

The slow time course of muscarinic action suggested its possible mediation by some internal messenger (such as intracellular Ca^{2+} (Krnjević, 1977) or a cyclic nucleotide (Greengaard, 1976; Woody et al., 1978; Cole and Nicoll, 1984). More recently, the emphasis has been on phospholipid metabolites, especially diacylglycerol and protein kinase C (PKC) (Malenka et al., 1986; Nicoll, 1988; El-Fakahany et al., 1988), as well as inositol trisphosphate (IP_3) (Dutar and Nicoll, 1988). In the experiments reported here briefly, we tested some possible mechanisms of ACh action.

Involvement of cyclic GMP in muscarinic actions

A selective inhibitor of cyclic AMP- and GMP-dependent kinases N-[2-(methylamino)ethyl]-5-isoquinolinesulfonamide (H-8) (Hidaka et al., 1984), was applied to investigate the putative role of cyclic GMP in muscarinic effects.

These experiments were performed on the hippocampus *in situ*, in urethane-anaesthetized Sprague-Dawley rats, recording field potentials at the somatic or apical dendritic level in CA1 and CA3, applying locally ACh and other agents by iontophoresis from multibarrelled micropipettes (Krnjević et al., 1981; Rovira et al., 1983).

As illustrated in Figure 1, the typical enhancement of a CA1 population spike by ACh (A), often associated with the appearance of a second population spike (Krnjević et al., 1981), was consistently unaffected by a prolonged (2-min) application of H-8 (B). Similarly, the depression of the EPSP field caused by ACh release at the apical dendritic level (C) was not changed when ACh was applied against a background of H-8 release (D).

On the other hand, H-8 clearly reduced the enhancement of population spikes by 8-Br-cyclic AMP, showing that such applications of H-8 are indeed capable of blocking a cyclic nucleotide-mediated effect (presumably due to depression of the slow AHP (Madison and Nicoll, 1986)). Repeated tests of dibutyryl cyclic GMP (by local, extracellular iontophoresis) also had very little effect on population spikes. Hence

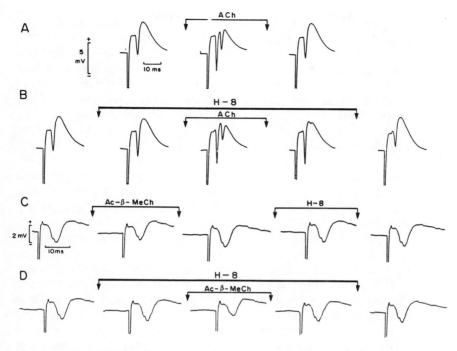

Figure 1. Field responses of CA1, evoked by fimbrial stimulation in rat under urethane, show that typical cholinergic effects are not blocked by H-8—a selective inhibitor of cyclic nucleotide-dependent kinase. *A, B* Recording from pyramidal cell layer; facilitation of population spikes by ACh (56 nA iontophoretic release) is not depressed by H-8 (84 nA), applied for 10 min. *C, D* Recording from stratum radiatum: release of H-8 (84 nA for 10 min) had little effect by itself and it did not prevent depressant action of acetyl-β-methylcholine (54 nA, for 1 min).

there was no evidence that cyclic GMP plays a major role as a mediator of muscarinic (or other) excitatory actions in the hippocampus. Moreover, even a prolonged release of H-8 produced little change in population spikes; if anything they were somewhat enhanced and the positive component of the field depressed, contrary to expectations if cyclic nucleotides exerted a significant on-going modulation of excitability.

Involvement of protein kinase C (PKC) in muscarinic actions

Phorbol ester actions

We confirmed Malenka et al.'s (1986) finding that, like ACh, phorbol esters enhance the CA1/CA3 population spike when applied in the pyramidal cell layer; unlike ACh, however, phorbol esters did not evoke a second population spike. An even greater difference was seen at the apical dendritic level, where, in contrast to ACh (Hounsgaard, 1978; Rovira et al., 1983), phorbol diacetate strongly *potentiated* the EPSP field. There are two possible explanations for these results: either that PKC and ACh can produce significantly different effects, and therefore PKC is not a major muscarinic second messenger; alternatively, phorbol esters may have actions that are unrelated to activation of PKC (Nishizuka, 1986).

Effects of PKC antagonists

Another, less selective antagonist of cyclic nucleotide-dependent kinases, 1-(5-isoquinolinesulfonyl)-2-methylpiperazine dihydrochloride (H-7), is also an effective inhibitor of PKC activation (Hidaka et al., 1984). The release of H-7 powerfully facilitated CA1/CA3 population spikes, resulting in bursts of 3–4 spikes (Fig. 2A). This prolonged (but reversible) effect was comparable to that of bicuculline. In further tests, however, there was no major reduction of GABA's inhibitory effect, though IPSPs (recorded intracellularly in hippocampal *slices*) were markedly depressed. A block of late IPSPs may account for the prolongation of EPSP fields produced by H-7 (Fig. 2B). This unexpected strong excitatory effect (apparently due to disinhibition) greatly reduces the usefulness of H-7 as PKC inhibitor. The absence of a comparable excitation in some previous experiments may be due to the absence of

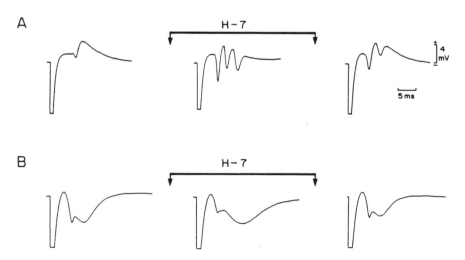

Figure 2. In another rat under urethane, H-7 release (56 nA for 10 min) caused a marked increase in size and number of population spikes evoked in CA1 pyramidal layer by constant fimbrial stimulation (*A*) and greatly prolonged the corresponding EPSP field in stratum radiatum (*B*). Both effects were reversible, as shown by traces at right in *A* and *B*, obtained respectively 5 and 15 min after the end of H-7 release.

inhibitory tone (Mochida and Kobayashi, 1988) or the use of much smaller applications (Lovinger et al., 1987).

Sphinganine, another inhibitor of PKC activation (Hannun et al., 1986), proved to be more useful. When applied alone, it did not produce any major alteration of field responses, but there was a consistent reduction of the effects of concomitant applications of ACh. This gives some support for previous evidence that some muscarinic effects are mediated via PKC (Malenka et al., 1986; Nicoll, 1988; Mochida and Kobayashi, 1988).

Blockage of membrane currents by ACh

These experiments were performed on CA1 neurons in rat hippocampal slices, using a single-electrode voltage clamp technique (Krnjević and Leblond, 1987) and 0.5 μM tetrodotoxin to inactivate Na currents. When Ca-inward currents were studied, 3 mM Cs^+ and 10 mM tetraethylammonium were also supplied.

Outward currents

In many cells depolarizing pulses evoked both an early, fast decaying outward current, clearly seen only at a relatively negative holding potential (V_H) (I_A-type; cf. Gustafsson et al., 1982)—and a delayed, non-inactivating outward current (I_c-type; Brown and Griffith, 1983). Both currents are visible at $V_H - 70$ mV in Figure 3, in the first inset trace. The corresponding upper graph (open triangles) plots the amplitude of the early current evoked by hyperpolarizing and depolarizing pulses. During the application of carbachol, both early and delayed outward currents vanished, as shown by the inset traces and also, for the early (A-type) currents, by the closed triangles. These observations are in keeping with previous reports that muscarinic agents can block both I_A (Nakajima et al., 1986) and the Ca-dependent I_c (Belluzzi et al., 1985).

Inward current

In a small number of cells, carbachol was also tested on TTX-insensitive, presumably Ca^{2+}-mediated inward currents (at $V_H - 40$ or $- 50$ mV) in CA1 neurons in slices. In these preliminary experiments, the results were partly in accordance with Gähwiler and Brown's (1987) observations in hippocampal monolayers, carbachol depressed both fast and slowly-inactivating Ca currents in some, but not all cells. Thus, the

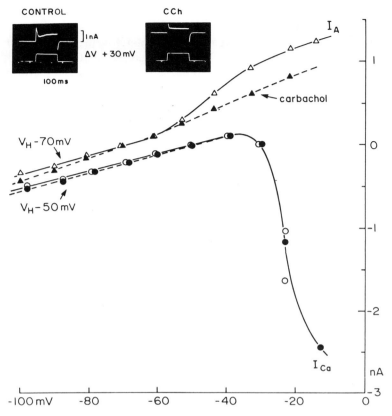

Figure 3. Current-voltage data from CA1 pyramidal cell in hippocampal slices showing block of fast-inactivating outward current (I_A) by carbachol (100 μM), but no significant change in Ca-inward current (I_{Ca}). Recording was by single-electrode voltage clamp (SEVC) with 3 M KCl electrode at 33.5°: for I_A in presence of 0.5 μM tetrodotoxin (TTX) only and at holding potential (V_H) -70 mV; for I_{Ca}, with further additon of 3 mM Cs and 10 mM tetraethylammonium (TEA), V_H -50 mV. Open symbols are control values, and closed symbols currents recorded during superfusion with carbachol (CCh).

negative slope region in the lower current-voltage plot in Figure 3—obtained from a CA1 cell—was not significantly altered by carbachol. On the other hand, when the inward current was evoked at regular 5-s intervals, by a fixed depolarizing pulse (Fig. 4), carbachol first caused a small enhancement, and then, after 2 min, a rapidly progressive and maintained depression. But when the regular series of pulses were interrupted for 30 s (point at asterisk), the inward current temporarily regained its full amplitude, suggesting that carbachol's main effect (on this cell at any rate) was either an open channel type of block or enhanced Ca inactivation.

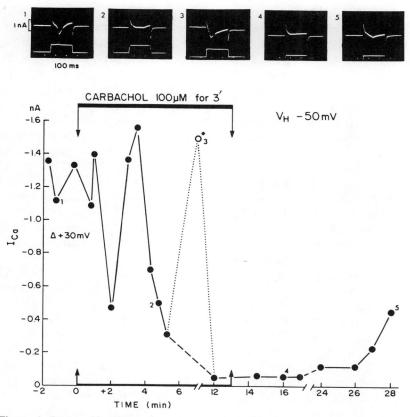

Figure 4. Comparable SEVC data showing apparent inactivation of I_{Ca} by carbachol (100 μM). Inward currents, evoked by constant 30 mV depolarizing pulses delivered at frequency of 0.2 Hz, were recorded in presence of TTX, Cs and TEA, at 33.5° and V_H −50 mV (as in previous figure); not all responses are shown. Marked depression was evident after 4 min of carbachol; but when regular series of pulses was interrupted for 30 s a large inward current was again evoked (at asterisk). The depression rapidly returned on resuming regular 0.2 Hz stimulation, and there was only partial recovery 20 min after washing out CCh. Traces illustrate currents observed at various times, as indicated by numbers.

Discussion

Clearly, the present picture is far more complicated than the relatively simple scheme generally accepted even 2–3 years ago, according to which the muscarinic excitatory effects had a single basic mechanism, blockage of most K conductances.

Two discoveries have been especially illuminating. The first is that muscarinic ACh suppresses Ca^{2+} currents in a variety of nerve cells, notably in sympathetic ganglia (Akasu and Koketsu, 1982; Belluzzi et al., 1985; Wanke et al., 1987), as well as the hippocampus (Gähwiler and

Brown, 1987). A block of postsynaptic Ca currents could also explain the muscarinic presynaptic action (Yamamoto and Kawai, 1967; Hounsgaard, 1978; Krnjević et al., 1981; Ben Ari et al., 1981; Haas, 1982; Segal, 1983). At the same time, it provides an explanation for the depression of Ca^{2+} dependent G_Ks, such as I_c and I_{AHP} (and perhaps even I_A if this is more Ca-dependent than is usually believed (cf. Bourque, 1988)).

The second important finding is that ACh causes a rise in free Ca^{2+} level in isolated hippocampal neurons (Kudo et al., 1988): at low concentrations, ACh releases Ca^{2+} from intracellular storage sites and at higher concentrations, Ca^{2+} influx is generated. Clearly, the raised $[Ca^{2+}]$, could inactivate Ca^{2+} current (Eckert and Chad, 1984; Chad and Eckert, 1986), thus causing most (and perhaps all) of the features of muscarinic excitation. The initial 'trigger' may well be acceleration of phosphoinositol turnover, and therefore increased formation of both diacylglycerol (and thus activation of PKC), and IP_3, which triggers Ca^{2+} release from internal stores (Berridge, 1986).

It has been argued that IP_3 mediates the muscarinic block of M current in hippocampal cells by a *Ca-independent* mechanism—because intracellular release of Ca-chelators in sufficient amount to block the slow AHP does not prevent the IP_3 or ACh-evoked depression of I_M (Dutar and Nicoll, 1988). This argument is based on the assumption that I_{AHP} is a true K outward current; but in view of many similarities between the slow AHPs of hippocampal and bursting pacemaker neurons of Aplysia, it is perhaps more likely to reflect a Ca^{2+}-influx-dependent inactivation of a steady inward current (Kramer and Zucker, 1985).

Conclusion

According to our observations on the hippocampus *in situ*, neither cyclic GMP nor H-8 (an antagonist of cyclic nucleotide-dependent kinases) have much effect on CA1/CA3 population spikes or on the excitatory action of ACh. This is further evidence against the idea that cyclic nucleotides play a major role as cholinergic second messengers. On the other hand, the results of test with a PKC antagonist sphinganine are in keeping with some involvement of PKC in cholinergic actions. (Another PKC antagonist, H-7, proved to be a very powerful excitant, probably via *disinhibition*). Preliminary experiments confirmed previous reports that carbachol depresses A- and C-type K currents, as well as inward Ca^{2+} current; though the latter effect was sometimes mainly due to frequency-dependent inactivation of Ca currents. It is suggested that a single, primary muscarinic action, the acceleration of phosphinositide turnover, may account for a variety of secondary effects: on the one hand, via activation of PKC, a number of possible PKC-mediated actions, such as block of the slow AHP (Nicoll, 1988) and perhaps block of an adenosine-generated leak current (Madison et al., 1987; El Fakahany et al., 1988); on the other hand, via IP_3 formation, a block of I_M (Dutar and Nicoll, 1988) and a rise in cytoplasmic free Ca^{2+} (Kudo et al., 1988) that may cause inactivation of both Ca^{2+} inward currents, and Ca^{2+}-dependent G_Ks (Belluzzi et al., 1985; Wanke et al., 1987; Gähwiler and Brown, 1987).

Acknowledgements. We are grateful to the Medical Research Council of Canada for its financial support and to the Savoy Foundation for a Studentship Award to Nadia Agopyan.

Akasu, T., and Koketsu, K. (1982) Modulation of voltage-dependent currents of muscarinic receptor in sympathetic neurones of bullfrog. Neurosci. Lett. 29: 41–45.

Belluzzi, O., Sacchi, O., and Wanke, E. (1985) Identification of delayed potassium and calcium currents in the rat sympathetic neurone' under voltage clamp. J. Physiol. 358: 109–129.

Ben-Ari, Y., Krnjević, K., Reiffenstein, R. J., and Reinhardt, W. (1981) Inhibitory conductance changes and action of γ-aminobutyrate in rat hippocampus. Neuroscience 6: 2445–2463.

Benardo, L. S., and Prince, D. A. (1982) Cholinergic excitation of mammalian hippocampal pyramidal cells. Brain Res. 249: 315–331.

Berridge, M. J. (1986) Regulation of ion channels by inositol trisphosphate and diacylglycerol. J. exp. Biol. 124: 323–335.

Bourque, C. W. (1988) Transient calcium-dependent potassium current in magnocellular neurosecretory cells of the rat supraoptic nucleus. J. Physiol. 397: 331–347.

Brown, D. A. (1984) Muscarinic excitation of sympathetic and central neurones. In: Subtypes of Muscarinic Receptors. Trends Pharmac. Sci., Suppl. Jan. 84, pp. 32–34.

Brown, D. A. and Griffith, W. H. (1983) Calcium-activated outward current in voltage-clamped hippocampal neurones of the guinea-pig. J. Physiol. 337: 287–301.

Chad, J. E., and Eckert, R. (1986) An enzymatic mechanism for calcium current inactivation in dialysed Helix neurones. J. Physiol. 378: 31–51.

Cole, A. E., and Nicoll, R. A. (1984) The pharmacology of cholinergic excitatory responses in hippocampal pyramidal cells. Brain Res. 305: 283–290.

De La Garza, R., McGuire, T. J., Freeman, R., Hoffer, B. J. (1987) Selective antagonism of nicotine actions in the rat cerebellum with alpha-bungarotoxin. Neuroscience 23: 887–891.

Dutar, P., and Nicoll, R. A. (1988) Stimulation of phosphatidylinositol (PI) turnover may mediate the muscarinic suppression of the M-current in hippocampal pyramidal cells. Neurosci. Lett. 85: 89–94.

Eccles, J. C., Fatt, P., and Koketsu, K. (1954) Cholinergic and inhibitory synapses in a pathway from motor-axon collaterals to motoneurones. J. Physiol. 126: 524–562.

Eckert, R., Chad, J. E. (1984) Inactivation of Ca channels. Prog. Biophys. molec. Biol. 44: 215–267.

El-Fakahany, E. E., Alger, B. E., Lai Wi, S., Pitler, T. A., Worley, P. F., and Baraban, J. M. (1988) Neuronal muscarinic responses: role of protein kinase C. FASEB J. 2: 2575–2583.

Gähwiler, B. H., and Brown, D. A. (1987) Muscarine affects calcium-currents in rat hippocampal pyramidal cells in vitro. Neurosci, Lett. 76: 301–306.

Greengard, P. (1976) Possible role for cyclic nucleotides and phosphorylated membrane proteins in postsynaptic actions of neurotransmitters. Nature 260: 101–108.

Gustafsson, B., Galvan, M., Grafe, P., and Wigström, H. (1982) A transient outward current in a mammalian central neurone blocked by 4-aminopyridine. Nature 299: 252–254.

Haas, H. L. (1982) Cholinergic disinhibition in hippocampal slices of the rat. Brain Res. 233: 200–204.

Halliwell, J. V., and Adams, P. R. (1982) Voltage-clamp analysis of muscarinic excitation in hippocampal neurons. Brain Res. 250: 71–92.

Hannun, Y. A., Loomis, C. R., Merrill, A. H., and Bell, R. M. (1986) Sphingosine inhibition of protein kinase C activity and of phorbol dibutyrate binding *in vitro* and in human platelets. J. biol. Chem. 261: 12604–12609.

Hidaka, H., Inagaki, M., Kawamoto, S., and Sasaki, Y. (1984) Isoquinolinesulfonamides, novel and potent inhibitors of cyclic nucleotide dependent protein kinase and protein kinase C. Biochemistry 23: 5036–5041.

Hounsgaard, J. (1978) Presynaptic inhibitory action of acetylcholine in area CA1 of the hippocampus. Exp. Neurol. 62: 787–797.

Kramer, R. H., and Zucker, R. S. (1985) Calcium-induced inactivation of calcium current causes the inter-burst hyperpolarization of *Aplysia* bursting neurones. J. Physiol. 362: 131–160.

Krnjević, K. (1975) Acetylcholine receptors in vertebrate CNS. In: Iversen L. L., and Snyder, S. H. (eds), Handbook of Psychopharmacology. Plenum, New York, pp. 97–125.

Krnjević, K. (1977) Control of neuronal excitability by intracellular divalent cations: A possible target for neurotransmitter actions. In: Fields, W. S. (ed.), Neurotransmitter Function. Basic and Clinical Aspects. Intercontinental Med. Book. Corp., New York, pp. 11–26.

Krnjević, and Leblond, J. (1987) Anoxia reversibly suppresses neuronal Ca-currents in rat hippocampal slices. Can. J. Physiol. Pharmac. 65: 2157–2161.

Krnjević, K., and Phillis, J. W. (1963a) Acetylcholine-sensitive cells in the cerebral cortex. J. Physiol. 166: 296–327.

Krnjević, K., and Phillis, J. W. (1963b) Pharmacological properties of acetylcholine-sensitive cells in the cerebral cortex. J. Physiol. 166: 328–350.

Krnjević, K., Pumain, R., and Renaud, L. (1971) The mechanism of excitation by acetylcholine in the cerebral cortex. J. Physiol. 215: 247–268.

Krnjević, K., Reiffenstein, R. J., and Ropert, N. (1981) Disinhibitory action of acetycholine in the rat's hippocampus: Extracellular observations. Neuroscience 12: 2465–2474.

Kudo, Y., Ogura, A., and Iijima, T. (1988) Stimulation of muscarinic receptors in hippocampal neuron induces characteristic increase in cytosolic free Ca^{2+} concentration. Neurosci. Lett. 85: 345–350.

Lovinger, D. M., Wong K. L., Murakami, K., and Routtenberg, A. (1987) Protein kinase C inhibitors eliminate hippocampal long-term potentiation. Brain Res. 436: 177–183.

Madison, D. V., and Nicoll, R. A. (1986) Cyclic adenosine 3',5'-monophosphate mediates β-receptor actions of noradrenaline in rat hippocampal pyramidal cells. J. Physiol. 372: 245–259.

Madison, D. V., Lancaster, B., and Nicoll, R. A. (1987) Voltage clamp analysis of cholinergic action in the hippocampus. J. Neurosci. 7: 733–741.

Melenka, R. C., Madison, D. V., Andrade, R., and Nicoll, R. A. (1986) Phorbol esters mimic some cholinergic actions in hippocampal pyramidal neurons. J. Neurosci. 6: 475–480.

Masland, R. H., Mills, J. W., and Cassidy, C. (1984) The functions of acetylcholine in the rabbit retina. Proc. R. Soc. Lond. B 223: 121–139.

McCormick, D. A., and Prince, D. A. (1988) Postsynaptic actions of acetylcholine in the mammalian brain *in vitro*. In: Avoli, M., Reader, T. A., Dykes, R. W., and Gloor, P. (eds), Neurotransmitters and Cortical Functions. Plenum, New York, pp. 287–302.

Mochida, S., and Kobayaski, H. (1988) Protein kinase C activators mimic the M_2-muscarinic receptor-mediated effects on the aciton potential in isolated sympathetic neurons of rabbits. Neurosci. Lett. 86: 201–206.

Nakajima, Y., Nakajima, S., Leonard, R. J., and Yamaguchi, K. (1986) Acetylcholine raises excitability by inhibiting the fast transient potassium current in cultured hippocampal neurons. Proc. natl. Acad. Sci. USA 83: 3022–3026.

Nicoll, R. A. (1988) The coupling of neurotransmitter receptors to ion channels in the brain. Science 241: 545–551.

Nishizuka, Y. (1986) Studies and perspectives of protein kinase C. Science 233: 305–312.

Phillis, J. W. (1971) The pharmacology of thalamic and geniculate neurons. Int. Rev. Neurobiol. 14: 1–48.

Rovira, C., Ben-Ari, Y., Cherubini, E., Krnjević, K., and Ropert, N. (1983) Pharmacology of the dendritic action of acetylcholine and further observations on the somatic disinhibition in the rat hippocampus *in situ*. Neuroscience 8: 97–106.

Segal, M. (1983) Rat hippocampal neurons in culture: Responses to electrical and chemical stimuli. J. Neurophysiol. 50: 1249–1264.

Wanke, E., Ferroni, A., Malgaroli, A., Ambrosini, A., Pozzan, T., and Meldolesi, J. (1987) Activation of a muscarinic receptor selectively inhibits a rapidly inactivated Ca^{2+} current in rat sympathetic neurons. Proc. natl Acad. Sci. USA 84: 4313–4317.

Woody, C. D., Swartz, B. E., and Gruen, E. (1978) Effects of acetylcholine and cyclic GMP on input resistance of cortical neurons in awake cats. Brain Res. 158: 373–395.

Yamamoto, C., and Kawai, N. (1967) Presynaptic action of acetylcholine in thin sections from the guinea pig dentate gyrus *in vitro*. Exp. Neurol. 19: 176–187.

Presynaptic cholinergic action in the hippocampus

Menahem Segal, Varda Greenberger and Henry Markram

Center for Neurosciences, The Weizmann Institute of Science, Rehovot 76100, Israel

Summary. The hippocampus is among the regions in the brain richest in M1 cholinergic receptors. Topical application of acetylcholine (ACh) onto hippocampal slices produces a characteristic complex response consisting of a depolarization, an increase in input resistance especially upon depolarization and a blockade of a slow afterhyperpolarization (AHP). The first two of these responses can be recorded also in non-cholinergic septal neurons in an area which contains about 8% of the M1 muscarinic receptors found in the hippocampus. The responses of hippocampal but not septal neurons to ACh involve an increase in the spontaneous synaptic activity and a decrease in evoked responses to afferent stimulation. The dissociated hippocampal culture was used to study these presynaptic effects. The neurons in culture possess muscarinic receptors which develop gradually over a period of several weeks after plating. ACh rarely depolarizes hippocampal neurons in culture. Instead, it causes an increase in spontaneous discharge of small postsynaptic currents (PSC's) and a marked decrease of large, evoked PSC's. In some cultured hippocampal cells ACh reduced I_{Ca} without affecting any of several outward K currents studied. It is suggested that ACh reduces evoked activity by reducing Ca currents at presynaptic terminals.

The extensive utilization of the in vitro preparation in recent years have yielded a description of an array of effects of acetylcholine (ACh) in the brain (Benardo and Prince, 1982a; Benardo and Prince, 1982b; Cole and Nicoll, 1983; Cole and Nicoll, 1984a; Cole and Nicoll, 1984b; Constanti and Sim, 1987; Dodd et al., 1981; Egan and North, 1985; Gähwiler and Brown, 1985; Gähwiler and Dreifuss, 1982; Halliwell and Adams, 1982; Krnjevic and Phillis, 1963; Krnjevic et al., 1981).The main effects were recorded in hippocampal slices and include 1) a blockade of a sustained outward potassium current activated by depolarization (Im) (Halliwell and Adams, 1982); 2) a blockade of Ca-dependent K current underlying the afterhyperpolarization (AHP) which follows a burst discharge (Benardo and Prince, 1982a; Benardo and Prince, 1982b; Cole and Nicoll, 1984b; Madison et al., 1987); 3) a slow depolarization associated with little conductance changes (Segal, 1982) which might be caused by a blockade of a sustained K current activated at rest (Madison et al., 1987); and 4) a blockade of a transient outward K current (Nakajima et al., 1986). In addition to these effects on K currents, there are recent reports on muscarinic blockade of Ca current (Gähwiler and Brown, 1985; Wanke et al., 1987).

The muscarinic receptor in the brain has been subdivided into one M1 with a low affinity for agonist and a high affinity for the antagonist pirenzepine and a second one with a high affinity for the agonist, called M2 (Messer and Hoss, 1987). The two receptor subtypes are heterogenously distributed in the brain with M1 receptor more abundant in the neocortex and hippocampus and M2 receptors present primarily in subcortical structures. The two subtypes have been associated with different physiological actions and second messengers; M1 receptors are associated with the blockade of Im whereas M2 receptors with the blockade of AHP (Müller and Misgeld, 1986). A different physiological association might be present in other brain regions (Constanti and Sim, 1987; Egan and North, 1985).

The differential distribution of M1 and M2 receptors in the brain (Spencer et al., 1986) poses several questions as to the functional significance of these receptors. In the hippocampus M1 receptors are heavily concentrated in stratum radiatum of region CA1, an area with a scarce cholinergic innervation. M2 receptors, on the other hand, are concentrated near the pyramidal-oriens levels where the cholinergic fibers arising from the septal nuclei terminate. If indeed M1 receptors have low affinity for the agonist, what functional significance might those receptors have which can rarely be exposed to the agonist released from remote terminals?

In comparing cases where severe physiological cholinergic deficits are reported (as in aging or in TMT-poisoned rats, Fig. 1), there is no clear correlation between the two; only a marginal receptor loss in the hippocampus is seen in aged rats which sustain marked functional cholinergic deficits (Biegon et al., 1988). This brings up the question of how many receptors can a region lose before a functional loss can be registered.

In comparing M1 responses in cells of different brain regions, a striking observation can be made; a qualitatively similar depolarizing response associated with a decreased conductance can be registered in CA1 neurons and in non-cholinergic neurons of the vertical sector of the diagonal band of Broca (Fig. 2). The responses of the latter neurons to ACh share a similar sensitivity to M1 antagonist pirenzepine. This region contains only about 8.4% of the M1 receptor density seen in CA1 region (Table 1). Can one then assume that 91.6% of CA1 M1 receptors are redundant, or do they serve another function, not detected with standard intracellular methodologies?

There are several indications of heterosynaptic effects of ACh in the hippocampus; a depolarizing response to topical application of ACh or to stimulation of local cholinergic fibers (Segal, 1988b) is normally associated with an increase in synaptic noise indicating an increase in synaptic release (Fig. 1). This effect is not seen in septal neurons (Fig. 2) and in hippocampus treated with drugs which reduce cholinergic receptor density and function (Fig. 1).

Finally, our earlier work (Segal, 1983) indicates that in tissue-cultured hippocampal neurons the predominant effect of ACh is on neurons which are presynaptic to the recorded one.

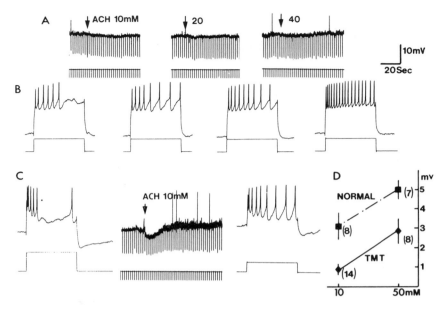

Figure 1. A comparison of responses to topical application of ACh in trimethyltin (TMT) treated (*A* and *B*) and control (*C*) hippocampal neurons in a slice preparation (from Segal, 1988a). *A* Deficient response to ACh, in TMT-treated cell. Application of microdrops containing increasing concentrations of ACh produced little changes in resting potential and input resistance of the cell as estimated by passage of hyperpolarizing current pulses. In response to depolarizing current pulses 0.5 s in duration (*B*) the cell discharges several action potentials and accommodates, i.e. slows down the discharge rate. Following increasing concentrations of ACh the cell discharges more action potentials than seen normally. *C* In normal rat hippocampal slices the application of ACh produces a complex response consisting of an initial hyperpolarization followed by a late depolarization associated with an increase in spontaneous postsynaptic discharges and a marked decrease in accommodation. *D* Magnitude of depolarization evoked by two different concentrations of ACh in normal and TMT-treated cells. The scale in *A* is also valid for the middle of Panel *C*. All depolarizing current pulses are of 0.5-s duration.

We have now extended our work with tissue-cultured hippocampal neurons and studied effects of ACh on interactions among neurons in culture.

Experiments were conducted in 3–6-week-old hippocampal cultures. We found that, like in the brain, the density of muscarinic receptors in hippocampal cultures develops gradually over a period of 2–3 weeks (Fernandez-Tome and Segal, 1987) to reach comparable levels to those of adult hippocampus *in vivo*—about 2000 fmoles/mg protein (Table 1). Topical application of ACh from pressure micropipette or by local perfusion from wide-bore pipette produced little changes in resting membrane properties of the majority of cultured neurons studied. The main consistent effect was an increase in small, spontaneous, postsynaptic current (PSC) discharges. In order to obtain a systematic control of

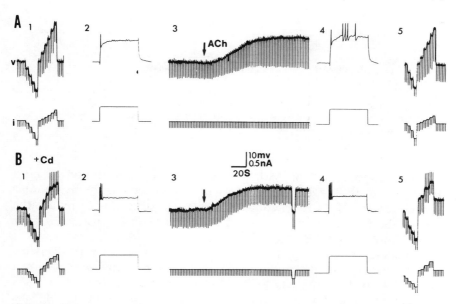

Figure 2. Responses to ACh of a putative non-cholinergic neuron recorded in a slice of the vertical limb of the diagonal band. *A* In the normal condition (1 and 5), a series of 100-ms hyperpolarizing constant current pulses was applied while the membrane was held at different potentials. A marked rectification is seen in the hyperpolarizing region and an apparent resistance increase in the depolarizing region. 1 = before, 5 = after application of ACh. Note the marked increase in resistance especially in the depolarizing region. 2 and 4 = a response to a 0.5-s depolarizing current pulse before (2) and after (4) ACh. While the depolarizing current pulse caused the generation of a single action potential before ACh, it produced several discharges after topical application of ACh. 3 = Topical application of ACh produced a slow 16-mV depolarization associated with a marked increase in input resistance. *B* Same cell after microperfusion of the slice with a 1-mM Cd-containing Krebs solution. Cd reduced considerably the depolarization-induced apparent resistance increase (1) and changed the pattern of response to a depolarizing command (2), did not prevent the depolarizing action of ACh (16 mV, # 3) but eliminated the conductance changes associated with a response to ACh. Note that in both *A* and *B* the marked depolarization and resistance change seen in response to ACh are not accompanied by an increase in spontaneous PSP discharge (as seen in Fig. 1C).

synaptic discharge we used a low Ca:Mg ratio (1:5) to reduce spontaneous and polysynaptic activity. We could then activate presynaptic neurons with short (10 ms) pulses of glutamate, applied via a pressure micropipette (Fig. 3). Using low density plating we selected fields of view of 0.9 mm diameter which contained 5–10 neurons. These were normally interconnected such that pulse activation of one produced a train of PSC's in a follower cell. The PSC's were either excitatory or inhibitory and there was no clear morphological distinction between the two types of presynaptic neurons. The IPSC's had a reversal potential near rest (−50 to −60 mv) and normally had a slower decay time constant (5–20 ms) than the EPSC's which had a more positive reversal potential (about zero mv) and a faster decay time constant (2–3 ms).

92

Table 1

Tissue	Receptor	Assay	Amount (fmoles/mg protein)
Tissue culture 7 days	M	homogenate	1,200
Tissue culture 14 days	M	homogenate	2,030
Adult CA1 brain	M1	autoradiography	1,910
Adult CA1 brain	M2	autoradiography	250
Adult Vertical DB	M1	autoradiography	160
Adult Vertical DB	M2	autoradiography	340

Topical application of ACh caused a marked suppression of both EPSC's and IPSC's (Fig. 3). This effect was seen with micromolar concentrations of ACh. The main effect of the drug was to reduce the size of the PSC's without changing their decay time constant. Higher concentrations of ACh caused synaptic failures. A total blockade of PSC's could be seen with 10–50 μM of ACh. The effect was reversible and full restoration of the PSC's was seen some 5–10 min after removal of ACh. In these same cells ACh did not affect the responses to topical application of glutamate onto the recorded neurons (Fig. 4). This indicates that the reduction in PSC's was not due to a reduced

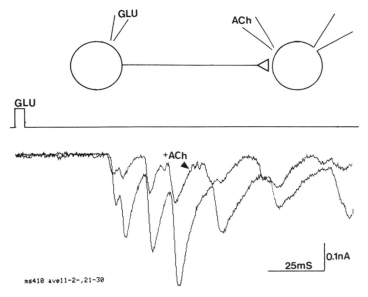

Figure 3. Pressure application of glutamate (10 mM solution) onto a presynaptic neuron (top left) produces excitatory postsynaptic currents (EPSC's) in the follower cell in a hippocampal culture (top right). Bottom: Averages of EPSC's recorded in the follower cell to 10 successive pulses of glutamate applied to a presynaptic cell at a rate of 0.1 Hz before and during (arrowhead) exposure of the cell to ACh. The cell was bathed in a solution containing 1 mM Ca and 5 mM Mg, and clamped at resting potential (-55 mV).

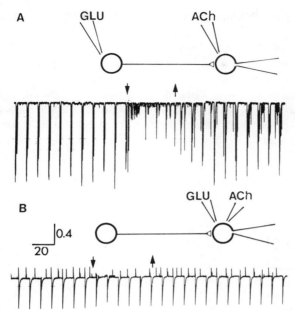

Figure 4. *A* Effects of ACh on successive responses to applications of glutamate to a presynaptic cell. Glutamate pulses are applied at a rate of 0.1 Hz to a presynaptic neuron and the PSC's are registered. Application of ACh near the recorded neuron (downward arrowhead) produced an increase in discharge of small EPSC's and a suppression of the evoked PSCs. Removal of ACh (upward arrowhead) restored the responses. *B* Repetitive applications of glutamate to the recorded cell produced large inward currents. These were not modified by topical application of ACh. Scale: 20 s, 0.4 nA. Cells were voltage clamped at resting potential (-55 mV).

postsynaptic sensitivity to the putative neurotransmitter substance but to a reduction of transmitter release. Occasionally, a neuron that was initially stimulated with glutamate to evoke PSC's in other cells was recorded from while other neurons were excited with glutamate to produce PSC's on it. The same picture emerged; there were no changes in the resting properties of the recorded neuron but a marked decrease in evoked PSC's in the presence of ACh. These experiments indicate that the effects of ACh on PSC's are not due to changes in membrane properties of somata of the pre- or postsynaptic neurons but probably to a change in properties of synaptic terminals.

In trying to analyze the ionic conductance affected by ACh to produce the observed PSC changes we initially searched for a possible effect of ACh on several species of K currents. We could not evoke in these cells consistently the Ca-dependent K current underlying the slow AHP. Likewise, we could not evoke Im in most of the cultured cells examined. The transient outward K current that could be evoked consistently in most of the cells studied was not modified by

ACh in any clear manner, unlike in a previous report (Nakajima et al., 1986).

Calcium currents were studied in the presence of tetrodotoxin, tetraethylammonium, 4-aminopyridine and Cs to block Na and K currents. Cells were recorded with pipettes containing Cs^+ and ATP. Depolarizing commands evoked Ca-dependent inward currents. The currents consisted of a transient and a persistent inward component. In most cells studied both components were present and shared the same voltage dependency (Fig. 5).

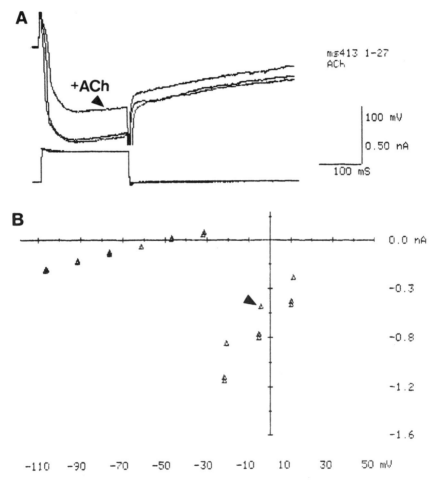

Figure 5. Calcium currents recorded in a cultured hippocampal neuron before, during (arrowhead) and after removal of ACh. *A* A specimen record of the current changes and the voltage commands. *B* A summary curve of the current-voltage relations before, during and after exposure to ACh. The cell was bathed in a medium containing 10 mM TEA, 5 mM Mg, 1 mM Ca, 1 μM TTX and recorded with a pipette containing ATP.

The presence of ACh reduced markedly I_{Ca} in many of the cells studied. Both the transient and the sustained components of the inward current responses were reduced by 30–50% during exposure to ACh. These effects were reversible and were blocked by the addition of atropine to the incubation medium.

It appears that ACh can block at least partially an inward Ca current without really affecting K currents which dominate the behavior of the cell in the depolarizing potential range. The reduction in Ca current can underly the reduction in evoked PSC's seen in these cells.

Assuming that a similar effect of ACh is exerted also *in vivo*, it should be seen in the *in vitro* slice preparation. Indeed, previous reports indicate that ACh suppresses synaptic responses in hippocampal slices (Dodd et al., 1981; Segal, 1982). It is pertinent to examine which of the afferent systems impinging upon the hippocampal pyramidal neurons are modulated by ACh. In our experiments we could not find major changes in muscarinic receptor density after fornix transection which deprives the hippocampus of septal and some brainstem afferents (Greenberger and Segal, unpublished observations). That leaves the perforant path and local association pathways as prime candidates for being regulated by ACh. These possibilities are being currently examined. In a preliminary study we found a 20% reduction in CA1 QNB binding following destruction of region CA3, a major source of innervation of CA1 pyramidal cells. The fact that a major fraction of the muscarinic receptors in the hippocampus of the rat survives after killing most of the hippocampal neurons with kainic acid (Greenberger and Segal, unpublished observations), supports the assertion that a major fraction of the muscarinic receptors do not reside on postsynaptic neurons, but on afferents to them, which are not destroyed by kainic acid.

The possibility that a major part of the muscarinic receptor population is associated with a decrease in evoked transmitter release somehow contradicts the general conception of ACh as being primarily an excitatory neurotransmitter. The question is therefore, is ACh an excitatory transmitter because it depolarizes neurons or is it an inhibitory one because it reduces evoked transmitter release? One can of course argue that it does both, depending on the brain region, and the current physiological status of the tissue involved.

Benardo, L. S., and Prince, D. A. (1982a) Cholinergic excitation of mammalian hippocampal pyramidal cells. Brain Res. 249: 315–331.

Benardo, L. S., and Prince, D. A. (1982b) Ionic mechanisms of cholinergic excitation in mammalian hippocampal pyramidal cells. Brain Res. 249: 333–344.

Biegon, A., Duvdevani, R., Greenberger, V., and Segal, M. (1988) Aging and brain cholinergic muscarinic receptors: An autoradiographic study in the rat. J. Neurochem. 51: 1381–1385.

Cole, A. E., and Nicoll, R. A. (1983) Acetylcholine mediates a slow synaptic potential in hippocampal pyramidal cells. Science 221: 1299–1301.

Cole, A. E., and Nicoll, R. A. (1984a) Characterization of a slow cholinergic postsynaptic potential recorded *in vitro* from rat hippocampal pyramidal cells. J. Physiol. (Lond.) 353: 173–188.

Cole, A. E., and Nicoll, R. A. (1984b) The pharmacology of cholinergic excitatory responses in hippocampal pyramidal cells. Brain Research 305: 285–290.

Constanti, A., and Sim, J. A. (1987) Muscarinic receptors mediating suppression of the M-current in guinea pig olfactory cortex neurons may be of the M2 subtype. Br. J. Pharmac. 90: 3–5.

Dodd, J., Dingledine, R., and Kelly, J. S. (1981) The excitatory action of acetylcholine on hippocampal neurons of the guinea-pig and rat maintained *in vitro*. Brain Res. 207: 109–127.

Egan, T. M., and North, R. A. (1985) Acetylcholine acts on M2-muscarinic receptors to excite rat locus coeruleus neurones. Br. J. Pharmac. 85: 733–735.

Fernandez-Tome, P., and Segal, M. (1987) Ontogenesis of muscarinic receptors in cultured rat hippocampal cells. Dev. Brain Res. 35: 158–160.

Gähwiler, B., and Brown, D. A. (1985) Functional innervation of cultured hippocampus neurones by cholinergic afferents from cocultured septal explants. Nature (Lond.) 313: 577–579.

Gähwiler, B. H., and Dreifuss, J. J. (1982) Multiple actions of acetylcholine on hippocampal pyramidal cells in organotypic explant cultures. Neuroscience 8: 1243–1256.

Halliwell, J. V., and Adams, P. R. (1982) Voltage clamp analysis of muscarinic excitation in hippocampal neurons. Brain Res. 250: 71–92.

Krnjevic, K., and Phillis, J. W. (1963) Acetylcholine sensitive cells in the cerebral cortex. J. Physiol. (Lond.) 166: 296–327.

Krnjevic, K., Reiffenstein, R. J., and Ropert, N. (1981) Disinhibitory action of acetylcholine in the rats' hippocampus: extracellular observations. Neuroscience 6: 2465–2474.

Madison, D. V., Lancaster, B., and Nicoll, R. A. (1987) Voltage clamp analysis of cholinergic action in the hippocampus. J. Neurosci. 7: 733–741.

Messer, W. S. Jr., and Hoss, W. (1987) Selectivity of pirenzepine in the central nervous system. 1. Direct autoradiographic comparison of the regional distribution of pirenzepine and carbamylcholine binding sites. Brain Res. 407: 27–36.

Müller, W., and Misgeld, U. (1986) Slow cholinergic excitation of guinea pig hippocampal neurons is mediated by two muscarinic receptor subtypes. Neurosci. Lett. 67: 107–112.

Nakajima, Y., Nakajima, S., Leonard, R. J., and Yamaguchi, K. (1986) Acetylcholine raises the excitability by inhibiting the fast transient potassium current in cultured hippocampus neurons. Proc. natl Acad. Sci. USA 83: 3022–3026.

Segal, M. (1982) Multiple actions of acetylcholine at a muscarinic receptor studied in the rat hippocampal slice. Brain Res. 246: 77–87.

Segal, M. (1983) Rat hippocampal neurons in culture: responses to electrical and chemical stimuli. J. Neurophysiol. 50: 1249–1264.

Segal, M. (1988a) Behavioral and physiological effects of trimethyltin in the rat hippocampus. Neurotoxicology 9: 481–490.

Segal, M. (1988b) Synaptic activation of a muscarinic receptor in the rat hippocampus. Brain Res. 42: 79–86.

Spencer, D. G., Horvath, E., and Traber, J. (1986) Direct autoradiographic determination of M1 and M2 muscarinic acetylcholine receptor distribution in the rat brain: relation to cholinergic nuclei and projections. Brain Res. 380: 59–68.

Wanke, E., Ferroni, A., Malgaroli, A., Ambrosini, A., Pozzan, T., and Meldolesi, J. (1987) Activation of a muscarinic receptor selectivity inhibits a rapidly inactivating Ca current in rat sympathetic neurons. Proc. natl Acad. Sci. 84: 4313–4317.

Opposing effects of acetylcholine on the two classes of voltage-dependent calcium channels in hippocampal neurons

M. Toselli* and H. D. Lux

*Max-Planck-Institut für Psychiatrie, Abteilung Neurophysiologie, Am Klopferspitz 18A,
D-8033 Planegg, Federal Republic of Germany*

Summary. Acetylcholine (Ach) was tested for its effect on calcium currents in primary cultures of embryonic rat hippocampal neurons. Ach reversibly depressed, in a dose-dependent way, the high voltage activated (HVA) Ca currents. The effect was antagonized by atropine. Our results suggest that a pertussis toxin (PTX)-sensitive GTP binding protein (G-protein) is involved in the signal transduction mechanism between the Ach receptor and the HVA Ca channel. Activating rather than depressive effects of Ach were observed on the low voltage-activated component of Ca currents. This effect was also antagonized by atropine but is not mediated by a PTX-sensitive G-protein.

Introduction

Activation of muscarinic receptors can alter the properties of voltage-dependent calcium (Ca) conductances: in neostriatal neurons Ca-dependent plateau potentials are modulated by muscarine (Misgeld et al., 1986). Wanke et al. (1987) have shown that acetylcholine (Ach) selectively inhibits a Ca current in sympathetic neurons. A depressive action of muscarine on Ca currents was also observed in hippocampal cells (Gähwiler and Brown, 1987). Hippocampal pyramidal cells receive a clearly defined cholinergic input from neurons in the medial septal nucleus (Lewis and Shute, 1967). We investigated the effect of Ach on Ca currents from primary cultures of hippocampal neurons. The use of the whole cell patch-clamp technique with cultured neurons allows a more precise control of membrane voltage and Ca current isolation (Hamill et al., 1981). Two types of neuronal voltage-dependent Ca currents have been identified which differ with respect to their time course of inactivation and pharmacological properties (Llinas and Yarom, 1981; Carbone and Lux, 1987; Fox et al., 1987; Yaari et al., 1987). One type, high voltage-activated (HVA), inactivates slowly and incompletely, is fully depressed by ω-conotoxin (McCleskey et al, 1987) and partially blocked by $10\,\mu M$ verapamil. The second type, low voltage-activated (LVA), activates between -50 and $-30\,mV$, inactivates quickly and in a voltage-dependent manner. It is completely

*Present address: Istituto di Fisiologia Generale, Via Forlanini 6, I-27100 Pavia, Italy.

blocked by 100 μM nickel, but is insensitive to ω-conotoxin and verapamil (10 μM). Here we report on opposite effects of Ach on the two
types of Ca channels and analyze the regulatory mechanisms for this
dual effect. Tentatively, implications of this effect on the control of cell
excitability and transmitter release will be also discussed.

Methods

Hippocampal neurons obtained from 18- to 19-day-old rat embryos
and maintained in culture for 3 to 10 days were used. Whole cell Ca
currents were isolated using pipettes of 3 to 5 MΩ resistance filled with
(in mM): 120 CsCl, 20 TEA-Cl, 10 EGTA, 2 MgCl$_2$, 0.1 GTP,
10 Hepes/CsOH (pH 7.4), 4 Mg-ATP, 5 creatine phosphate, 20 U/ml
creatine phosphokinase. The bath solution contained (in mM):
120 NaCl, 5 CaCl$_2$, 2 MgCl$_2$, 10 glucose, 10 Hepes/NaOH (pH 7.4), 5
4-aminopyridine, 5 TEA-Cl, 0.3 μM TTX.

Results

Dual effect of Ach on the HVA and LVA Ca currents

Recently, we have found that Ach (1–100 μM) reversibly depressed,
in a dose-dependent manner, the HVA Ca current by about 40% (see
Fig. 1A). Atropine can antagonize this effect (Toselli and Lux, 1988).
Activating rather than depressing effects of Ach were observed on the
LVA Ca current: 10 μM Ach reversibly increased the peak of this
inward current by about 30% (fig. 1B). This effect was antagonized by
atropine (200 nM), indicating that muscarinic cholinergic receptors are

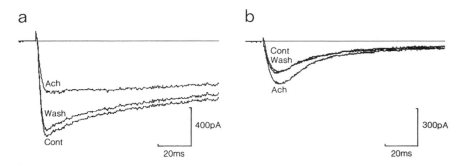

Figure 1. Effects of 10 μM Ach on voltage-dependent Ca currents in hippocampal neurons
before (Cont), during (Ach) and after (Wash) Ach application. *a* HVA Ca currents recorded
at 0 mV. *b* LVA Ca currents recorded at −20 mV. The cell in *b* was pretreated with PTX.
Holding potential in *a* and *b*: −80 mV.

involved in the modulation of the two classes of Ca channels. The inward currents were completely isolated from outward potassium currents by internally perfusing the cells either with CsCl or with the impermeant cation N-methyl-D-glucamine and adding the K channels blockers tetraetylammonium and 4-aminopyridine both internally and externally. Under these conditions, neither the zero current potential nor the maximum of the I/V curve changed during Ach application. In addition, no changes in the input resistance were observed.

Mechanism of action of Ach on the two classes of Ca channels

Guanine nucleotide binding proteins (G-proteins) seem to link the mAch receptor to the HVA Ca channel as suggested by the following observations (Toselli and Lux, 1988): 1) When the neurons were intracellularly perfused with $100 \mu M$ GTPγS, which irreversibly activates G-proteins, Ach irreversibly inhibited the HVA Ca current. 2) The transmitter turned out to be inactive in cells preincubated for 6–18 h with 50 ng/ml of pertussis toxin (PTX), which catalyzes the ADP-ribosylation of different kinds of G-proteins, making them unable to transduce receptor signals.

However, the Ach effect on the LVA Ca current was still present in PTX-preincubated cells, suggesting that a PTX-sensitive G-protein is not involved in the cholinergic responses of this type of Ca channel.

In the brain, muscarinic agonists cause inhibition of adenylate cyclase activity, leading to a decrease of cAMP levels (for a review see Nathanson, 1987). We did not observe any significant effect of intracellularly perfused cAMP on the Ach-mediated depression of the HVA Ca current (Toselli and Lux, 1988).

In the hippocampus, activation of mAch receptors causes a large increase in phosphatidylinositol (PI) turnover (Fisher and Bartus, 1985), leading to the formation of diacetylglycerol, an activator of protein kinase-C (PKC). The phorbol ester 12-tetradecanoylphorbol-13 acetate (TPA) as well as the synthetic diacylglycerol 1-oleoyl-2-acetyl glycerol (OAG) are potent activators of the enzyme PKC (Nishizuka, 1984; Ganong et al., 1986). Application of $5 \mu M$ OAG reversibly reduced both the LVA and HVA Ca currents (Fig. 2). Similar results were obtained with TPA ($10 \mu M$). We have shown that Ach has the opposite effect on the LVA Ca current. In contrast, both Ach and PKC activators have a similar inhibitory effect on the HVA Ca currents. Using a perfusion system which enables complete exchange of solutions around the cell in about 50 ms (see Konnerth et al., 1988), we measured the onset of the Ca current depression during drug delivery. Half maximal effect occurred in about 200 ms with either OAG or TPA. The action of Ach was much slower, being half maximal in about 1.5 s. The

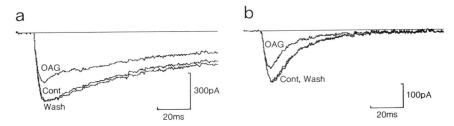

Figure 2. Effects of $5\,\mu M$ OAG on voltage-dependent Ca currents before (Cont), during (OAG) and after (Wash) OAG application. *a* HVA Ca currents recorded at 0 mV. *b* LVA Ca currents recorded from the same cell as in *a* at -30 mV. Holding potential in *a* and *b*: -80 mV.

fast kinetics of OAG and TPA suggests that these drugs may be acting on the outside of the membrane, rather than through an enzymatic pathway on the inner membrane. Furthermore, when the cells were internally perfused with 50 μM H-7, an inhibitor of PKC (Hidaka et al., 1984), no significant effect was observed on the OAG responses.

Discussion

We have shown that activation of mAch receptors has opposite effects on the two classes of Ca channels. The conditions used in this study for recording Ca currents ensure that what we observed was a genuine effect of Ach on Ca currents and not secondary to any activation of K-conductances. Although our results do not rule out the involvement of PKC in the cholinergic modulation of voltage-dependent Ca currents, they indicate that caution should be used when interpreting the effects of PKC activators in intact cell preparations because of a possible direct action of these drugs on the outside of the membrane. In addition to diacylglycerol, another product of the mAch receptor-induced PI breakdown is inositol triphosphate (IP_3), which causes an increase in $[Ca^{2+}]_i$ by redistribution from intracellular stores (Pozzan et al., 1986). It would seem that the Ach-induced decrease in the HVA Ca currents is not a consequence of an increase in intracellular Ca released from internal stores, since in the interior of the cell Ca was buffered with the Ca chelator EGTA (10 mM). In the hippocampus the cholinergic modulation of the HVA Ca currents is also independent of the cAMP cascade since cAMP, when present in the patch pipette, did not significantly alter the Ach response. On the other hand, a PTX-sensitive G-protein is clearly involved in these responses. Different transmitters have been shown to modulate Ca currents in other cell types with a similar mechanism (for a review see Dunlap et al., 1987). Furthermore, two recent reports show that G-proteins can directly modify the activity of voltage-dependent channels (Yatani et al., 1987a, b). We propose, as a working hypothesis, that in hippocampal neurons a G-protein directly couples mAch receptors to the HVA Ca channel.

We have shown that the cholinergic activation of the LVA Ca currents is not mediated by PTX-sensitive G-proteins. Although receptor-generated signals mediated by PTX-insensitive G-proteins have been reported (Evans et al., 1985; Isom et al., 1987; Nakajima et al., 1988), further experiments are necessary to clarify the mechanism of action of Ach on this class of Ca channels.

Ca-dependent control of cell excitability

In hippocampal pyramidal cells a Ca-activated K conductance generating a slow afterhyperpolarization (AHP) is responsible for spike frequency adaptation (Madison and Nicoll, 1984). The slow AHP is blocked by Ach and phorbol esters can mimick this effect, suggesting that a PKC could be involved in this cholinergic action (Malenka et al., 1986). But the muscarinic analog oxotremorine, which only weakly activates PI turnover (Fisher and Bartus, 1985), is as potent as carbachol in blocking the slow AHP (Dutar and Nicoll, 1988). We have shown that the site of action of the PKC activators OAG and TPA in depressing Ca currents is most likely on the exterior of the membrane and independent of PKC activation. We propose that modulation of neuronal excitability may be triggered by a cholinergic-induced depression of the HVA Ca channel which, in turn, controls the Ca-dependent slow AHP. Our proposal does not exclude a parallel, direct action of Ach also on the slow AHP.

Ca-dependent control of transmitter release

Transmitter release is triggered by a rise in intracellular Ca. Activation of muscarinic receptor M_2 causes inhibition of synaptic transmission (Weiler et al., 1984), while activation of muscarinic receptor M_1 facilitates transmitter release (Kilbinger and Nafziger, 1985). In some preliminary experiments we observed that pirenzepine (200 nM) antagonizes the cholinergic activation of the LVA Ca current. On the contrary, the muscarinic inhibition of the HVA Ca currents was not suppressed by the same antagonist in separate recordings. We tentatively propose that Ca-influx into the neurons and, consequently, transmitter release could be increased or decreased through a dual action of Ach on the LVA or on the HVA Ca channels respectively, depending on the voltage state of the cell membrane, the two classes of Ca channels being selectively activated at different voltage ranges.

Carbone, E., and Lux, H. D. (1987) Kinetics and selectivity of a low voltage activated calcium current in chick and rat sensory neurones. J. Physiol. 386: 547–570.

Dunlap, K., Holz, G. G., and Rane, S. G. (1987) G proteins as regulators of ion channel function. Trends Neurosci. 10: 241–244.

Dutar, P., and Nicoll, R. A. (1988) Stimulation of phosphatidyl inositol turnover may mediate the muscarinic suppression of the M current in hippocampal pyramidal cells. Neurosci. Lett. 85: 89–94.

Evans, T., Martin, M., Hughes, A. R., and Harden, T. K. (1985) Guanine nucleotide sensitive, high affinity binding of carbachol to muscarinic cholinergic receptors of 1321N1 astrocytoma cells is insensitive to pertussis toxin. Molec. Pharmac. 27: 32–37.

Fisher, S. K., and Bartus, R. T. (1985) Regional differences in the coupling of muscarinic receptors to inositol phospholipid hydrolysis in guinea pig brain. J. Neurochem. 45: 1085–1095.

Fox, A. P., Nowycky, M. C., and Tsien, R. W. (1987) Kinetic and pharmacological properties distinguishing three types of calcium currents in chick sensory neurons. J. Physiol. 394: 149–172.

Gähwiler, B. H., and Brown, D. A. (1987) Muscarine affects calcium currents in rat hippocampal pyramidal cells in vitro. Neurosci. Lett. 76: 301–306.

Ganong, B. R., Loomis, C. R., Hannun, Ya., and Bell, R. M. (1986) Specificity and mechanism of protein kinase C activation by sn-1,2 diacylglycerols. Proc. natl Acad. Sci. 83: 1184–1188.

Hamill, O. P., Marty, A., Neher, E., Sakmann, B., and Sigworth, F. J. (1981) Improved patch clamp techniques for high-resolution current recording from cells and cell-free membrane patches. Pflügers Arch. 391: 85–100.

Hidaka, H., Inagaki, M., Kawamoto, S., and Sasaki, Y. (1984) Isoquinolinesulfonamides, novel and potent inhibitors of cyclic nucleotide dependent protein kinase and protein kinase C. Biochemistry 23: 5036–5041.

Isom, L. L., Cragoe, E. J., and Limbird, L. E. (1987) Multiple receptors linked to inhibition of adenilate cyclase accelerate Na^+/H^+ exchange in neuroblastoma × glioma cells via a mechanism other than decreased cAMP accumulation. J. biol. Chem. 262: 17504–17509.

Kilbinger, H., and Nafziger, M. (1985) Two types of neuronal muscarine receptors modulating acetylcholine release from guinea pig myenteric plexus. Naunyn-Schmiedeberg Arch. Pharmak. 328: 304–309.

Konnerth, A., Lux, H. D., and Morad, M. (1988) Proton induced transformation of calcium channel in chick dorsal root ganglion cells. J. Physiol. 386: 603–633.

Lewis, P. R., and Shute, C. C. D. (1967) The cholinergic limbic system: projections to hippocampal formation, medial cortex, nuclei of the ascending cholinergic reticular system, and the subformical organ and supra-optic crest. Brain 90: 521–540.

Llinas, R., and Yarom, Y. (1981) Properties and distribution of ionic conductances generating electroresponsiveness of mammalian inferior olivary neurons in vitro. J. Physiol. 315: 569–584.

Malenka, R. C., Madison, D. V., Andrade, R., and Nicoll, R. A. (1986) Phorbol esters mimick some cholinergic actions in hippocampal pyramidal neurons. J. Neurosci. 6: 475–480.

McCleskey, E. W., Fox, A. P., Feldman, D. H., Cruz, L. J., Olivera, B. M., Tsien, R. W., and Yoshikami, D. (1987) ω-Conotoxin: direct and persistent blockade of specific types of calcium channels in neurons but not muscle. Proc natl Acad. Sci. 84: 4327–4331.

Madison, D. V., and Nicoll, R. A. (1984) Control of the repetitive discharge of rat CA1 pyramidal neurons in vitro. J. Physiol. 354: 319–331.

Misgeld, U., Galabresi, P., and Dodt, U. (1986) Muscarinic modulation of calcium-dependent plateau potentials in rat neostriatal neurons. Pflügers Arch. 407: 482–487.

Nakajima, Y., Nakajima, S., and Inoue, M. (1988) Pertussis toxin insensitive G protein mediates substance P-induced inhibition of potassium channels in brain neurons. Proc. natl Acad. Sci. 85: 3643–3647.

Nathanson, N. M. (1987) Molecular properties of the muscarinic acetylcholine receptor. A. Rev. Neurosci. 10: 195–236.

Nishizuka, Y. (1984) The role of protein kinase C in cell surface signal transduction and tumor promotion. Nature 308: 693–698.

Pozzan, T., DiVirgilio, F., Vincentini, L. M., and Meldolesi, J. (1986) Activation of muscarinic receptors in PC12 cells. Biochem. J. 234: 547–553.

Toselli, M., and Lux, H. D. (1989) GTP binding proteins mediate acetylcholine inhibition of voltage-dependent calcium channels in hippocampal neurons. Pflügers Arch. 413: 319–321.

Wanke, E., Ferroni, A., Malgaroli, A., Ambrosini, A., Pozzan, T., and Meldolesi, J. (1987) Activation of a muscarinic receptor selectively inhibits a rapidly inactivated Ca current in rat sympathetic neurons. Proc. natl Acad. Sci. USA 84: 4313–4317.

Weiler, M. H., Misgeld, U., and Cheong, D. K. (1984) Presynaptic muscarinic modulation of nicotinic excitation in the rat neostriatum. Brain Res. 296: 111–120.

Yaari, Y., Hamon, B., and Lux, H. D. (1987) Development of two types of calcium channels in cultured mammalian hippocampal neurons. Science 235: 680–682.

Yatani, A., Codina, J., Brown, A. M., and Birnbaumer, L. (1987) Direct activation of mammalian atrial muscarinic potassium channels by GTP regulatory protein G_k. Science 235: 207–211.

Yatani, A., Codina, J., Imoto, Y., Reeves, J. P., Birnbaumer, L., and Brown, A. M. (1987) A G protein directly regulates mammalian cardiac calcium channels. Science 238: 1288–1292.

Muscarinic slow EPSPs in neostriatal and hippocampal neurons *in vitro*

U. Misgeld

Max-Planck-Institut für Psychiatrie, Abteilung Neurophysiologie, Am Klopferspitz 18A, D-8033 Planegg-Martinsried, Federal Republic of Germany

Summary. Cholinergic slow excitatory postsynaptic potentials (slow EPSPs) can be elicited by presynaptic tetanic stimulation in brain slices obtained from rat neostriatum or guinea pig hippocampus. Slow EPSPs are generated by a reduction of a K-leak-conductance. In hippocampal neurons slow EPSPs are amplified by the reduction of an outward current termed I_{AHP} through the activation of a second muscarinic receptor subtype. While hippocampal slow EPSPs might be involved in information processing across hippocampal pathways, muscarinic modulation in the neostriatum consists of a presynaptic tuning of nicotinic fast synaptic transmission.

The muscarinic slow EPSP as a marker for synaptic release of ACh

Muscarinic slow EPSPs can be elicited by repetitive presynaptic stimulation in neurons of sliced brain areas in which other cholinergic markers such as choline acetyltransferase and acetylcholinesterase (AChE) activity, choline uptake and receptor binding have indicated a dense cholinergic innervation. The neostriatal (Misgeld et al., 1980), the interpeduncular (Sastry, 1980) and the hippocampal slice (Cole and Nicoll, 1983) may serve as examples. Slow EPSPs of central neurons have the same characteristic features as the muscarinic slow EPSPs of autonomic ganglion cells (Kuffler, 1980). In addition to the slow time course (s instead of ms for fast EPSPs), these characteristic features are: A decrease in size if the cell membrane is hyperpolarized, an accompanying increase in input resistance, enhancement by AChEs and blockage by muscarinic antagonists such as atropine and pirenzepine. Thus, the muscarinic slow EPSP indicates that cholinergic terminals have been activated from the site chosen for stimulation. This, in turn, allows the study of the effects of synaptically released ACh as opposed to the study of application of exogenous ACh or agonists.

In cells in which a slow EPSP can be elicited by repetitive presynaptic stimulation, single shock stimulation elicits a fast EPSP (Fig. 1). The size of the fast EPSP allows it to reach spike threshold, while in most instances this is not so for the slow EPSP. Several possibilities exist for the association of fast synaptic events and slow EPSPs at neuro-neuronal synapses (Fig. 1). The simplest possibility is diagrammed in Figure 1A. Single stimulation of the cholinergic fiber elicits a fast EPSP, while

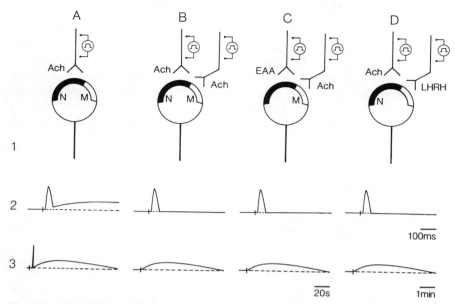

Figure 1. Schematic presentation of various possibilities for the association of fast and slow EPSPs. *A* Stimulation of the ACh-releasing fiber induces a fast and a slow EPSP. *B* Stimulation of one cholinergic fiber induces a fast EPSP, stimulation of the other cholinergic fiber induces a slow EPSP. *C* Stimulation of a fiber releasing an excitatory amino acid (EAA) induces a fast EPSP, while stimulation of a cholinergic fiber induces a slow EPSP. *D* Stimulation of a peptidergic (e.g. LHRH) fiber induces a slow EPSP as actually observed in some autonomic ganglion cells. Please note the difference in time calibration for the slow EPSP.

repetitive stimulation of the same fiber additionally elicits a slow EPSP. Alternatively, the fast EPSP and the slow EPSP may result from the activation of separate fibers (Fig. 1B). It has been suggested that activation of Renshaw cells by antidromic stimulation of the cholinergic motoraxons elicits a nicotinic fast response, while activation of descending supraspinal, albeit unidentified cholinergic fibers results in a muscarinic slow response (Ryall and Haas, 1975). Hippocampal neurons which receive a dense cholinergic innervation are equipped with muscarinic receptors, but results as to the function of nicotinic receptors are ambiguous (Segal, 1978; Krnjevic et al., 1981; Cole and Nicoll, 1984; Aracava et al., 1987). Here, cholinergic fibers may be associated with fibers conveying fast information by the release of an excitatory amino acid (Fig. 1C). Finally, slow EPSPs are generated by the release of peptides, however these slow EPSPs last longer than muscarinic slow EPSPs and, hence, are termed ultraslow EPSPs (Fig. 1D).

The present review focuses on findings obtained in neostriatal and hippocampal slices of the rodent brain and compares the properties of the slow EPSP evoked by repetitive stimulation and effects of ACh release following single stimuli or occurring spontaneously. In terms of the organization of their cholinergic inputs, the neostriatum and hippocampus share some common properties, but also differ considerably. As far as the differences are concerned, the major source of cholinergic synapses in the neostriatum comes from intrinsic neurons, while in the hippocampus cholinergic synapses originate predominantly from septohippocampal afferents. In both areas ultrastructural features of the synapses are similar (Frotscher, this volume) and presumed cholinergic synaptic profiles form 'classical' point-to-point synapses.

Properties of slow EPSPs in neostriatal and hippocampal slices

With intracellular recording techniques in slices from rat neostriatum or guinea pig hippocampus a slow EPSP can be observed which follows repetitive stimulation in the presence of anticholinesterases, e.g. physostigmine ($0.1-10 \mu M$). The slow EPSP is blocked by atropine ($0.1-1 \mu M$) or pirenzepine ($0.1-1 \mu M$), hence it is cholinergic in origin. Detailed descriptions of slow EPSPs in neostriatal and hippocampal neurons have been given elsewhere (Dodt and Misgeld, 1986; Müller and Misgeld, 1986; Misgeld et al., 1989). In this article, slow EPSPs of neostriatal and hippocampal neurons will be described separately before differences are addressed.

To elicit a slow EPSP in neostriatal neurons rather robust stimulation at frequencies of 20–100 Hz (Fig. 2A) is required, which can be applied anywhere in the neostriatal tissue in a distance of up to 1 mm from the recording site. Most effective are several trains of stimuli of 20 Hz which can evoke a slow EPSP lasting up to 2 min. Each individual stimulus in the train is followed by a fast EPSP, the first triggering one or two action potentials due to the high stimulus intensity necessary for evoking the slow EPSP. With normal acetyl-cholinesterase activity in the tissue (Misgeld et al., 1980) no slow EPSP is seen, and even in the presence of physostigmine at the resting membrane potential (-70 mV), only an increase in apparent membrane resistance and sometimes a small depolarization can be seen (Fig. 2A, 1). The depolarization occurring at -60 to -50 mV membrane potentials is rarely of sufficient amplitude to generate spiking (Fig. 2A, 2). At membrane potentials negative to the K-equilibrium potential, the slow EPSP reverses in sign to a hyperpolarizing potential. The reversal potential depends on the K-concentration in the superfusate. Further, the slow EPSP is blocked by low concentrations of the K-channel blocker Ba ($100 \mu M$). Taken together, these findings indicate that the slow EPSP is generated by a reduction of K-leak-conductance (Dodt and Misgeld, 1986).

In hippocampal neurons, relatively short (20–50 Hz, 0.2–0.5 s) stimulus trains, if applied near the cell layers, elicit a slow depolarization

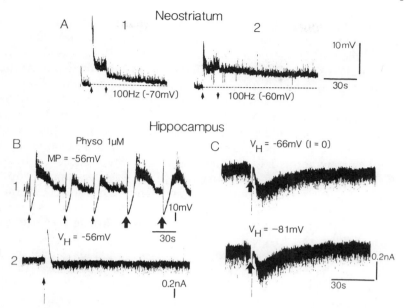

Figure 2. Slow EPSPs of neostriatal and hippocampal cells. *A* Neostriatal cell: About 1000 stimuli (between arrows) induce a small depolarization at the resting membrane potential of −70 mV (1), while the same stimulation elicits a long-lasting, yet subthreshold slow EPSP during depolarization to −60 mV by constant current injection (2), physostigmine (1 μM) present. *B* and *C* CA3 neuron: 1) Short stimulus trains (arrow; 20 Hz, 0.5 s) applied near the CA3 cell layer in the presence of physostigmine (1 μM) at a membrane potential of −56 mV (MP) elicited, following a hyperpolarization, a depolarization triggering a spike series. Repeating the stimulation elicited a much smaller slow EPSP. Only after doubling the stimulus intensity was a slow EPSP of the original size evoked. 2) Voltage clamp at the same holding potential (V_H) reveals the underlying small inward current following the outward current. Spikes truncated by the frequency characteristics of the recorder. *C* Reduction of the outward current by Cs in the impaling electrode reveals a small inward current even at holding potentials where I_M does not operate. Physostigmine not present. After a longer recording period with the Cs-filled electrode the inward current was blocked as well (not shown).

which is associated with an enhanced excitability (Cole and Nicoll, 1983; Müller and Misgeld, 1986; Misgeld et al., 1989). The enhancement of excitability is reinforced by a reduction of cell discharge accommodation. The associated increase in apparent input resistance is replaced by a conductance increase during periods of enhanced firing. The amplitude of the slow EPSP increases with membrane depolarization and decreases with membrane hyperpolarization. Only rarely can the slow EPSP be reversed in sign at membrane potentials negative to the K-equilibrium potential.

Weak stimulation may elicit a reduction of the afterhyperpolarization following a train of action potentials, while strong stimulation elicits a membrane depolarization in addition. Pirenzepine which has a higher affinity for M1 than for M2 receptor subtypes (Hammer et al., 1980) blocks, in the same cell, the membrane depolarization following the train stimulation in the presence of physostigmine, but not the reduction of the afterhyperpolarization. Atropine blocks both effects (Müller and Misgeld, 1986). Both effects, i.e. membrane depolarization and reduction of the afterhyperpolarization, can be induced by the application of exogenous ACh or carbachol (Müller et al., 1988). Voltage clamp analysis (Misgeld et al., 1989) reveals that a small inward current generated by a decrease in K-leak-conductance is amplified by inward rectifier currents reinforced by a reduction of an outward current termed I_{AHP}. The inward current induced by focally applied ACh is pirenzepine-sensitive and fades if ACh is applied in the presence of very low CCh concentrations. Thus the inward current is mediated by a desensitizing muscarinic receptor subtype, while the reduction of the I_{AHP} is mediated by a non-desensitizing receptor subtype (cf. Müller and Misgeld, this volume).

The coactivation of two receptor subtypes explains some of the peculiarities which slow EPSPs of hippocampal neurons have. In current clamp recording slow EPSPs rise rapidly when reaching a critical potential range (Fig. 2B), although voltage clamp recording at the same potential reveals only a small inward current. This is due to the fact that, in the presence of physostigmine, hippocampal neurons develop a region of negative slope conductance between -50 and -40 mV because the accumulated ACh blocks outward conductances (Misgeld et al., 1989) and, hence, a Ca-dependent inward current dominates in the voltage range of -50 to -40 mV. In the absence of physostigmine, stimulation elicits a small inward current dampened by an outward current. The outward current is partly due to the activation of Ca-dependent K-conductances and partly to the opening of K-channels by neurotransmitters. However, if this outward current is reduced by Cs in the recording pipette, a small slow inward current can be seen transiently even at holding potentials of -80 mV (Fig. 2C). Desensitization of the muscarinic receptor subtype mediating the slow inward current is responsible for fading of the slow EPSP if the train stimulation is repeated at time intervals of less than 3 min (Fig. 2B).

Taken together, there are differences in the stimulation conditions under which slow EPSPs can be elicited in neostriatal and hippocampal neurons. These are given by the organization of the cholinergic input. More importantly, two receptor subtypes cooperate in the generation of slow EPSPs in hippocampal neurons, which does not seem to be the case in neostriatal neurons. The result for hippocampal neurons is an amplification of the small inward current generated in both neostriatal and hippocampal neurons by a reduction of their K-leak-conductance.

Possible functions for slow EPSPs

In autonomic ganglia, the function of synaptically released ACh can simply be viewed as that of a messenger sustaining impulse traffic across the cholinergic synapse through nicotinic receptors. The information transfer may be tuned pre- and postsynaptically by muscarinic receptors, but also by other modulators (cf. Akasu, this volume). For neurons in the CNS it is generally assumed that muscarinic receptors predominate, suggesting that information transfer is not the primary issue for ACh released from central synapses (Krnjevic, 1974). Antimuscarinic, but not antinicotinic drugs exert powerful effects on behavioral, electrophysiological and clinical paradigms. Various suggestions have been made as to the function of muscarinic receptors in the brain. Three hypotheses shall be compared to the properties of slow EPSPs in hippocampal and neostriatal slices: i) muscarinic modulation of so-called braking currents (Brown, 1983), ii) slow information transfer through muscarinic receptors, and iii) presynaptic modulation of fast information transfer.

Activation of muscarinic receptors depresses a variety of K-currents: I_{AHP} (Benardo and Prince, 1982; Cole and Nicoll, 1983; Müller and Misgeld, 1986), I_M (Halliwell and Adams, 1982) and I_A (Nakajima et al., 1986), which all contribute to spike repolarization and discharge accommodation. Accommodation of discharge in hippocampal neurons is inhibited by muscarinic agonists (Fig. 3C). This effect is seen in all principal cells of the hippocampus after stimulation near the cell layer or after bath application of low agonist concentrations, even if no membrane depolarization occurs.

With prolonged application of agonists, this effect does not desensitize. On the contrary it is progressive and leads to burst discharges of the neuron under muscarinic receptor activation (Müller et al., 1988). Blocking of AChE in the tissue by physostigmine (Misgeld et al., 1989) and thereby increasing the lifetime of spontaneously released ACh in the tissue has the same effect. The effect is not pirenzepine ($1 \ \mu M$)-sensitive, but blocked by atropine ($1 \ \mu M$). Atropine, applied in the absence of muscarinic agonisits to displace endogenous ACh from muscarinic receptor sites, has no effect on discharge properties of hippocampal neurons, indicating that modulation of discharge in hippocampal neurons by endogenous ACh is prevented by AChE in the tissue.

Obviously, muscarinic depression of 'braking' currents alters input-output relations of hippocampal neurons. The tonic, modulatory effect could facilitate excitatory input coming from other synapses onto the same neuron. However, this concept does not seem applicable to neo-striatal neurons. In control neostriatal neurons discharge with little accommodation (Fig. 3A), which is not changed by high concentrations of muscarinic agonists (Fig. 3B), antagonists or AChE.

110

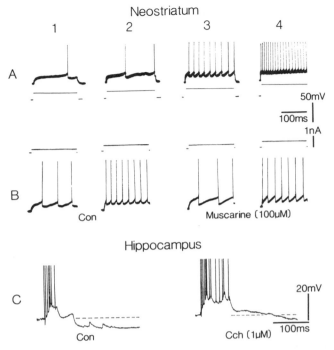

Neostriatum

Figure 3. Discharge patterns of neostriatal (*A*, *B*) and a hippocampal CA3 (*C*) neuron. *A*) A neostriatal cell exhibits little accommodation when stimulated by current injection of increasing strength. *B*) Same discharge pattern of another cell. The discharge pattern is retained by this cell in the presence of a high muscarine concentration. Only an increase in input resistance is evident. *C* Accommodation of discharge in a CA3 neuron stimulated by a 120-ms current pulse of 0.5 nA is removed by a low concentration of carbachol (Cch).

The proposed concept is largely based on the fact that voltage-dependent I_M does not depolarize a neuron at its resting membrane potential (Brown, 1983). Muscarinic receptor activation of hippocampal neurons, however, either by electrical tetanic stimulation or by exogenous agonists reduces a voltage-independent K-leak-conductance through activation of a pirenzepine-sensitive receptor. The resulting depolarization takes the neuron to a voltage range where inward currents dominate, reinforced by the reduction of the outwardly directed I_{AHP} through the activation of a pirenzepine-insensitive muscarinic receptor site. Hence, the neuron is driven to discharge (Fig. 2B). Because the cascade of events can be prevented by blockade of the pirenzepine-sensitive receptor subtype, desensitization of this receptor can ensure that discharges generated by muscarinic activation are time locked to an activation of cholinergic synapses. From this point of view, muscarinic processes, indeed, might be involved in information processing across hippocampal cholinergic pathways (Misgeld et al., 1989). Again, this suggestion does not apply to the neostriatal cholinergic system because there the slow EPSP rarely induces firing.

In addition to the depression of K-conductances, activation of muscarinic receptors reduces Ca-dependent plateau-potentials in neostriatal neurons (Misgeld et al., 1986) and reduces Ca-currents in hippocampal neurons (Gähwiler and Brown, 1987; Toselli and Lux, this volume). Reduction of Ca-currents, if occurring at presynaptic terminals, could reduce transmitter release which might be responsible for the reduction of the amplitudes of fast EPSPs in hippocampal (Hounsgaard, 1978; Valentino and Dingledine, 1981; Segal, 1982) as well as neostriatal slices (Misgeld et al., 1982; Dodt and Misgeld, 1986). In slices from both brain areas a reduction of stimulated ACh release has been observed as well (Hadházy and Szerb, 1977; Weiler et al., 1984; Weiler, this volume). However, there is no evidence that the fast EPSPs, which were found to be reduced by muscarinic agonists in hippocampal slices, are mediated by ACh. The suggested presynaptic action of ACh on non-cholinergic synapses may be a pharmacological by-product resulting from exogenous ACh application, because neither physostigmine nor atropine themselves seem to affect fast EPSPs (Valentino and Dingledine, 1981).

In contrast in neostriatal slices, physostigmine reduces and atropine enhances the fast EPSP which is evoked by intrastriatal stimulation (Fig. 4; Misgeld et al., 1982; Dodt and Misgeld, 1986). Nicotinic antagonists reduce the amplitude of the fast EPSP, as does nicotine itself in high concentrations (0.5–1 mM) (Misgeld et al., 1982) presumably by desensitizing nicotinic receptors. In addition, fast excitation elicited by intra-striatal stimulation resists deafferentation of the neostriatum (Misgeld et al., 1979). This suggests that, in the neostriatum, muscarinic receptors modulate nicotinic information transfer by intrinsic cholinergic neurons through a negative feedback control of ACh release (Weiler et al., 1984). Whether this mechanism applies also

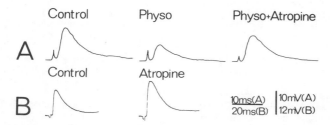

Figure 4. Muscarinic modulation by endogenous ACh of the fast EPSP evoked in neostriatal neurons by intrastriatal stimulation. *A* Application of physostigmine (1 μM) reduced the amplitude of a fast EPSP and the effect was antagonized by 10 μM atropine. *B*) Atropine (10 μM) itself enhanced the amplitude of a fast EPSP elicited in a slice not treated with muscarinic agonists or ACh-E inhibitors. Responses averaged from 8 traces each.

112

to some cholinergic synapses in the hippocampus, is an open question as long as the answer to a possible synaptic function for nicotinic receptors is not definite.

Acknowledgements. The author thanks A. Lewen and E. Schroeder for assistance in preparation of the manuscript and Figures. The study was supported by the Deutsche Forschungsgemeinschaft (SFB 220, C-4).

Aracava, Y., Deshpande, S. S., Swanson, K. L., Rapoport, H., Wonnacott, S., Lunt, G., and Albuquerque, E. X. (1987) Nicotinic acetylcholine receptors in cultured neurons from the hippocampus and brain stem of the rat characterized by single channel recording. Fedn Eur. Biochem. Soc. 222: 63–70.

Benardo, L. S., and Prince, D. A. (1982) Cholinergic excitation of mammalian hippocampal pyramidal cells. Brain Res. 249: 315–331.

Brown, D. A. (1983) Slow cholinergic excitation—a mechanism for increasing neuronal excitability. Trends neurol. Sci. 8: 302–307.

Cole, A. E., and Nicoll, R. A. (1983) Acetylcholine mediates a slow synaptic potential in hippocampal pyramidal cells. Science 221: 1299–1301.

Cole, A. E., and Nicoll, R. A. (1984) Characterization of a slow cholinergic post-synaptic potential recorded in vitro from rat hippocampal cells. J. Physiol. (Lond.) 352: 173–188.

Dodt, H. U., and Misgeld, U. (1986). Muscarinic slow excitation and muscarinic inhibition of synaptic transmission in the rat neostriatum. J. Physiol. (Lond.) 380: 593–608.

Gähwiler, B. H., and Brown, D. A. (1987). Muscarine affects calcium-currents in rat hippocampal pyramidal cells in vitro. Neurosci. Lett. 76: 301–306.

Hádhazy, P., and Szerb, J. C. (1977). The effect of cholinergic drugs on [^{3}H]-acetylcholine release from slices of rat hippocampus, striatum and cortex. Brain Res. 123: 311–322.

Halliwell, J. V., and Adams, P. R. (1982). Voltage-clamp analysis of muscarinic excitation in hippocampal neurons. Brain Res. 250: 71–92.

Hammer, R., Berrie, C. P., Birdsall, N. J. M., Burgen, A. S. V., and Hulme, E. C. (1980). Pirenzepine distinguishes between different subclasses of muscarinic receptors. Nature 283: 90–92.

Hounsgaard, J. (1978) Presynaptic inhibitory action of acetylcholine in area CA1 of the hippocampus. Exp. Neurol. 62: 787–797.

Krnjevic, K. (1974) Chemical nature of synaptic transmission in vertebrates. Physiol. Rev. 54: 418–540.

Krnjevic, K., Reiffenstein, R. J., and Ropert, N. (1981) Disinhibitory action of acetylcholine in the rat's hippocampus: Extracellular obervations. Neuroscience 6: 2465–2474.

Kuffler, S. W. (1980) Slow synaptic responses in autonomic ganglia and the pursuit of a peptidergic transmitter. J. exp. Biol. 89: 257–286.

Misgeld, U., Calabresi, P., and Dodt, H. U. (1986) Muscarinic modulation of calcium dependent plateau potentials in the rat neostriatal neurons. Pflügers Arch. 407: 482–487.

Misgeld, U., Müller, W., and Polder, H. R. (1989) Potentiation and suppression by eserine of muscarinic synaptic transmission in the guinea-pig hippocampal slice. J. Physiol. (Lond.) 409: 191–206.

Misgeld, U., Okada, Y., and Hassler, R. (1979) Locally evoked potentials in slices of rat neostriatum: A tool for the investigation of intrinsic excitatory processes. Exp. Brain Res. 34: 575–590.

Misgeld, U., Weiler, M. H., and Bak, I. J. (1980) Intrinsic cholinergic excitation in the rat neostriatum: Nicotinic and muscarinic receptors. Exp. Brain Res. 39: 401–409.

Misgeld, U., Weiler., M. H., and Cheong, D. K. (1982) Atropine enhances nicotinic cholinergic EPSPs in rat neostriatal slices. Brain Res. 253: 317–320.

Müller, W., and Misgeld, U. (1986) Slow cholinergic excitation of guinea-pig hippocampal neurons is mediated by two muscarinic receptor subtypes. Neurosci. Lett. 67: 107–112.

Müller, W., Misgeld, U., and Heinemann, U. (1988) Carbachol effects on hippocampal neurons in vitro: Dependence on the rate of rise of carbachol tissue concentration. Exp. Brain Res. 72: 287–298.

Nakajima, Y., Nakajima, S. , Leonard, R. J., and Yamaguchi, K. (1986) Acetylcholine raises excitability by inhibiting the fast transient potassium current in cultured hippocampal neurons. Proc. natl Acad. Sci. USA 83: 3022–3026.

Ryall, R. W., and Haas, H. L. (1975) On the physiological significance of muscarinic receptors on Renshaw cells: A hypothesis. In: Naser, P. G. (ed.), Cholinergic mechanisms. Raven Press, New York, pp. 335–341.

Sastry, B. R. (1980) Excitatory postsynaptic potentials in the mammalian central nervous system associated with an increase in the membrane resistance. Life Sci. 27: 1403–1407.

Segal, M. (1978) The acetylcholine receptor in the rat hippocampus; nicotinic, muscarinic or both? Neuropharmacology 17: 619–623.

Segal, M. (1982) Multiple actions of acetylcholine at a muscarinic receptor studied in the rat hippocampal slice. Brain Res. 246: 77–87.

Valentino, R. J., and Dingledine, R. (1981) Presynaptic inhibitory effect of acetylcholine in the hippocampus. J. Neurosci. 1: 784–792.

Weiler, M. H., Misgeld, U., and Cheong, D. K. (1984) Presynaptic muscarinic modulation of nicotinic excitation in the rat neostriatum. Brain Res. 296: 111–120.

Carbachol and pirenzepine discriminate effects mediated by two muscarinic receptor subtypes on hippocampal neurons *in vitro*

W. Müller and U. Misgeld

Max-Planck-Institut für Psychiatrie, Abteilung Neurophysiologie, Am Klopferspitz 18A, D-8033 Planegg-Martinsried, Federal Republic of Germany

Summary. Measurement of [Cch] in the bath and in slices demonstrated a considerable concentration discrepancy between the bath and the extracellular space. With fast (bolus) application of Cch this discrepancy is due to the speed of diffusion, while equilibration with continuous application is considerably impaired by cellular uptake of Cch (Creese and Taylor, 1967). Low concentrations ($\leq 1\ \mu$M) of Cch reduce the afterhyperpolarization following a train of action potentials and depolarize the membrane. Analysis of $[Cch]_0$ (t) and the effects of pirenzepine allowed these effects to be assigned to two different muscarinic receptor subtypes.

Introduction

Activation of muscarinic receptors on hippocampal neurons in slices reduces at least three separate potassium conductances:

1) a voltage-independent potassium conductance (Madison et al., 1987; Benson et al., 1988) to induce a membrane depolarization (Müller and Misgeld, 1986);

2) a calcium-dependent potassium conductance to suppress discharge accommodation and the afterhyperpolarization (AHP) following a train of action potentials (Benardo and Prince, 1982a; Benardo and Prince, 1982b; Cole and Nicoll, 1983; Cole and Nicoll, 1984; Müller and Misgeld, 1986);

3) a potassium conductance active only at membrane potentials positive to -60 mV to cause a membrane depolarization at membrane potentials positive to -60 mV (I_M, Halliwell and Adams, 1982).

The suggestion that there are several muscarinic receptor subtypes instead of only one, which is based on a considerable overall binding heterogeneity of muscarinic agents (Birdsall et al., 1980; Hammer et al., 1980), can explain this divergence of effects. Supporting the assumption of muscarinic receptor heterogeneity, muscarinic receptors couple to a variety of intracellular messengers, e.g. inositol(-tris)phosphate (IP_3), diacylglycerol, protein kinase C, cyclic GMP and cyclic AMP. Finally, different subtypes have indeed been cloned (Kubo et al., 1986).

Theory of discrimination of muscarinic receptors by carbachol and pirenzepine

Heterogeneity of binding exists not only for the antagonist pirenzepine but also for agonists. Receptors arbitrarily termed M1 bind pirenzepine with high affinity ($K_i = 20$ nM) and carbachol (Cch) with low affinity ($K_D = 1.7 \, \mu$M, Fig. 1) while receptors termed M2 bind pirenzepine with low affinity ($K_i = 0.2$–1 μM) and Cch with high affinity ($K_D = 77$ nM, Fig. 1, Birdsall et al., 1980; Birdsall and Hulme, 1983; Hammer et al., 1980). Due to the difference in Cch binding constants, a $[Cch]_0$ of 10–100 nM should selectively evoke M2-mediated effects (Fig. 1, solid lines) while a higher $[Cch]_0$ should activate M1 and M2 receptors. In this case the assumption was made that the response is proportional to the number of activated receptors. However, because coupling of receptors to channels is not always direct and may involve second messengers, activation of a fraction of receptors may be sufficient to evoke a maximal response ('receptor reserve'). Such a receptor reserve would shift effective agonist concentrations to lower values and

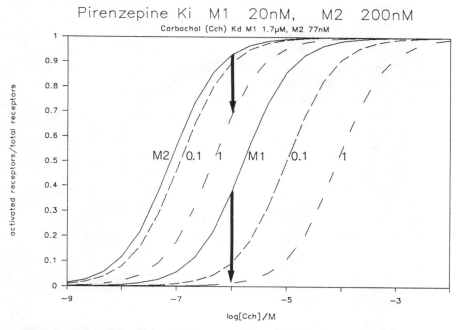

Figure 1. Theoretical binding of Cch to two muscarinic binding sites with a K_D of 77 nM (M2) and a K_D of 1.7 μM (M1), values taken from Birdsall et al. (1980). Cch binding is shifted to higher Cch concentrations by 0.1 or 1 μM pirenzepine (dashed lines 0.1 and 1, $K_i = 20$ nM for M1, $K_i = 200$ nM for M2, values taken from Hammer et al., 1980). In equilibrium with 1 μM Cch, pirenzepine 1 μM reduces Cch binding to M2 receptors by 27% but reduces Cch binding to M1 receptors by 97% (arrows).

would cause a steeper concentration-response relation than expected from the Langmuir adsorption isotherm (Fig. 1). Assuming a comparable receptor reserve for M1 and M2 receptors, Cch concentrations selective in evoking M2-mediated effects would be shifted simply to lower concentrations. If there is, however, a receptor reserve for the M1 receptor only, there would be no $[Cch]_0$ which can induce selectively M2-mediated effects. Actually, there could be a reverse situation, i.e. M1 receptors could be activated by lower concentrations of Cch than those necessary for M2 receptor activation, but this would require a more than tenfold receptor reserve for M1 receptors and no receptor reserve for M2 receptors. On the other hand, desensitization of a receptor shifts the ED_{50} to higher concentrations, depending on the rate of desensitization and the rate of increase of $[Cch]_0$ with time. Thus, if only the M1 receptor desensitizes, the $[Cch]_0$ range differentiating between M1 and M2 receptors will be larger than expected, as actually observed for 'slow' Cch application (see below).

Variation of pirenzepine binding is likely to be only tenfold for M1 and M2 receptors (Hammer et al., 1980). Nevertheless, since the affinity profiles of pirenzepine and Cch are opposite (Birdsall and Hulme, 1983; Hammer et al., 1980), pirenzepine (0.1–1 μM) is presumably quite effective in selectively blocking M1-mediated effects for a considerable range of Cch concentrations (0.3 < $[Cch]_0$ < 10 μM, Fig. 1). On the other hand, at a $[Cch]_0$ of 10 nM, pirenzepine (1 μM) will block 82% of Cch binding to M2 receptors (Fig. 1). Therefore, for the discrimination of muscarinic receptor subtypes by electrophysiological tests in slices, the actual Cch concentration in the tissue has to be known.

The variation of ligand receptor binding over some three decades of drug concentration (Fig. 1), the rather small difference between pirenzepine binding constants, a possibly nonlinear receptor-effector coupling (e.g. receptor reserve, desensitization) and a considerable uncertainty about extracellular drug concentration in slices (see below) make pharmacological determination of the muscarinic receptor subtype mediating a single muscarinic effect rather difficult. The discrimination of two simultaneously involved receptors is easier when a comparison of the two effects is possible in one neuron (Müller and Misgeld, 1986).

Determination of extracellular Cch concentration in brain slices after bath application of Cch

In most slice studies, extracellular drug concentration is deduced from the bath concentration. However, extracellular drug concentration will equal bath concentration only in an equilibrium state, i.e. after prolonged superfusion of the slice with a fixed drug concentration, the required time of superfusion depending on experimental factors chosen (e.g. the use of 'submerged' or 'interface' slices, cf. Müller et al., 1989). If the drug is, however, metabolized in the tissue by enzymatic degradation or taken up into the intracellular compartment, equilibrium may never be reached. A good example is demonstrated by acetylcholine (Ach) and Cch. The stable Ach analogue Cch, which is not degraded by Ach-esterase, is effective in inducing muscarinic effects at considerably lower bath concentrations than Ach. Nevertheless, effective $[Cch]_{bath}$ is still high in comparison to binding constants. Passive and carrier-mediated uptake of Cch (Creese and Taylor, 1967) can slow its equilibration. Therefore, in order to know the variation of extracellular Cch concentration over time ($[Cch]_0(t)$) and the time needed to reach drug equilibrium between bath and slice, we measured $[Cch]_0(t)$ and simultaneously $[Cch]_{bath}(t)$ with ion-sensitive microelectrodes during bath application of Cch (for methods see Müller et al., 1988). [Cch] was measured at a depth of 150 μm in the granule cell layer of the dentate gyrus of 300-μm thick guinea pig hippocampal slices, and simultaneously some 100 μm beneath the slice in the bath.

With continuous application of Cch, a fast increase of $[Cch]_{bath}$ was observed after some minutes latency (Fig. 2, left line) in our interface chamber. In the tissue, $[Cch]_0$ increased slowly at an approximately constant rate until, after some 60 min, an apparent plateau of $[Cch]_0$ was obtained. With bolus application of Cch to the perfusion line we obtained a fast increase of $[Cch]_{bath}$. In the tissue, increase of $[Cch]_0$ was delayed by about 1 min and only a peak $[Cch]_0$ of some 30% of bath [Cch] peak (cf. Müller et al., 1988) was reached. Variation of peak $[Cch]_0$ within the slice was less than 30% between a depth of 100 and 200 μm. Bolus application resembles bath application of drugs in submerged slices if insufficient time is allowed for equilibration.

The significance of factors impairing the equilibration of Cch between the bath and the extracellular space of the slice was evaluated by numeric simulation (cf. Müller et al., 1988). We assumed free diffusion of Cch in an extracellular space with a tortuosity $\lambda = 1.6$ and a 50-μm boundary layer of non-flowing perfusate beneath the slice (cf. Bingmann and Kolde, 1982). The measured variation of $[Cch]_{bath}$ over time ($[Cch]_{bath}(t)$) was taken as a boundary condition. Time course and peak of $[Cch]_0$ after bolus application were reproduced within narrow limits by the simulation. However, with continuous application of Cch (300 μM), the theoretical increase of $[Cch]_0(t)$ calculated under the specifications detailed above (Fig. 2, squares) was much faster than the measured one. This must be due to a Cch decrease in the tissue. Such a decrease could be a 1:1 passive uptake of Cch, which would slow the increase of $[Cch]_0$ in the extracellular space (Fig. 2, dots). Creese and Taylor (1967) reported that Cch is taken up by rat cortical slices over a time course of 7 h up to a 13-fold concentration by a combined passive and carrier-mediated uptake (half saturation constant = 41 μM), which

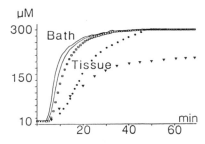

Figure 2. Theoretical time course of extracellular [Cch] in the tissue during continuous application of Cch (300 μM). Measured [Cch]$_{bath}$ (t) taken as a boundary condition. [Cch] below 10 μM were not calculated, since [Cch] below 10 μM did not generate a measurable signal in the ion-sensitive microelectrodes. Left line is measured [Cch]$_{bath}$ (t). Measured [Cch]$_0$ increased with time at an almost constant rate to reach a plateau at some 60 min. Without cellular uptake theoretical [Cch]$_0$ would increase almost as fast as [Cch]$_{bath}$ (squares). With the assumption of a 1:1 uptake into cells, theoretical [Cch]$_0$ (t) approximates the measured [Cch]$_0$ (t). With the inclusion of a passive and carrier-mediated uptake, as described by Creese and Taylor (1967), equilibrium is not obtained within one hour of continuous application of Cch (triangles).

is relatively fast for a low [Cch], e.g. 1 μM. With the inclusion of such an uptake, calculated [Cch]$_0$ reached only two thirds of [Cch]$_{bath}$ of 300 μM after continuous application for one hour (Fig. 2, triangles). Since measurement of [Cch]$_0$ over one hour with an ion-sensitive microelectrode can deviate by 30% (e.g. due to a 0.5 mV electrode drift), our measurements do not prove that the equilibrium was indeed reached. A realistic approximation of [Cch]$_0$ after one hour of continuous application of Cch (300 μM) might therefore be a value somewhere in between equilibrium of bath and tissue and the value calculated according to the Cch-uptake measurements of Creese and Taylor (triangles in Fig. 2).

Equilibration of a low [Cch] (1 μM) would be even more impaired since the carrier-mediated uptake is relatively fast for a low [Cch] (Creese and Taylor, 1967). In our simulation continuous application of 1 μM [Cch] for one hour resulted in a theoretical [Cch]$_0$ of only one third of the applied [Cch] (not shown). Thus, uptake of Cch might prevent activation of both M1 and M2 receptors, depending on the mode of Cch application chosen.

Discrimination of M1- and M2-mediated effects

To discriminate receptors by Cch application, a correct concentration-response relation is needed, i.e. Cch must be applied until equilibrium is reached, or [Cch]$_0$ has to be measured. Since it takes a long time to reach Cch equilibrium (see above), establishment of a correct concentration-response relation would be rather tedious. With a known

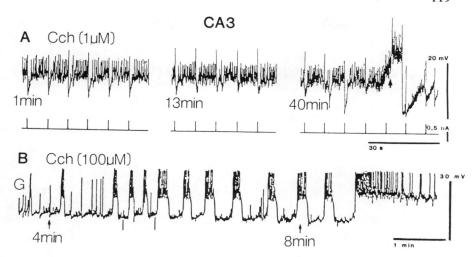

Figure 3. Effects of continuous application of Cch. *A* The afterhyperpolarization (AHP) following a train of action potentials (direct stimulation: 120 ms, 0.5 nA) decreased continuously with time of continuous application of Cch (1 μM). After 40 min paroxysmal depolarizations were observed (arrow). Membrane potential, which was −60 mV, did not change during Cch application. *B* During continuous application of Cch (100 μM), paroxysmal depolarizations were observed after a few minutes and neurons depolarized in a step-like manner after ca. 9 min. Neurons were not further depolarized during the further increase of [Cch]$_0$. Membrane potential: −60 mV.

time course of [Cch]$_0$, a complete concentration-response relation can be obtained during continuous application of Cch. During an increase of [Cch]$_0$ from 10 nM to 1 μM (cf. Fig. 1) by continuous application of Cch (1 μM), the M2-mediated effect should commence after some minutes and increase for an hour. Indeed, the reduction of the AHP followed this time course (Fig. 3A), indicating that this effect is mediated 1:1 by M2 receptors. Accordingly, during continuous application of 100 μM Cch, an M1-mediated effect should commence after a few minutes and increase for 20 min. As expected continuous application of 100 μM Cch rapidly blocked the AHP by activation of the Cch high affinity M2 receptors, but the neurons were depolarized after ca. 9 min in a step-like manner with no tendency toward further depolarization (Fig. 3B). The latter result indicates a more complex concentration-response relation for the depolarization than for the reduction of the AHP. The complexity of the concentration-response relation could result from M1 receptor desensitization (Misgeld et al., 1989).

Therefore, to study M1-mediated effects we applied Cch as a bolus. Thus we obtained a transient depolarization of some mV at extracellular Cch concentrations expected to activate M1 receptors (Fig. 5A). Reduction of the AHP preceded the depolarization due to the higher affinity of Cch to M2 receptors. However, because of a short-term desensitization, neurons repolarized before peak of [Cch]$_0$ was reached. In summary,

120

Cch discriminates M1 and M2 receptors by higher affinity for M2 receptors and desensitization of M1 receptors.

To reassure that the depolarization is mediated by M1 receptors and the reduction of the AHP by M2 receptors, we applied pirenzepine. Of course, pirenzepine may block all muscarinic effects, depending on the relation of $[Cch]_0$ to $[pirenzepine]_0$. For a $[Cch]_0$ of about 1 μM,

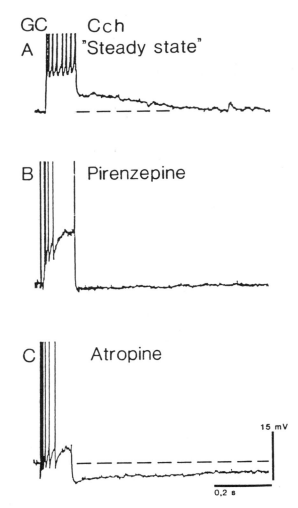

Figure 4. Effects of muscarinic antagonists in the presence of a high [Cch] (continuous application of Cch 300 μM for more than one hour). *A* In a steady state all neurons had low membrane potentials. At a holding potential of −60 mV, direct stimulation elicited a train of action potentials followed by an afterdepolarization. *B* Bolus application of pirenzepine (expected peak 6 μM) hyperpolarized neurons and blocked the afterdepolarization, but did not recover an AHP. *C* After wash out of pirenzepine, bolus application of atropine (6 μM) hyperpolarized neurons and recovered an AHP.

pirenzepine (0.1–1 μM) blocks Cch binding to M1 receptors almost completely but has almost no effect on Cch binding to M2 receptors (Fig. 1). Correspondingly, after continuous application of Cch (1 μM) for more than one hour, bolus application of pirenzepine with an expected peak concentration of 0.6 μM had no significant effect on the blockage of the AHP by Cch (Müller et al., 1988). Neither pirenzepine nor atropine (0.6 μM) had an effect on the membrane potential. Atropine (0.6 μM) restored an AHP (Müller et al., 1988). After depolarization of neurons by continuous application of Cch (100 μM) for more than one hour, both pirenzepine and atropine (6 μM) hyperpolarized neurons. The effect of pirenzepine concurs with the binding data for the M1 receptor, but may also be explained with a pirenzepine-sensitive, Cch-low affinity binding site ($K_D = 125$ μM, $K_i = 1$ μM). In high [Cch]$_0$ neurons not only lacked an AHP but also exhibited an afterdepolarization. Because of the high [Cch]$_0$ not even 10 μM pirenzepine displaces Cch from the M2 receptor. Therefore, the blockade of this afterdepolarization by pirenzepine 6 μM (Fig. 4B) is not likely to be due to a partial blockade of the M2 receptor. Atropine (6 μM) is expected to block the M2 receptor in the presence of a high [Cch]$_0$. Accordingly, atropine recovered an AHP (Fig. 4C).

Pirenzepine (1 μM) applied for ca. 20 min completely blocked depolarizations induced by Cch bolus applications (peak [Cch]$_0$ 0.3–0.6 μM, Fig. 5), in agreement with Cch binding to M1 receptors (Fig. 1). To avoid desensitization by repeated Cch application Cch was applied at intervals of more than 20 min. In control applications at this interval

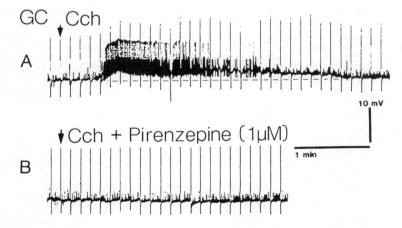

Figure 5. Blockade of the depolarization by binding of pirenzepine to M1 receptors. *A* Bolus application of Cch (arrow, expected peak concentration 0.3 μM at 2.5 min) induces a membrane depolarization with a peak already after some 40 s. *B* After continuous application of pirenzepine (1 μM) for 20 min, the Cch effect on the membrane potential is completely blocked. Reduction of the AHP was not significantly changed in comparison to control. Membrane potential: -60 mV.

Cch induced a second depolarization without fading. The AHP was almost completely blocked by Cch (peak $[Cch]_0$ 0.6 μM) in control as well as in the presence of pirenzepine (1 μM for 20 min).

Because of the desensitization of M1 receptors, it is easier to determine the relative insensitivity of M2 receptors to pirenzepine than the relative sensitivity of M1 receptors. Agonist and pirenzepine discrimination, when $[Cch]_0(t)$ is known, is therefore an appropriate way of discriminating these receptor subtypes in slices by electrophysiological techniques.

Acknowledgements. The authors thank A. Lewen and B. Muffler for excellent technical assistance, and E. Schroeder for editorial help. The gift of pirenzepine by Thomae GmbH, Biberach (FRG) is gratefully acknowledged. This study was supported by the Deutsche Forschungsgemeinschaft (SFB 220, C-4).

Benardo, L. S., and Prince, D. A. (1982a) Cholinergic excitation of mammalian hippocampal pyramidal cells. Brain Res. 249: 315–331.

Benardo, L. S., and Prince, D. A. (1982b) Cholinergic pharmacology of mammalian hippocampal pyramidal cells. Neuroscience 7: 1703–1712.

Benson, D. M., Blitzer, R. D., and Landau, E. M. (1988) An analysis of the depolarization produced in guinea-pig hippocampus by cholinergic receptor stimulation. J. Physiol. Lond. 404: 479–496.

Bingmann, D., and Kolde, G. (1982) PO_2-profiles in hippocampal slices of the guinea pig. Exp. Brain Res. 48: 89–96.

Birdsall, N. J. M., Hulme, E. C., and Burgen, A. (1980) The character of the muscarinic receptors in different regions of the rat brain. Proc. R. Soc. Lond. B 207: 1–12.

Birdsall, N. J. M., and Hulme, E. C. (1983) Muscarinic receptor subclasses. TIPS 4: 459–463.

Cole, A. E., and Nicoll, R. A. (1983) Acetylcholine mediates a slow synaptic potential in hippocampal pyramidal cells. Science 221: 1299–1301.

Cole, A. E., and Nicoll, R. A. (1984) Characterization of a slow cholinergic post-synaptic potential recorded in vitro from rat hippocampal pyramidal cells. J. Physiol. Lond. 352: 173–188.

Creese, R., and Taylor, D. B. (1967) Entry of labeled carbachol in brain slices of the rat and the action of d-tubocurarine and strychnine. J. Pharmac. exp. Ther. 157: 406–419.

Halliwell, J. V., and Adams, P. R. (1982) Voltage-clamp analysis of muscarinic excitation in hippocampal neurons. Brain Res. 250: 71–92.

Hammer, R., Berrie, C. P., Birdsall, N. J. M., Burgen, A. S. V., and Hulme, E. C. (1980) Pirenzepine distinguishes between different subclasses of muscarinic receptors. Nature 283: 90–92.

Kubo, T., Fududa, K., Mikami, A., Maeda, A., Takahashi, H., Mishina, M., Haga, T., Haga, K., Ichiyama, A., Kangawa, K., Kojima, M., Matsuo, H., Hirose, T., and Numa, S. (1986) Cloning, sequencing and expression of complementary DNA encoding the muscarinic acetylcholine receptor. Nature 323: 411–416.

Madison, D. V., Lancaster, B., and Nicoll, R. A. (1987) Voltage clamp analysis of cholinergic action in the hippocampus. J. Neurosci. 7(3): 733–741.

Misgeld, U., Müller, W., and Polder, H. R. (1989) Potentiation and suppression of muscarinic synaptic transmission in the guinea-pig hippocampal slice. J. Physiol. 409: 191–206.

Müller, W., and Misgeld, U. (1986) Slow cholinergic excitation of guinea pig hippocampal neurons is mediated by two muscarinic receptor subtypes. Neurosci. Lett. 67: 107–112.

Müller, W., Misgeld, U., and Heinemann, U. (1988) Carbachol effects on hippocampal neurons in vitro: dependence on the rate of rise of carbachol tissue concentration. Exp. Brain Res. 72: 287–298.

Müller, W., Misgeld, U., and Heinemann, U. (1989) Kinetics of carbachol effects in guinea pig hippocampal slices. In: Kessler, M., Höper, J., and Harrison, K. (eds), Theory and application of ion-selective electrodes in physiology and medicine. Springer, Berlin Heidelberg, in press.

Cholinergic activation of medial pontine reticular formation neurons *in vitro*

R. W. Greene, H. L. Haas, U. Gerber and R. W. McCarley

Harvard Medical School and VAMC, Brockton, MA, USA,
and Johannes Gutenberg-Universitat, Mainz, Federal Republic of Germany

Summary. In vivo microinjections of cholinergic compounds into the medial pontine reticular formation have produced some or depending on the injection site, all of the phenomena of REM, thus providing the only adequate pharmacological model of this behavioral state. The necessary anatomical substrate, a cholinergic projection to the mPRF was recently demonstrated, however the direct effect of cholinergic agonists on mPRF neurons is unknown. We have examined the effects of carbachol on mPRF neurons recorded *in vitro* from brainstem slices of Sprague-Dawley rats (8–10 days old). Three kinds of response to the application of carbachol (0.5–1 μM) were observed (n = 15) as follows: a depolarizing response (67%), a hyperpolarizing response (20%) and a biphasic response consisting of a hyperpolarizing response followed by a depolarizing response (13%). Under voltage clamp control, the depolarizing response was observed as an inward current resulting from a decrease in conductance which was constant over the membrane potential range of -100 to -50 mV. Reversal potential was negative to -80 mV. An increase in the excitability of neurons (as measured by responses to identical intracellularly applied depolarizing current pulses) during the depolarizing responses was due to the increase in steady state inward current. When intracellular DC current of equal amplitude but opposite polarity was applied, no increase in excitability was observed. This response was always blocked by the addition of atropine (0.5–1 μM) to the perfusate.

The hyperpolarizing response was observed as an increase in outward current due to an increase in conductance with marked voltage sensitivity (over the range of -100 to -50 mV) characteristic of the anomalous rectifier. Preliminary data indicated that the hyperpolarizing response was more sensitive to pirenzepine (complete blockade at 1.0 μM) than the depolarizing response (complete blockade at 2 μM) but neither response was affected by pirenzepine concentrations of 200 nM or less.

Cholinergic effects on evoked depolarizing PSPs were examined on neurons with depolarizing (n = 3) and biphasic (n = 1) responses and in all cases, the PSPs were enhanced. This enhancement was blocked by atropine.

In conclusion, it is suggested that activation of two different muscarinic receptors (neither of which is the M_1 receptor) on mPRF neurons results in two different responses, a decrease in a voltage-insensitive potassium conductance and an increase in the anomalous rectifier.

Introduction

The only phenomenologically adequate model of REM sleep is produced by microinjection of cholinergic compounds (both agonists and cholinesterase inhibitors) into site-specific locations within the medial pontine reticular formation (mPRF; Amatruda et al., 1975; Baghdoyan et al., 1985). This procedure can elicit a complete array of REM phenomena including muscle atonia (as in natural REM, the lower motor neurons are hyperpolarized), EEG desynchronization, PGO waves, rapid eye movements and hippocampal theta rhythm.

A number of important questions as to the physiological relevance of the cholinergic model of REM sleep remain. First, what is the cholinergic input to the mPRF? There are no cholinergic cells within the mPRF, however, in the dorsal pontine tegmentum are two groups of cholinergic neurons defined as Ch5 and Ch6 by Mesulam and coworkers (Mesulam et al., 1983) in the pedunculopontine (PPT) and lateral dorsal tegmental nucleus (LDT). Recently, coworkers in our lab (Mitani et al., 1988) have investigated the projections from these two nuclei to the mPRF. Employing double-labelling with WGA-HRP and choline acetyltransferase (ChAT) monoclonal antibodies, they identified 10.2% of ChAT positive cells in ipsilateral LDT and 3.7% in contralateral LDT as having projections to the mPRF. In PPT, 5.2% of the ChAT positive ipsilateral neurons and 1.3% of the contralateral demonstrated projections to the mPRF. These percentages are similar (slightly less for PPT) to those observed for double-labelled ChAT positive neurons with WGA-HRP injections in the thalamus of comparable size (personal communication with Dr Hallanger computed from data of her previously published study; Hallanger et al., 1987). Thus, the presence of a cholinergic projection to the mPRF was confirmed.

Although the appropriate anatomy is prerequisite for the validation of the cholinergic model of REM sleep, the mimicry of REM events at the cellular level by cholinergic agonists must also be established. *In vivo*, intracellular recordings from mPRF neurons of unanesthetized cats have been made in conjunction with monitoring of the behavioral state of the animal (Ito and McCarley, 1984). On passage from slow wave sleep to REM sleep, the membrane potential depolarized 7–10 mV, the excitability of the neuron increased and post-synaptic potentials were enhanced. All these phenomena reversed with the behavioral state transition back to waking. The presence of similar cholinergic actions on mPRF neurons at the cellular level are unknown. Extracellular microiontophoretic investigations have shown a cholinergically elicited increase in neuronal firing (Greene and Carpenter, 1985) but the mechanism of this action was beyond the technical scope of the study. In the work presented here, the actions of cholinergic agonists and antagonists on the membrane properties of mPRF neurons are examined in an *in vitro* slice preparation of the rat pontine brainstem.

Methods

Brainstem slices from Sprague-Dawley rats (8–10 days old) were perfused with a modified Ringers solution containing 124 mM NaCl, 26 mM $NaHCO_3$, 1.25 mM KH_2PO_4, 2 mM KCl, 2.4 mM $CaCl_2$ and

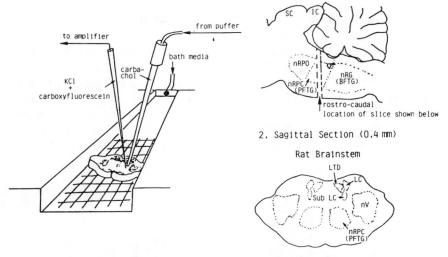

1. Schematic of Recording Setup

3. Coronal Section (interaural -0.3 mm)

Figure 1. Schematic of the recording chamber and the area to which recordings were restricted. *1* The recording chamber was a shallow (<1 mm) channel, the width of the slice. Bath media was gravity fed to the closed end and dripped off the open end. The slice rested on nylon netting and was anchored in place from above with wide mesh netting (not shown). The puffer pipette was submerged but above the surface of the slice. *2, 3* The location of the recording site was within the nucleus reticularis pontis caudalis. BFTG, bulbar frontal tegmental field; IC, inferior colliculus; LC, locus coeruleus; subLC, sublocus coeruleus; LTD, laterodorsal tegmental nucleus; nRG, nucleus reticularis gigantocellularis; nRPC, nucleus reticularis pontis caudalis; nRPO, nucleus reticularis pontis oralis; PFTG pontine frontal tegmental field; SC, superior colliculus.

1.3 mM $MgCl_2$, and maintained at 30°C. Anatomical localization of recording and stimulating electrodes to the mPRF was confirmed by marker lesions followed by Nissl staining of the slice (see Fig. 1). Drugs were bath applied except for 5 experiments in which carbachol was applied by puffer, employing glass pipettes with tip diameters of 10–50 μM placed submerged in media but above the surface of the slice. Pressurized nitrogen (1–6 p.s.i.) controlled by a solenoid valve was used to eject the drug. The only difference in response between these techniques was in a shorter time to onset of carbachol action with the puffer, a difference consonant with closer proximity to the neuron. Current clamp recordings were made as previously described with glass electrodes filled with 2M KCl (Greene et al., 1986). Voltage clamp recordings were obtained with a sample and hold circuit (electrode resistance 50–70 MΩ). Only neurons which maintained robust and stable electrophysiological properties throughout the wash-in (2–5 min) and wash-out (10-min) periods were used in this study.

126

Results and discussion

We exposed 21 neurons in the nucleus reticularis pontis caudalis to carbachol at concentrations of 0.5–1.0 μM in the bath or 1–10 mM in puffer electrodes. Thirteen neurons responded to carbachol with a depolarization of 16 ± 7 mV SD associated with an increase of input resistance of $21 \pm 18\%$ SD. In over half the neurons, the depolarization was of sufficient amplitude to reach action potential threshold (Fig. 2A, B). Carbachol evoked a hyperpolarization associated with a decrease in input resistance in 4 neurons (Fig. 5A), and a biphasic response (a

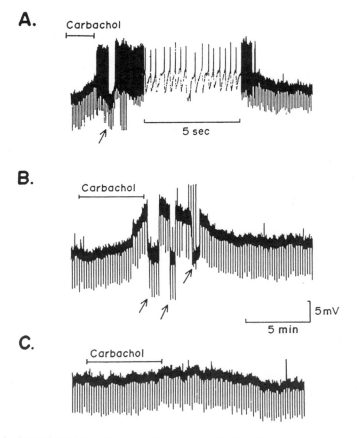

Figure 2. Carbachol elicits a depolarization associated with a decrease in conductance that is blocked by atropine. *A, B* Chart records of a typical depolarizing response of an mPRF neuron to bath application of carbachol (0.5 μM during time indicated). Downward deflections were due to intracellular current pulses (400 ms, 200 pA) applied to assess input resistance. At arrows, membrane potential was returned to the baseline potential by D.C. hyperpolarizing current to avoid voltage-sensitive changes of the membrane resistance not specific to carbachol. *C* Atropine (0.5 μM) blocks the depolarizing response to carbachol (same neuron as in *B*).

shorter duration hyperpolarization followed by depolarization) in 3 neurons. Only one neuron did not respond to carbachol.

These were direct, non-synaptically mediated effects, as indicated by their presence with bath application of tetrodotoxin (1 μM) which prevents sodium action potential-dependent synaptic activity (n = 12). These carbachol effects were seen on each of the main types of mPRF neuron, the low threshold burst (LTB) neurons and the non-burst (NB) neurons (Greene et al., 1986). Four of the 6 LTB neurons exposed to carbachol responded with a depolarization, one with a hyperpolarization and the other with a biphasic response; in the NB neuron sample, depolarization was seen in 9, a biphasic response in 2 and hyperpolarization in 3.

The depolarizing response was further analyzed as to its pharmacological and voltage sensitivities. The addition of atropine (0.5 μM) to the perfusate (n = 4) resulted in complete blockade of the carbachol-evoked depolarization (Fig. 2C). Under voltage clamp control of the membrane potential, carbachol evoked a net inward current associated with decrease in chord conductance, as indicated by a decrease in the amount of current required to hyperpolarize the membrane potential 10 mV (Fig. 3A). The voltage sensitivity of the evoked conductance decrease was examined by changing the membrane potential at a constant rate of 1 mV/200 ms from -100 to -50 mV while recording the resulting current before and during exposure to carbachol (Fig. 3B). The current evoked by carbachol was measured by subtraction of current recorded during carbachol from that during control (Fig. 3C); the slope of the I/V plot of this current was constant over the measured range and thus indicative of a voltage-insensitive conductance change. The reversal potential was $-98 +/- 17$ mV SD ($[K^+]_0 = 3.25$ mM; n = 4, derived by extrapolation in 2 cases).

The depolarization response evoked by carbachol was accompanied by an increase in neuronal excitability, since identical amplitude and duration depolarizing current pulses (Fig. 4A, B) elicited from 1.3 to fourfold as many action potentials during carbachol application as compared with control (n = 3). To determine if the steady-state inward current elicited by carbachol was alone sufficient to account for the increased excitability we combined carbachol application with intracellular injection of a hyperpolarizing direct current to return the membrane potential to control level. In this condition, excitability was not increased (Fig. 4C).

Finally, we examined the effects of carbachol on post-synaptic potentials (PSPs) elicited by stimulation of the contralateral pontine reticular formation, since it is known that enhanced PSPs are obtained by reticular stimulation during REM as contrasted with slow wave sleep (Ito and McCarley, 1984). Stimulation of the contralateral mPRF (10–100 μA amplitude, 100 μs duration) with bipolar double barrel

128

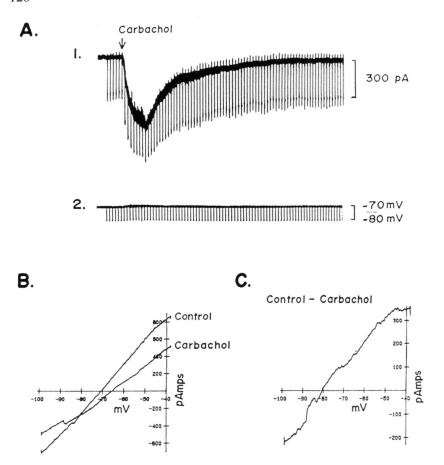

Figure 3. The depolarization elicited by carbachol is mediated by a voltage-insensitive decrease in conductance. *A* Decreased membrane conductance during puffer application (arrow, 1 s, 3 p.s.i.) of carbachol (1 mM) is indicated by the decreased amplitude of downward deflections in the upper current record in response to 10 mV, 400 ms membrane potential shift commands (lower record; applied every 10 s) in a neuron under voltage clamp control. *B* I/V plot generated by a constant depolarization of the membrane potential (1 mV/200 ms) from −100 mV to −50 mV during control conditions and during exposure to 0.5 μM carbachol in the perfusate. *C* I/V plot obtained by digital subtraction of the carbachol I/V plot from that of control shown in *B*. Note the voltage insensitivity of the carbachol current.

glass electrodes filled with perfusate, evoked depolarizing PSPs (Fig. 4D). In the presence of carbachol which depolarized the neurons of this sample (n = 3), the evoked PSPs were enhanced (when examined from the same baseline membrane potential) and this enhancement was blocked in the presence of atropine.

An analysis of the hyperpolarizing response showed a non-linear current-voltage relationship (examined between −100 and −50 mV) of

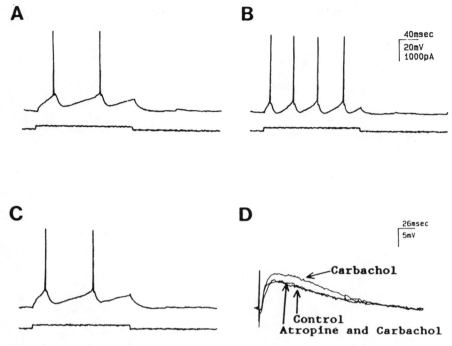

Figure 4. Carbachol elicits an increase in neuronal excitability and in the amplitude of excitatory post-synaptic potentials (PSPs). *A* Control conditions. Two oscilloscope traces of neuronal membrane potential response (upper) to intracellular depolarizing current injection of 400 ms duration and 150 pA (lower). Baseline membrane potential is −65 mV. *B* Same stimulus conditions and neuron as in *A*, but with puffer application of carbachol (10 mM, 0.5 s, 1.5 p.s.i.). The membrane potential depolarized 6 mV and the number of spikes increased. *C* Same neuron with same stimulus and carbachol application parameters as in *B*, but with D.C. hyperpolarizing current injection to return the membrane potential to the control level. Note the number of action potentials is the same as in *A*. *D* Three superimposed oscilloscope traces from an mPRF neuron of a PSP elicited by stimulation of the contralateral mPRF (stimulation artifact is the first biphasic positive-negative deflection). Topmost trace is during bath perfusion with carbachol (0.5 μM), and bottom traces are during the control condition and during bath perfusion with both carbachol and atropine (0.5 μM). Note the 20% increase of PSP amplitude in the presence of carbachol compared to control and carbachol/atropine conditions.

the carbachol evoked current in the presence of 1 μM TTX (Fig. 5, methods as described for depolarizing response). It was characterized by the presence of inward rectification (characteristic of the anomalous rectifier current; Katz, 1949; Hagiwara and Takahashi, 1974), i.e. slope conductance was greater at membrane potential levels negative to the reversal potential. At membrane potentials between −65 mV and −50 mV (currents at potentials depolarized to −50 mV were not examined) the slope conductance was less than 2.5 nS, so that the outward current evoked by carbachol was nearly constant over this range. This

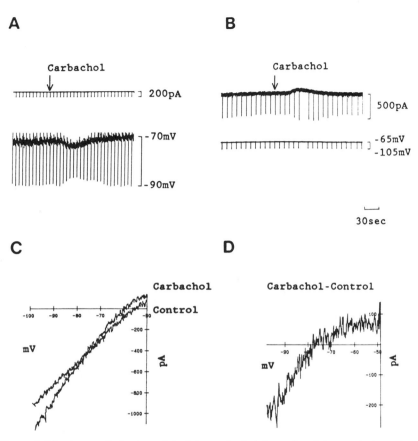

Figure 5. The hyperpolarization elicited by carbachol is mediated by an increase in an inwardly rectifying conductance. *A, B* Records of current (upper record) and membrane potential (lower record) obtained in current clamp (*A*) and voltage clamp (*B*) conditions during puffer application of carbachol at arrow (1.0 mM carbachol, 10 s, 5 p.s.i.). In *A*, downward deflections result from current injection (400 ms) and in *B* from a membrane potential shift command (400 ms). Following carbachol ejection there is an increase in chord conductance observable in *A* and *B*. *C* I/V plot from another neuron generated by a constant depolarization of the membrane potential from -100 to -50 mV (1 mV/400 ms) before and during exposure to carbachol ($[K^+]_0 = 5.0$ mM). *D* I/V plot of the current elicited by carbachol constructed by subtraction of the control plot from the carbachol plot. Note that the slope conductance of the outward (positive) current is less than that of the inward current indicative of inward rectification.

was in contrast to a slope conductance of 12 nS over the range of -100 to -80 mV.

Inwardly rectifying currents in invertebrates (Hagiwara et al., 1976) vertebrate somatic muscle (Standen and Stanfield, 1980) heart muscle (Noble, 1983) and vertebrate central neurons (Halliwell and Adams, 1982) can be blocked by externally applied CsCl. We have begun an examination of the sensitivity of the inwardly rectifying carbachol current to CsCl and its relationship with the CsCl sensitive intrinsic current of mPRF neurons. I/V plots were generated as previously described. Preliminary data indicate CsCl (5 mM) antagonized a powerful inward current which showed inward rectification at about -60 mV (Fig. 6). Carbachol-elicited current was also reduced by CsCl but to a lesser extent (58% reduction at -50 mV, Fig. 7). Comparison of the two currents showed a marked difference in reversal potentials (Fig. 7). The intrinsic inward rectifier had a more depolarized E_{rev} suggestive of a greater sodium permeability, as seen in hippocampal neurons (Halliwell and Adams, 1982). The E_{rev} of the carbachol current was estimated by extrapolation to the x intercept of the portion of the I/V plot with constant slope conductance (negative to -80 mV) in bath

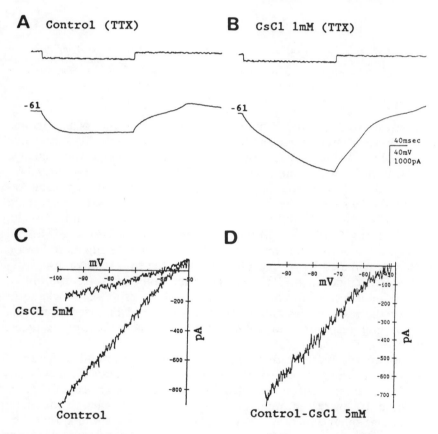

Figure 6. CsCl antagonizes an inwardly rectifying current. *A, B* are records of injected current (upper traces) and membrane potential (lower traces) in a mPRF neuron before (*A*) and during (*B*) perfusion with 1 mM CsCl. *C* has two I/V plots taken under voltage clamp control before (control) and during perfusion with 5 mM CsCl. *D* is an I/V plot resulting from the digital subtraction of the CsCl 5 mM plot from that of control shown in *C*. Note the change in slope beginning at about -65 mV.

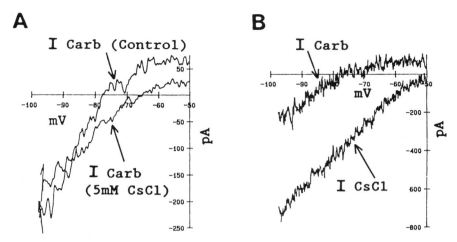

Figure 7. CsCl antagonizes a carbachol-elicited and an intrinsic inwardly rectifying current. *A* is two I/V plots of the carbachol evoked current before (control) and during perfusion with 5 mM CsCl. *B* is two I/V plots of carbachol and CsCl sensitive current. Note the more hyperpolarized reversal potential of the carbachol sensitive current.

concentrations of K^+ of 2.5, 5 and 10 mM (Fig. 8). A plot of E_{rev} against $\ln[K]_0$ gave a straight line with a slope of 22 mV/$\ln$[mM] (the value predicted by the Nernst equation at 30°C is 26), consistent with a change primarily in P_K with a small contribution from Na ($\alpha = 0.02$ to 0.04).

Based on sensitivity to pirenzepine, cholinergic depolarizing responses mediated by a decrease in conductance are mediated by M_1 receptors in enteric (North et al., 1985) and hippocampal (Muller and Misgeld, 1986) neurons. Cholinergic hyperpolarizing responses associated with an increase in conductance in parabrachial neurons (Egan and North, 1986) are mediated by M_2 receptors as indicated by a relative insensitivity to pirenzepine. In mPRF neurons, pirenzepine at concentrations of 200 nM or less had no effect on either type of response. Blockade of 30% of the response was accomplished with 1 μM pirenzepine for the depolarizing response and 0.4 μM for the hyperpolarizing response (Fig. 9).

In summary, over two-thirds of mPRF neurons responded to carbachol with a strong depolarizing response that was associated with an increase in input resistance. There was an increase in excitability as measured by increased number of spikes to the same depolarizing current and enhancement of PSPs elicited by electrical stimulation of the contralateral pontine reticular formation. The remaining third responded with either a biphasic hyperpolarization-depolarization or hyperpolarization alone, consistent with microiontophoretic ACh effects on spike rates of identified reticulospinal mPRF neurons (Greene and Carpenter, 1985). Pirenzepine and atropine blockade of carbachol

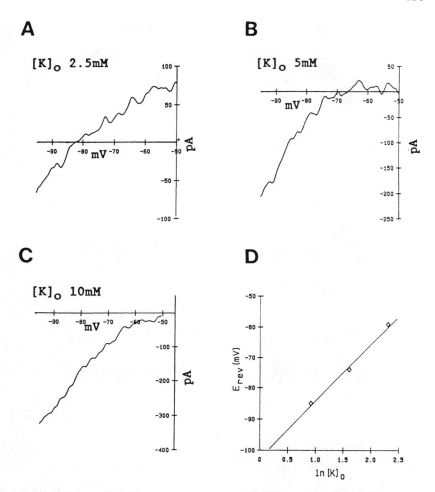

Figure 8. The hyperpolarizing current elicited by carbachol is mediated primarily by an increase in potassium conductance. *A*, *B* and *C* are I/V plots of current evoked by carbachol during perfusion with media containing 2.5, 5.0 and 10.0 mM K respectively. *D* is a plot of the reversal potentials (see text for methodological details) with respect to $\ln[K]_0$ using data illustrated in *A*, *B* and *C*. The slope of the plot is 22 mV/ln mM.

effects suggests mediation by a muscarinic receptor. The insensitivity to pirenzepine concentrations of less than 200 nM suggests mediation by other than a M_1 receptor.

With respect to the mechanism of depolarization, we suggest the most likely cause is an inward current induced by a decrease in a voltage insensitive membrane conductance; the reversal potential suggests a reduction primarily in potassium permeability, although ion substitution studies will be required for a definitive answer. A similar depolarizing response to muscarinic activation has been reported in neurons of

134

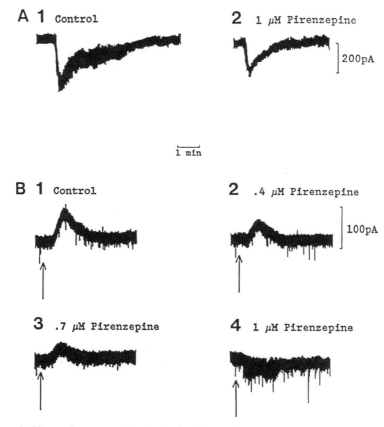

Figure 9. Pirenzepine antagonizes both the depolarizing and hyperpolarizing responses to carbachol but with different sensitivities. *A* and *B* are chart recordings of current during voltage clamp control of 2 mPRF neurons, held at −60 mV. In *A*, carbachol elicited an inward current that was reduced by 30% in the presence of 1 μM pirenzepine. The neuron illustrated in *B*, responded to carbachol with an outward current which was reduced 30% by 0.4 μM pirenzepine and abolished by 1 μM pirenzepine.

the hippocampus and the medial and lateral geniculate nuclei; in both cases this was mediated via reduction of a potassium conductance active at all membrane potentials, with an apparent lack of inactivation and a reversal potential more negative than −90 mV (Madison et al., 1987; McCormick and Prince, 1987). These are also characteristics of the S current modulated by neurotransmitters in invertebrate neurons (Siegelbaum et al., 1982; Pollock et al., 1985; Brezina et al., 1987).

Cholinergically evoked hyperpolarizations have been observed in both the brainstem (Egan and North, 1986) and thalamus (McCormick and Prince, 1986) although voltage sensitivity was not analyzed. Hyperpolarizations resulting from an enhancement of an inwardly rectifying potassium conductance were reported in locus coeruleus and submucosal neurons in response to opioid agonists (North et al., 1987), and in striatal and hippocampal neurons in response to adenosine (Trussell and Jackson, 1985, 1987) or serotonin (Yakel et al., 1988). In cardiac cells, muscarinic cholinergic agonists elicit an increase in potassium conductance (Noma and Trautwein, 1978; Sakmann et al., 1983) similar in its inwardly rectifying voltage sensitivity to the responses observed in the present study. Single channel analysis showed the muscarinic current in cardiac cells to be distinct from the intrinsic inward rectifier (G_{kl}) of atrial and ventricular cells (Sakmann et al., 1983). In mPRF neurons the intrinsic inward rectifier appears to have a different ionic selectivity from that of the cholinergic inward rectifier current based on differences of reversal potential. Our data indicate the latter has greater selectivity for K ions. Furthermore, the intrinsic current was more readily antagonized by CsCl and is thus likely to be distinct from the carbachol sensitive current.

The depolarizing effects evoked by carbachol on mPRF neurons are indistinguishable from the neuronal alterations which occur with the onset of REM. These include a 7–10 mV depolarization of the membrane potential, an increase in excitability and an enhancement of PSP's evoked from the pontine reticular formation (Ito and McCarley, 1984). Furthermore the effects on firing pattern of a carbachol-induced depolarization also parallel those occurring during natural REM sleep in that there is the absence of a 'burst' discharge pattern (McCarley and Hobson, 1975), due to depolarization-dependent inactivation of the low threshold calcium spike responsible for the burst discharge (Greene et al., 1986) as demonstrated in thalamic neurons (McCormick and Prince, 1987).

The direct effects of cholinergic muscarinic activation of mPRF neurons *in vitro* are consistent with the presence of physiological mechanisms underlying cholinergic activation of the mPRF during naturally induced REM sleep. At the single cell level these results provide a mechanism for the action of microinjected muscarinic cholinergic agents in pharmacological production of a REM-like state.

Acknowledgements. We thank Thea Pickering for expert assistance. Supported by the Veterans Administration, National Institute of Mental Health R37 MH39, 683, Army Defense Grant 87PP7811 and the Swiss National Science Foundation to U.G.

Amatruda, T. T., Black, D. A., McKenna, R. M., McCarley, R. W., and Hobson, J. A. (1975) Sleep cycle control and cholinergic mechanisms: differential effects of carbachol injections at pontine brain stem sites. Brain Res. 98: 510–515.

Baghdoyan, H. G., McCarley, R. W., and Hobson, J. A. (1985) Cholinergic manipulation of brainstem reticular systems: effects on desynchronized sleep generation. In: A. Waquier, J. Monti, J. P. Gillard, and M. Raulovacki (eds), Sleep: Neurotransmitters and Neuromodulators. Raven Press, New York, pp. 15–27.

Brezina, V., Eckert, R., and Erxleben, C. (1987) Modulation of potassium conductances by an endogenous neuropeptide in neurones of aplysia californica. J. Physiol. (Lond.) 382: 267–290.

136

Egan, T. M., and North, A. R. (1986) Acetycholine hyperpolarizes central neurones by acting on an m2 muscarinic receptor. Nature (Lond.) 319(30): 405–407.

Greene, R. W., and Carpenter, D. O (1985) Actions of neurotransmitters on pontine medial reticular formation neurons of the cat. J. Neurophysiol. 54: 520–531.

Greene, R. W., Haas, H. L., and McCarley, R. W. (1986) A low threshold calcium spike mediates firing pattern alterations in pontine reticular neurons. Science 234: 738–740.

Hagiwara, S., and Takahashi, K. (1974) The anomalous rectification and cation selectivity of the membrane of a starfish egg cell. J. Membr. Biol. 18: 61–80.

Hagiwara, S., Miyazaki, S., and Rosenthal, N. P. (1976) Potassium current and the effect of cesium on this current during anomalous rectification of the egg cell membrane of a starfish. J. gen. Physiol. 67: 621–638.

Hallanger, A. E., Levey, A. I., Lee, H. J., Rye, D. B., and Wainer, B. H. (1987) The origins of cholinergic and other subcortical afferents to the thalamus in the rat. J. comp. Neurol. 262: 105–124.

Halliwell, J. V., and Adams, P. R. (1982) Voltage-clamp analysis of muscarinic excitation in hippocampal neurons. Brain Res. 250: 71–92.

Ito, K., and McCarley, R. W. (1984) Alterations in membrane potential and excitability of cat medial pontine reticular formation neurons during changes in naturally occurring sleep-wake states. Brain Res. 292: 169–175.

Katz, B. (1949) Les constantes électriques de la membrane du muscle. Arch. Sci. physiol. 3: 285–299.

Madison, D. V., Lancaster, B., and Nicoll, R. A. (1987) Voltage clamp analysis of cholinergic action in the hippocampus. J. Neurosci. 7: 733–741.

McCarley, R. W., and Hobson, J. A. (1975) Discharge patterns of cat pontine brain stem neurons during desynchronized sleep. J. Neurophysiol. 38: 751–766.

McCormick, D. A., and Prince, D. A. (1986) Acetylcholine induces burst firing in thalamic reticular neurones by activating a potassium conductance. Nature (Lond.) 319(30): 402–405.

McCormick, D. A., and Prince. D. A. (1987) Actions of acetylcholine in the guinea-pig and cat medial and lateral geniculate nuclei, in vitro. J. Physiol. (Lond.) 392: 147–165.

Mesulam, M.-M., Mufson, E. J., Wainer, B. H., and Levey, A. I. (1983) Central cholinergic pathways in the rat: An overview based on an alternative nomenclature (Ch1–Ch6). Neuroscience 10: 1185–1201.

Mitani, A., Ito, K., Hallanger, A. E., Wainer, B. H., Kataoka, K., and McCarley, R. W. (1988) Cholinergic projections from the laterodorsal and pedunculopontine tegmental nuclei to the pontine gigantocellular tegmental field in the cat. Brain Res. 451: 397–402.

Muller, W., and Misgeld, U. (1986) Slow cholinergic excitation of guinea pig hippocampal neurons is mediated by two muscarinic receptor subtypes. Neurosci. Lett. 67: 107–112.

Noble, D. (1983) The surprising heart: a review of recent progress in cardiac electrophysiology. 353: 1–50.

Noma, A., and Trautwein, W. (1978) Relaxation of the acetylcholine-induced potassium current in the rabbit sinoatrial node cell. Pflügers Arch. Ges. Physiol. 377: 193–200.

North, R. A., Slack, B. E., and Surprenant, A. (1985) Muscarinic M1 and M2 receptors mediate depolarization and presynaptic inhibition in guinea-pig enteric nervous system. J. Physiol. 368: 435–452.

North, R. A., Williams, J. T., Surprenant, A., and Christie, M. J. (1987) Mu and delta receptors both belong to a family of receptors which couple to a potassium conductance. Proc. natl Acad. Sci. USA 84: 5487.

Pollock, J. D., Bernier, L., and Camardo, J. S. (1985) Serotonin and cyclic adenosine 3′:5′-monophosphate modulate the potassium current in tail sensory neurons in the pleural ganglion of Aplysia. J. Neurosci. 5: 1862–1871.

Sakmann, B., Noma, A., and Trautwein, W. (1983) Acetylcholine activation of single muscarinic K + channels in isolated pacemaker cells of the mammalian heart. Nature (Lond.) 303: 250–253.

Siegelbaum, S. A., Camardo, J. S., and Kandel, E. R. (1982) Serotonin and cyclic AMP close single K + channels in Aplysia sensory neurones. Nature (Lond.) 299: 413–417.

Standen, N. B., and Stanfield, P. R. (1980) Rubidium block and rubidium permeability of the inward rectifier of frog skeletal muscle fibres. J. Physiol. 304: 415–435.

Trussell, L. O., and Jackson, M. B. (1985) Adenosine-activated potassium conductance in cultured striatal neurons. Proc. natl Acad. Sci. USA 82: 4857–4861.

Trussell, L. O., and Jackson, M. B. (1987) Dependence of an adenosine-activated potassium current on a GTP-binding protein in mammalian central neurons. J. Neurosci. 7: 3306–3316.

Yakel, J. L., Trussell, L. O., and Jackson, M. B. (1988) Three serotonin responses in cultured mouse hippocampal and striatal neurons. J. Neurosci. 8: 1273–1285.

Cholinergic responses in human neocortical neurones

J. V. Halliwell*

MRC Neuropharmacology Research Group, Department of Pharmacology, University College London, Gower Street, London WCIE 6BT, England

Summary. Neurones in deeper layers of slices of temporal or frontal human neocortex maintained *in vitro* were impaled with microelectrodes and responses to cholinergic agonists were studied under current and voltage clamp conditions. A range of membrane currents were identifiable: inactivating and persistent Na^+-conductances, inactivating and persistent Ca^{2+}-conductances, two types of inward currents activated by hyperpolarization (I_Q and $I_{f.i.r.}$) and voltage and Ca^{2+}-activated K^+-conductances, which were distinguished on the grounds of their characteristic voltage or pharmacological specificity. The cholinergic agonists muscarine or carbachol were applied in the medium superfusing the slices. Two major effects were observed: consistently, the time and voltage-dependent noninactivating K^+-conductance I_M was suppressed and, when Ca^{2+}-influx was permitted (in the absence of Ca^{2+}-channel blockers), a Ca^{2+}-activated K^+-conductance was transiently or persistently potentiated. Consistent with a suppression of I_M, muscarine excited human neocortical neurones only when applied during a period of membrane depolarization to a potential at which I_M would be expected to exert a braking effect on excitability. Applied at a potential negative to the M-current activation range, muscarine had no excitatory or even an inhibitory effect on the cell. Collectively, these results demonstrate that in the human, I_M can be a target for cholinergic regulation and, in addition, complex effects of ACh on other conductances could modulate cell firing patterns.

Introduction

Acetylcholine (ACh) is present in appreciable quantities in the human cortex and originates in the cell groups of the ventral forebrain, notably the nucleus basalis of Meynert. The transmitter is implicated in the maintenance of many important behavioral states such as memory and attention. Depletion of ACh and presumed cholinergic dysfunction is a prime feature of neurodegenerative disease like Alzheimer's disease. Therefore from both a functional and pathological viewpoint it is important to know the basic mechanisms of cholinergic action in the human brain. Experiments with brain slices have provided much information about the actions of ACh and its analogues in various regions of the brains of nonhuman mammals. These studies are undertaken in the belief that the mechanisms so elucidated are good models and can be extrapolated to the human species, but without at least *some* investigation of human tissue the validity of this position remains open to

* Present address: Department of Physiology, Royal Free Hospital School of Medicine, University of London, Rowland Hill Street, London NW3 2PF, England.

question. This chapter, therefore, presents some of the results obtained in this laboratory from experiments with slices prepared from neocortical tissue samples obtained from patients undergoing necessary surgery for the removal of deep-lying tumours. Although the results are from a relatively small sample of neurones, the range of membrane currents encountered, together with the consistent observation of widespread synaptic activity indicate the viability of the experimental preparation and give credence to the conclusions about ACh action. As will be seen, ACh can have an excitatory action on human neurones, by means of suppressing an identifiable K^+-conductance, the M-current, as in other mammals; in addition, it appears that cholinergic receptor activation can simultaneously enhance another (Ca^{2+}-activated) K^+-conductance.

Methods

Samples of neocortical tissue were received within 1 h from 10 patients undergoing surgery for the removal of subcortical tumours and were free from epileptic symptoms (6 craniopharyngeoma, 2 glioma, 1 meningeoma and 1 pituitary tumour); 7 specimens were from the anterior temporal cortex and the remainder from the lateral part of the frontal lobe. The tissue samples were trimmed into blocks (5 × 10 × 10 mm) which were chopped normal to the cortical surface to yield slices 500–600 μm thick that contain all the cortical layers and some subjacent white matter. The slices were kept submerged in oxygenated artificial cerebrospinal fluid (ACSF) in a holding chamber at room temperature for up to 16 h until required for recording. They were transferred singly to a recording chamber (Halliwell and Adams, 1982) where they were superfused with preoxygenated ACSF (5 ml/min) at 28 °C. The ACSF had the following composition (mM): Na^+ 145, K^+ 3, Ca^{2+} 2.5, Mg^{2+} 1.2, Cl^- 128, $H_2PO_4^-$ 1.2, HCO_3^- 25, glucose 11; pH 7.4 when saturated with 5% CO_2, 95% O_2. Neurones lying 800–1800 μm beneath the pial surface were impaled with microelectrodes containing 3 M KCl or 4 M K acetate and their properties could be studied under conditions of current or voltage clamp (Halliwell and Adams, 1982).

Properties of human neocortical neurones *in vitro*

Stable impalements lasting between 0.5 h and 4 h were achieved in 17 neurones in 10 tissue samples of both cortical areas. The criteria for selecting neurones deemed to be 'healthy' were observation of spontaneous or evoked action potentials >80 mV (92 ± 6 mV; mean + SEM), resting potentials negative to −60 mV (−67 ± 2 mV), and resting input resistances >30 MΩ (58 ± 4 MΩ). Spontaneous depolarizing synaptic

140

potentials occurred particularly in neurones in which the Cl⁻-gradient had been disturbed by recording with a KCl-filled electrode. Focal electrical stimulation (100 μs, 10–100 μA) of the grey matter 1–2 mm from the recording site evoked an EPSP/IPSP sequence in all of 6 neurones so tested (Fig. 1A); IPSP reversal was 20–30 mV positive to resting potential in cells impaled with KCl-containing electrodes,

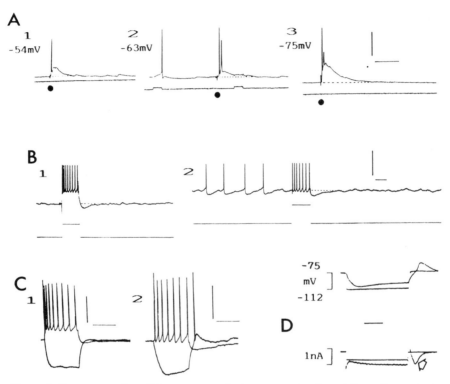

Figure 1. Some properties of human neocortical neurones *in vitro*. *A* Synaptic responses recorded in human temporal neocortical cell impaled with a KCl-filled electrode and evoked by a 20 μA, 100 μs electrical stimulus to the nearby grey matter (●); the cell was polarized to different potentials by the passage of different amounts of direct current (lower trace in pair) (1–3). In trace 2 at resting potential a short current pulse (0.3 nA, 30 ms) was delivered before, and just after the synaptic stimulus; note the inhibition of the direct action potential during the later inhibitory phase of the PSP; cal: 40 mV, 100 ms. *B* Action potential trains evoked by a 400 ms current pulse (1) and after >30 s of 0.45 nA depolarizing current (2). During the long current injection (2) an additional 0.65 nA pulse of current was delivered. Note the ensuing AHP and the consequent inhibition of the spike train. Resting potential, −73 mV; spikes truncated; cal: 40 mV, 200 ms. *C* Responses two different neurones from frontal neocortex to ±1.0 nA (1) and ±0.5 nA (2) showing different forms of ETP. Resting potentials: −72 mV (1), −64 mV (2); cal: 40 mV, 200 ms. *D* Superimposed current and voltage clamp records from a frontal neocortical cell bathed in TTX-containing medium (0.5 μM) showing that the sag of the ETP (voltage, upper trace) is due to a slowly developing inward current upon hyperpolarization and that the rebound depolarization results from a transient inward current (arrowed) inactivated at rest (−70 mV); cal: 100 ms.

suggesting the involvement of Cl^--conductance underlying the IPSP. Collectively, these are indications that a good deal of neuronal circuitry survived the drastic treatment sustained during surgery and the subsequent slicing procedure. Other groups have observed similar responses in slices of epileptic human cortex (Schwartzkroin and Prince, 1976; Avoli and Olivier, 1987) but in contrast to these previous reports, during the present experiments no spontaneous or evoked bursting behaviour was encountered at all. This reflected the absence of seizure activity in the patients from whom the tissue samples came. When stimulated by the passage of depolarizing current through the impaling electrode, the majority of the present sample of human neurones fired a train of action potentials that was characterized by an initial firing frequency that was higher than the steady rate observed towards the termination of a long (>1 s) current pulse. In other words, the cells displayed frequency adaptation or accommodation. Termination of a discharge that displayed adaptation led to an afterhyperpolarization (AHP) of the neurone which had a time course of up to 0.5–1 s. This AHP was seen in cells impaled with both acetate and Cl^- containing electrodes and therefore reflects an underlying K^+-conductance. Reports of similar AHPs in neocortical neurones from a range of mammalian species including the human (Connors et al., 1982; Schwindt et al., 1988a, b; Avoli, 1986). A very much shorter AHP (~ 20 ms) followed individual action potentials (Fig. 1A$_2$).

Inward currents

Employing the single electrode voltage clamp facility of the pre-amplifier, it was possible to study some of the membrane currents in the subthreshold voltage range. As reported by Stafstrom et al. (1985) in feline neocortical neurones, there existed in human cells a TTX-sensitive persistent inward current, spanning a voltage range from ~ -70 mV to well positive to spike threshold; this was identified by slowly ramping the membrane potential from ~ -90 mV in a positive direction whereupon the current so driven possessed a net inward component that was abolished by 0.6 μm TTX. Hyperpolarization of human neurones activated one of two separate inward rectifier currents. Both were blocked by 1–2 mM Cs^+, but a differential sensitivity to Ba^{2+} identified the currents as the Ba^{2+}-insensitive mixed Na^+/K^+-conductance, I_Q/I_h, (Halliwell and Adams, 1982; Mayer and Westbrook, 1983; Crepel and Penit-Soria, 1986; Spain et al., 1987), occurring in about 60% of cells and the Ba^{2+}-sensitive pure K^+-conductance, $I_{f.i.r.}$/anomalous rectifier, (Constanti and Galvan, 1983a; Stanfield et al., 1985), present in about 40%. Possession of one or other of these conductances conferred a characteristic form to the electrotonic potential (ETP; not defined in

142

Fig. 1 legend) elicited from a human neocortical neurone by a > 300 ms hyperpolarizing current pulse (Fig. 1C). Hyperpolarization of human neurones had a further consequence: it relieved, in about 80% of cases inactivation of a Ca^{2+}-dependent inward current that reactivated at around -75 mV. This current (Fig. 1D) was responsible for rebound excitation that followed a hyperpolarizing event in the human neurones (Halliwell, 1986a); it was resistant to TTX and was only partially reduced by the Ca^{2+}-blocker Cd^{2+}. A more positively activating TTX-resistant inward current was also observed which was blocked by 200 μM Cd^{2+} (Halliwell, 1986b); Cd^{2+} concomitantly eliminated a high threshold slow TTX-resistant spike. It appears from these results that human neurones possess at least 2 types of Ca^{2+}-current with different sensitivity to Cd^{2+}. Both Ca^{2+}-conductances were observed in neurones possessing either type of inward rectifier.

Outward currents

Outward currents could be identified in human neocortical cells after blocking Na^+-conductances with TTX. The steady-state current/voltage (I/V) relation of these neurones displayed marked outward rectification in the voltage range positive to -60 mV whether determined from slow ramp protocols or from stepwise jumps to different potentials. This rectification is illustrated in Figure 2: membrane current responses triggered by standard rectangular voltage clamp steps were much larger

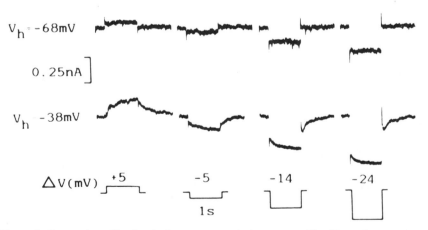

Figure 2. Outward rectification in human neocortical neurones. The figure shows currents elicited by the clamp steps (bottom row) from a holding potential of -68 mV (top row) and -38 mV (middle row). Note the enhanced standing membrane conductance at positive holding potential and the time-dependent conductance decrease resulting from the hyperpolarizing clamp steps. Temporal cortical neurone exposed to 0.5 μM TTX and possessing $I_{r.i.r.}$ at negative potentials.

when evoked from a holding potential of $\sim -40\,\text{mV}$ than those evoked from a holding potential close to rest. There are two notable features of the larger response at positive potentials: first, the instantaneous current driven by a negative clamp step is larger, signifying a greater standing conductance; second, an additional inward current develops slowly during the step, which reflects a conductance decrease, since the instantaneous current driven by the symmetrical return jump at the end of the command is smaller. The interpretation of this membrane behaviour is that it reflects the existence of the M-current (Brown and Adams, 1980; Halliwell and Adams, 1982) in human neurones. Further evidence in favour of this conclusion is that the underlying conductance does not inactivate, has a reversal potential dependent upon extracellular $[K^+]$, activates around $-60\,\text{mV}$ and has voltage-dependent kinetics that speed up at negative potentials (Halliwell, 1986b). In addition, the current was not dependent on Ca^{2+}-influx and as described below was reduced by muscarinic agonists. A less well characterized additional outward current was observed in human neurones; this could be triggered by positive steps from $-40\,\text{mV}$ and appeared to develop secondarily to an inward current that masked the instantaneous or 'leak' conductance (Fig. 2). The outward current deactivated slowly at positive potentials giving rise to an outward tail; hyperpolarizing steps during the tail revealed a reversal close to E_k and an acceleration of the current decline with hyperpolarization. This current and the preceding inward current were both sensitive to $300\,\mu\text{M}\;Cd^{2+}$, suggesting a Ca^{2+} and Ca^{2+}-activated K-conductance (I_C) as a generating mechanism (Brown and Griffith, 1983). Voltage clamp of the AHP following a current-induced train of action potentials revealed an outward tail current with similar properties to those of the presumed I_C. A similar current exists amongst a range of Ca^{2+}-dependent conductances responsible for AHPs in feline neocortical neurones (Schwindt et al., 1988a, b).

Actions of muscarine and carbachol on human neurones

Muscarinic receptor activation excites a range of cortical neurones (neocortical, allocortical and archicortical) in various mammalian species (e.g. Dodd et al., 1981; McCormick and Prince, 1986b; Constanti and Galvan, 1983b). With the exception of presumed interneurones in cingulate cortex (McCormick and Prince, 1986b), the excitatory effects result from or are accompanied by an increase in input resistance; this is thought to reflect a reduction in K^+-conductance as originally suggested by Krnjević and co-workers (1971). Potassium currents that restrain the excitability of neurones and which have been shown to be reduced by ACh or its analogues working through muscarinic (atropine-sensitive) receptors include the M-current (Brown and Adams, 1980;

Halliwell and Adams, 1982; Constanti and Galvan, 1983b) the slow Ca^{2+}-activated K-conductance (I_{AHP}) (Lancaster and Adams, 1986; Cole and Nicoll, 1984; Madison et al., 1987; Constanti and Sim, 1987), the transient K^+-conductance (I_A) (Nakajima et al., 1986) and a voltage-insensitive 'leak' K^+-conductance (Madison et al., 1987; Muller and Misgeld, 1986; Benson et al., 1988). Suppression of any of these listed K^+-conductances could underly an excitatory effect of muscarinic agonists in human cortex.

Bath application of 10–20 μm muscarine to 2 neurones with a negative membrane potential (< -65 mV) produced no change in resting potential or input resistance (Fig. 3A); however, when depolarized to near threshold, one of these cells depolarized further to generate steady firing when muscarine was reapplied (Fig. 3B). An increased input resistance accompanied the excitatory response at the depolarized potential. A small reversible effect of muscarine was seen in the cell in Figure 3 at the resting potential. A slight *reduction* of neuronal excitability ensued, which is not consistent with any of the cholinergic effects listed above. Since this occurred without any observable passive membrane changes, it is unlikely that a K^+-conductance, similar to that

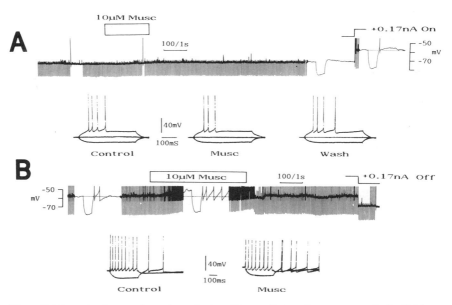

Figure 3. Muscarine has a voltage-dependent excitatory effect on human neocortical cell firing. *A* 10 μM muscarine was bath-applied as indicated at resting potential. Voltage trace; downward deflections are potentials resulting from regular 0.2 nA current pulses. Oscilloscopic records show cell responses elicited by ± 0.2 nA current injections before, during and following recovery from the muscarine application. *B* similar to *A*, but the continuous voltage trace was expanded before and during muscarine, which was applied while the cell was being constantly depolarized to -58 mV with 0.17 nA of outward current.

activated by ACh in thalamic or brainstem neurones (McCormick and Prince, 1986a; Egan and North, 1986) is involved.

The voltage-sensitive action of muscarine described above points strongly to an action on the M-current which a proportion of human neurones possesses (Halliwell, 1986b). The voltage clamp protocol which best demonstrates I_M is one where the neurone is clamped at a positive holding potential (~ -40 mV) to activate the conductance continuously; test hyperpolarizing steps of 10 mV or so are imposed at regular intervals to assess the standing conductance and observe the slow current relaxations that reflect deactivation and reactivation of I_M. When 300 μM Cd^{2+} was included in the bathing medium to prevent Ca^{2+}-influx, muscarine (10–20 μM) caused a mean inward deflection of the holding current at around 400 pA (n = 3), a reduction of the standing conductance at -40 mV and an inhibition of the relaxing component of the triggered current response (Fig. 4). This response pattern was accounted for by the sole effect of muscarine being M-current suppression: when steady-state I/V curves were determined in the presence and absence of muscarine, the two curves coalesced at negative potentials, excluding the possibility of other (leak?) conductances being involved (data not shown). Effects of muscarine or carbachol (up to 50 μM) were more complicated when the drugs were applied under conditions of free Ca^{2+}-entry. The holding current change was not consistent at -40 mV holding potential, but reduction of M-current relaxations were always encountered. In addition, further outward currents triggered by depolarization were *increased* and prolonged (Fig. 5). Figure 5 compares the effects under these conditions of carbachol, Cd^{2+} (to block Ca^{2+}-activated conductance) and another M-current blocker,

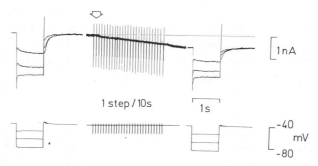

Figure 4. The voltage-dependent persistent outward current is reduced by muscarine and is identified as I_M. The figure depicts the effect of 20 μM muscarine on a temporal neocortical neurone bathed in ACSF containing, additionally, 0.5 μM TTX, 300 μM Cd^{2+} and 1 mM Cs^+. Superimposed traces of current (upper row) initiated by hyperpolarizing steps from a holding potential of -40 mV (voltage: lower row) are displayed and were recorded before (left) and after (right) exposing the cell to 20 μM muscarine. The middle section of the trace is a slower record and shows the moment, marked with the arrow, of switching to muscarine-containing medium and the onset of drug action. From Halliwell (1986b).

146

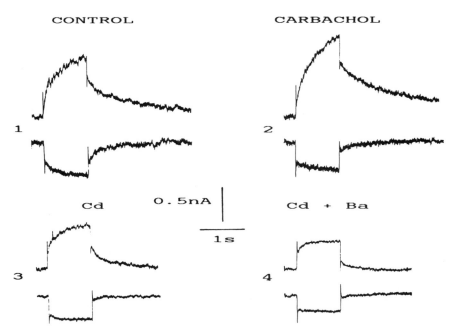

CONTROL   CARBACHOL

1

2

Cd   0.5nA

1 s

Cd + Ba

3

4

Figure 5. In normal (Cd^{2+}-free) medium carbachol enhances a Ca^{2+}-activated K^+-conductance. The figure shows currents driven from a holding potential of -40 mV by ± 10 mV clamp steps in control medium (1), in the presence of 20 μM carbachol (2), 300 μm Cd^{2+} (3) and 2 mM Ba^{2+} (which blocks M-current) in the continuing presence of Cd^{2+} (4).

Ba^{2+} (Brown and Adams, 1980) applied in the continuing presence of Cd^{2+} on currents evoked by both hyperpolarizing and depolarizing steps from -40 mV; the enhancement of the outward current evoked by depolarization is clear, as is the fact that the enhanced current is Cd^{2+}-sensitive and that the Cd^{2+}-insensitive component (I_M) is blocked by Ba. The conclusion that is reached is that a Ca^{2+}-activated K^+-current (probably I_C) can be enhanced by cholinergic activation at the *same* time that the M-current is being depressed.

Discussion

By any current electrophysiological criteria the sample of cortical cells that provided the results presented above can be judged a 'healthy' one. A full complement of membrane currents could be described and furthermore, significant differences between subgroups of neurones, for example, with respect to the type of anomalous or inward rectifier, were noted. One expects when sampling a very large population to encounter both common features and patterns of diversity. When comparing the results from human neurones with those from other mammalian species it is clear that, superficially, similar ionic mechanisms operate in the human and that there is no reason to doubt the validity of the animal models, at least at a cellular level. It is interesting that a proportion of human neocortical neurones possesses a low threshold transient Ca^{2+}-carrying conductance that has been described in other neuronal populations in different species (see Fox et al., 1987) but not yet in mammalian neocortex. Furthermore, it would seem that this conductance activates at a more negative potential in human cortex (e.g. Fig. 1D). As has been discussed by others (e.g. McCormick and Prince, 1986a) this makes IPSPs and other hyperpolarizing influences paradoxically excitatory (Fig. $1C_2$) since the current becomes available for activation upon repolarization. In addition to this (very) low threshold presumed Ca^{2+}-current, at least three other inward currents can be operating in the voltage range just negative to the activation point for the Na^+-conductance responsible for the fast action potential: I_Q, $I_{f.i.r.}$ and the persistent Na^+-current. There is a strong inbuilt tendency, therefore, for human (and other mammalian) neocortical neurones to fire spikes, which must be held in check by inhibitory mechanisms, since these cells are characterized by a low spontaneous discharge rate. Inhibitory synaptic input is one mechanism, but others are intrinsic to the individual cell and obviously include I_M and Ca^{2+}-activated K^+-conductance that are presently identified in human neurones. Cholinergic suppression of the M-current is not excitatory *per se* (Fig. 3) (McCormick and Prince, 1986b), but makes the cell more responsive to depolarizing stimuli by removing the hyperpolarizing restraint of I_M. Cholinergic enhancement of Ca^{2+}-activated K^+-conductance (probably I_C; Brown and Griffith, 1983) would make excitation more phasic, with or without simultaneous M-current inhibition; the response of most other mammalian cortical neurones is different from that of the human in that Ca^{2+}-activated K^+-conductances are unaffected or reduced by ACh or muscarinic agonists (see recent review by Nicoll (1988) for refs). Muscarine and carbachol had similar effects on both these identified K^+-currents in human neurones, so that the effects were probably mediated through muscarinic receptor subtypes (see this volume); further experiments with selective muscarinic antagonists would be required to confirm this but were not feasible with the present small sample of neurones. On the basis of the present results, little comment can be made about the mechanism whereby the conductances are modified; however, both M-current suppression and potentiation of I_C could arise from the muscarinic stimulation of phosphatidyl inositol breakdown and the subsequent production of inositol trisphosphate and liberation of intracellular Ca^+ (Nicoll, 1988).

The physiological consequences of these effects on K^+-currents would be to change the responsiveness of cortical neurones to other excitatory inputs when cholinergic afferents were active: suppression of the

M-current would lower the threshold for initiated firing whereas an increase in I_C might be expected to break a steady train into phasic groupings of action potentials interspersed with pauses.

Avoli, M. (1986) Inhibitory potentials in neurons of the deep layers of the *in vitro* neocortical slice. Brain Res. 370: 165–170.

Avoli, M., and Olivier, A. (1987) Bursting in human epileptogenic neocortex is depressed by an N-methyl-D-aspartate antagonist. Neurosci. Lett. 76: 249–254.

Benson, D. H., Blitzer, R. D., and Landau, E. M. (1988) An analysis of the depolarization produced in guinea-pig hippocampus by cholinergic receptor stimulation. J. Physiol. (Lond.) 404: 479–496.

Brown, D. A., and Adams, P. R. (1980) Muscarinic suppression of a novel voltage-sensitive K^+-current in a vertebrate neurone. Nature (Lond.) 283: 673–676.

Brown, D. A., and Griffith, W. H. (1983) Calcium-activated outward current in voltage-clamped hippocampal neurones of the guinea-pig. J. Physiol. (Lond.) 337: 287–301.

Cole, A. E., and Nicoll, R. A. (1984) Characterization of a slow cholinergic postsynaptic potential recorded *in vitro* from rat hippocampal pyramidal cells. J. Physiol. (Lond.) 352: 173–188.

Connors, B. W., Gutnick, M. J., and Prince, D. A. (1982) Electrophysiological properties of neocortical neurones *in vitro*. J. Neurophysiol. 48: 1302–1320.

Constanti, A., and Galvan, M. (1983a) Fast inward-rectifying current accounts for anomalous rectification in olfactory cortex neurones. J. Physiol. (Lond.) 335: 153–178.

Constanti, A., and Galvan, M. (1983b) M-current in voltage-clamped olfactory cortex neurones. Neurosci. Lett. 39: 65–70.

Constanti, A., and Sim, J. A. (1987) Calcium-dependent potassium conductance in guinea pig olfactory cortex neurones *in vitro*. J. Physiol. (Lond.) 387: 173–194.

Crepel, F., and Penit-Soria, J. (1986) Inward rectification and low threshold calcium conductance in rat cerebellar Purkinje cells. An *in vitro* study. J. Physiol. (Lond.) 372: 1–23.

Dodd, J., Dingledine, R., and Kelly, J. S. (1981) The excitatory action of acetylcholine on hippocampal neurones of the guinea-pig and rat maintained *in vitro*. Brain Res. 207: 109–127.

Egan, T. M., and North, R. A. (1986) Acetylcholine hyperpolarizes central neurones by acting on an M_2 muscarinic receptor. Nature (Lond.) 319: 405–407.

Fox, A. P., Nowycky, M. C., and Tsien, R. W. (1987) Single-channel recordings of three types of calcium channels in chick sensory neurones. J. Physiol. (Lond.) 394: 173–200.

Halliwell, J. V. (1986a) A low threshold Ca-current in human neocortical neurones. Neurosci. Lett. Suppl. S40.

Halliwell, J. V. (1986b) M-current in human neocortical neurones. Neurosci. Lett. 67: 1–6.

Halliwell, J. V., and Adams, P. R. (1982) Voltage-clamp analysis of muscarinic excitation in hippocampal neurones. Brain Res. 250: 71–92.

Krnjević, K., Pumain, R., and Renaud, L. (1971) The mechanism of excitation by acetylcholine in the cerebral cortex. J. Physiol. (Lond.) 215: 447–465.

Lancaster, B., and Adams, P. R. (1986) Calcium-dependent current generating the after hyperpolarization of hippocampal neurons. J. Neurophysiol. 55: 1268–1282.

Madison, D. V., Lancaster, B., and Nicoll, R. A. (1987) Voltage-clamp analysis of cholinergic action in the hippocampus. J. Neurosci. 7: 733–741.

Mayer, M. L., and Westbrook, G. L. (1983) A voltage-clamp analysis of inward (anomalous) rectification in mouse spinal sensory ganglion neurones. J. Physiol. 340: 19–45.

McCormick, D. A., and Prince, D. A. (1986a) Acetylcholine indices burst firing in thalamic reticular neurones by activating a potassium conductance. Nature (Lond.) 319: 402–405.

McCormick, D. A., and Prince, D. A. (1986b) Mechanisms of the action of acetylcholine in the guinea-pig cerebral cortex *in vitro*. J. Physiol. (Lond.) 375: 169–194.

Muller, W., and Misgeld, U. (1986) Slow cholinergic excitation of guinea-pig hippocampal neurons is mediated by two muscarinic receptor subtypes. Neurosci. Lett. 67: 107–112.

Nakajima, Y., Nakajima, S., Leonard, R. J., and Yamaguchi, K. (1986) Acetylcholine raises excitability by inhibiting the fast transient potassium current in cultured hippocamal neurones. Proc. natl Acad. Sci. USA 83: 3022–3026.

Nicoll, R. A. (1988) The coupling of neurotransmitter receptors to ion channels in the brain. Science 241: 545–551.

Schwartzkrion, P. A., and Prince, D. A. (1976) Microphysiology of human cerebral cortex studied *in vitro*. Brain Res. 115: 497–500.

Schwindt, P. C., Spain, W. J., Foehring, R. C., Stafstrom, C. E., Chubb, M. C., and Crill, W. E. (1988a) Multiple potassium conductances and their functions in neurons from cat sensorimotor cortex *in vitro*. J. Neurophysiol. 59: 424–449.

Schwindt, P. C., Spain, W. J., Foehring, R. C., Stafstrom, C. E., Chubb, M. C., and Crill, W. E. (1988b) Slow conductances in neurons from cat sensorimotor cortex and their role in slow excitability changes. J. Neurophysiol. 59: 450–467.

Spain, W. J., Schwindt, P. C., and Crill, W. E. (1987) Anomalous rectification in neurons from cat sensorimotor cortex *in vitro*. J. Neurophysiol. 57: 1555–1576.

Stafstrom, C. E., Schwindt, P. C., Chubb, M. C., and Crill, W. E. (1985) Properties of persistent sodium conductance and calcium conductance of layer V neurons from cat sensorimotor cortex *in vitro*. J. Neurophysiol. 53: 153–170.

Stanfield, P. R., Nakajima, Y., and Yamaguchi, K. (1985) Substance P raises neuronal excitability by reducing inward rectification. Nature (Lond.) 315: 498–501.

Cholinergic modulation of hippocampal epileptic activity *in vitro*

Yoel Yaari[a] and Morten S. Jensen[b]

[a]*Department of Physiology, Hebrew University School of Medicine, Jerusalem, Israel, and*
[b]*Pharma Biotec, Institute of Physiology, Aarhus University, Aarhus, Denmark*

Introduction

The hippocampus is an epilepsy-prone cortical structure (Green, 1964). It is frequently involved in clinical seizure disorders and it readily generates paroxysmal discharges when subjected to repetitive electric stimulation or when exposed to K^+-enriched solutions or to convulsant drugs (Kandel and Spencer, 1961; Zuckermann and Glaser, 1968). The factors responsible for this propensity are not clear. One putative factor is the massive cholinergic innervation of the hippocampus (Lewis et al., 1967). Repetitive activation of cholinergic fibers results in sustained excitation of principal hippocampal neurons (Cole and Nicoll, 1984). Likewise, topical application of acetylcholine (ACh) or analogue drugs to the hippocampus produces sustained paroxysms (Baker and Benedict, 1968). Systemic application of blood-brain-barrier permeant cholinergic drugs produces profound convulsant activity resembling limbic seizures (Turski et al., 1983).

Thus cholinergic-receptor activation is potentially epileptogenic, but does this process play a role in hippocampal seizure activity induced by other insults to the brain? We have investigated this question in an *in vitro* model of focal hippocampal epilepsy, produced by exposing rat hippocampal slices to solutions containing elevated concentrations of K^+ (Traynelis and Dingledine, 1988; Yaari and Jensen, 1988; Jensen and Yaari, 1988). Although this is an artificial method, the recordings obtained from this preparation strongly resemble naturally occurring and experimentally induced epileptic foci *in vivo*.

Potassium induces focal epilepsy in hippocampal slices

Rat hippocampal slices perfused with solutions containing 6–8.5 mM K^+ (instead of the normal 3.5 mM K^+) generate brief (duration 50–100 ms) bursts of synchronized neuronal discharge at a regular low frequency (once every 1 to 2 s) (Fig. 1). These events resemble in their

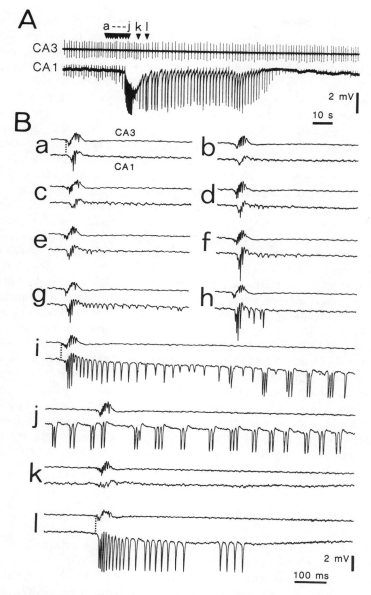

Figure 1. Paroxysmal activity induced by raising extracellular [K⁺] in rat hippocampal slice. *A* Simultaneous extracellular recordings from CA3 (upper trace) and CA1 (lower trace) in a slice perfused with solution containing 7 mM K⁺. Note the concurrent interictal paroxysms in the two fields and the tonic-clonic ictal episode in CA1. *B* Temporal relationship between CA3 and CA1 paroxysms. Selected parts from the traces shown in *A* (marked with triangles, a-l) are shown on an expanded time scale. Note that brief paroxysms in CA3 slightly precede the corresponding interictal and clonic bursts in CA1.

configuration the *interictal* paroxysms (the EEG 'spikes') that intervene between *ictal* (or *seizure*) episodes in brains harboring an active epileptic focus (Ajmone Marsan, 1969). Indeed, these brief paroxysms commence in a 'spiking' focus in the CA3 field and spread from there to neighboring hippocampal regions (Rutecki et al., 1985; Korn et al., 1987; Jensen and Yaari, 1988). Therefore, cutting the pathways connecting CA3 to CA1 eliminates this activity from area CA1.

Many preparations exposed to elevated extracellular $[K^+]$ also display spontaneous episodes of intense and prolonged (many seconds up to a few minutes) synchronized discharge (Traynelis and Dingledine, 1988; Jensen and Yaari, 1988). Unlike the intervening brief paroxysms (which we refer to as *interictal* events), these *ictal* paroxysms commence every few minutes in a 'seizure' focus in CA1. The ictal discharges in CA1 do not invade CA3, which continues uninterruptedly to generate brief paroxysms during the CA1 seizures.

As in animal models of focal epilepsy, the ictal episodes in this *in vitro* model consist of a tonic-clonic sequence. During the *tonic* phase, which endures for a few seconds, CA1 pyramidal cells fire repetitively and in synchrony, as indicated by the appearance of long bursts of population spikes in extracellular recordings from the CA1 pyramidal layer (Fig. 1). Concurrently, the d.c. potential in this layer shifts negatively, presumably due to activity-dependent rise in interstitial $[K^+]$ above baseline (Yaari et al., 1986; Yaari and Jensen, 1989). The continuous synchronized discharge subsides gradually and is followed by a series of intermittent giant bursts, which constitute the *clonic* phase of seizure. Each clonic paroxysm is triggered in CA1 by a burst propagating from CA3. Clonic activity may last 15 s up to several minutes. It is often succeeded by a period of relative depression, during which interictal activity in CA1 is temporarily reduced. This is invariably followed by a new ictal episode.

Different mechanisms underlie brief and sustained paroxysms

It is not yet clear why moderately elevating baseline extracellular $[K^+]$ induces 'spiking' and 'seizure' foci in hippocampal slices. An important factor in the development of this activity may be a K^+-induced attenuation of inhibitory postsynaptic potentials (Korn et al., 1987), which normally thwart the development of positive feedback in the densely packed neuronal network of the hippocampus. The positive feedback mechanisms underlying the brief (interictal and clonic) and sustained (tonic) paroxysms are probably different (Jensen and Yaari, 1988). Excitatory synaptic transmission mediated by glutamate is crucial for the genesis of brief paroxysms, since specific glutamate antagonists potently suppress interictal and clonic bursts (Yaari and Jensen, 1989).

The disposition of CA3 to generate brief paroxysms in elevated extracellular $[K^+]$, may thus be due to strong recurrent excitatory connections in this field (Christian and Dudek, 1988).

By contrast, sustained paroxysms are generated nonsynaptically (e.g. through the regenerative accumulation of interstitial K^+; Yaari et al., 1986), since they are not abolished by treatments which block chemical synaptic transmission (Konnerth et al., 1986; Jensen and Yaari, 1988). The high density of neuronal packaging in the CA1 field (Green and Maxwell, 1961) would favor the development of these nonsynaptic excitatory interactions, and thus may account for the propensity of CA1 neurons to generate sustained paroxysms.

Cholinergic agents differentially effect brief and sustained paroxysms

Atropine and hexamethonium (1 μM), which block, respectively, muscarinic and nicotinic cholinergic receptors, had no effect on K^+-induced paroxysms. Thus neither interictal nor ictal activity in elevated extracellular $[K^+]$ is initiated by intrinsic cholinergic excitation. However, exogenous cholinergic agents strongly affected this activity. Carbachol (1–5 μM) invariably halted the genesis of brief paroxysms in all hippocampal fields, including the clonic phase of CA1 seizures (Fig. 2). Glutamate-ergic synaptic transmission in area CA1 was concurrently reduced by carbachol, suggesting that depression of brief paroxysms (interictal and clonic) may involve partial blockage of excitatory synaptic transmission. Muscarine mimicked these effects of carbachol, while atropine and pirenzepine, but not hexamethonium (all drugs applied at 1 μM), completely reversed them. This pharmacological profile indicates that activation of muscarinic receptors mediates the cholinergic suppression of brief paroxysms.

While suppressing interictal and clonic bursts, carbachol enhanced the frequency of ictal episodes (Fig. 2). Moreover, carbachol invariably induced recurring sustained (tonic) paroxysms in slices that did not manifest any ictal activity in elevated extracellular $[K^+]$ (Fig. 3). These epileptogenic effects were associated with excitation of CA1 pyramidal cells, characterized by slow depolarization, loss of spike accommodation and block of the slow afterhyperpolarization (Benardo and Prince, 1982; Madison and Nicoll, 1984). Atropine and pirenzepine (1 μM) suppressed the carbachol-induced seizures, implicating muscarinic receptors in their generation.

Application of the acetylcholinesterase inhibitor neostigmine mimicked all effects of carbachol, presumably by elevating the concentration of free interstitial ACh. However, a direct action of neostigmine at muscarinic receptors cannot be excluded. Neostigmine (1 μM) suppressed interictal and clonic bursts and enhanced or provoked the

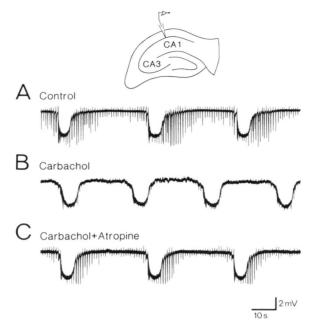

Figure 2. Differential effects of carbachol on K^+-induced interictal and ictal paroxysms in rat hippocampal slice. Extracellular recording from the CA1 pyramidal layer. *A* Paroxysmal activity induced by perfusing the preparation with 7.5 mM K^+. *B* Paroxysmal activity after a 10-min exposure to 5 μM carbachol. Note abolition of brief (interictal and clonic) bursts. *C* Paroxysmal activity 10 min after adding 1 μM atropine to the carbachol-containing solution. Note reappearance of brief paroxysms.

appearance of tonic paroxysms (Fig. 4). All these effects were reversed by atropine or pirenzepine.

Discussion

The data show that also in the hippocampus *in vitro*, cholinergic-receptor activation is potentially epileptogenic, i.e. carbachol (or neo-stigmine) can induce recurring sustained paroxysms when extracellular $[K^+]$ is elevated (Fig. 3). However, the paroxysmal activity generated in elevated extracellular $[K^+]$ in the absence of extraneous cholinergic stimulation most probably does not involve intrinsic cholinergic excitation, since it persists in the presence of cholinergic receptor blockers (atropine or hexamethonium).

Notwithstanding, cholinergic agonists acting at muscarinic receptors strongly modulate K^+-induced paroxysms. Interestingly, muscarinic-receptor activation exerted opposite effects on brief and sustained paroxysms. This may reflect the different mechanisms underlying the

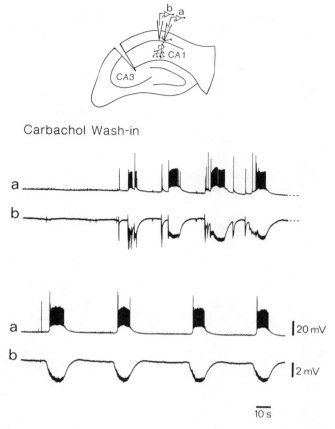

Figure 3. Epileptogenic effect of carbachol in an isolated CA1 perfused with high extracellular [K⁺]. Simultaneous intracellular (*a*, top traces) and extracellular (*b*, lower traces) recordings from the CA1 pyramidal layer in a cut preparation perfused with 7.5 mM. This preparation did not display paroxysmal activity until it was exposed to 2 μM carbachol. The upper pair of traces begins 5 min after introducing carbachol into the slice chamber. The two pairs of traces are continuous.

two forms of epileptic discharge. Muscarinic-receptor activation produces presynaptic inhibition of excitatory synapses in the hippocampus (Hounsgaard, 1978; Valentino and Dingledine, 1981). This effect would tend to suppress the synaptically-generated interictal and clonic paroxysms. Alternatively, suppression of brief paroxysms may be due to muscarinic-mediated reduction of calcium currents in the soma-dendritic membrane (Misgeld et al., 1986; Gahwiler and Brown, 1987), since these currents facilitate intrinsic bursting of hippocampal neurons (Wong and Prince, 1978).

The muscarinic enhancement or induction of tonic paroxysms in elevated extracellular [K⁺] probably results from facilitation of

156

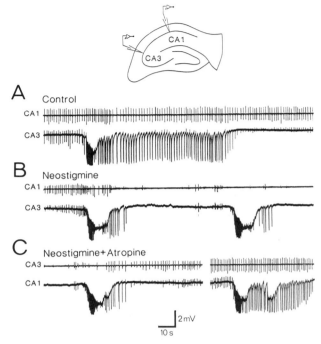

Figure 4. Differential effects of neostigmine on K^+-induced interictal and ictal paroxysms in rat hippocampal slice. Simultaneous extracellular recordings from CA1 and CA3 pyramidal layers. *A* Paroxysmal activity induced by perfusing the slice with 7 mM K^+. *B* Paroxysmal activity 5 min after adding 1 μM neostigmine. All interictal and clonic paroxysms disappeared after additional 5 min in neostigmine. *C* Paroxysmal activity 5 min (left) and 10 min (right) after adding 1 μM atropine to the neostigmine-containing solution.

non-synaptic excitatory interactions that generate these paroxysms. By directly exciting hippocampal neurons, muscarinic agonists may facilitate interstitial K^+ accumulation, which, in turn, would excite the neurons even further. Muscarinic receptors are also present on glial cells (Murphy et al., 1986). Their activation may interfere in some way (e.g. blockage of K^+ channels) with the normal clearance of excess interstitial K^+ through the glial network and facilitate its accumulation (Somjen, 1984).

There is only meagre evidence to date suggesting that cholinergic mechanisms may influence paroxysmal activity in epileptic brains of experimental animals and of humans in ways compatible with those observed *in vitro*. Raising brain ACh concentration by different mechanisms reduced, whereas applying atropine enhanced interictal activity in kindled rats (Fitz and McNamara, 1979). In human epileptics, raising brain ACh levels shortened the duration, but increased the frequency of complex partial seizures (McNamara et al., 1980). Whether these effects reflect differential muscarinic modulation of synaptic and nonsynaptic epileptogenic mechanisms by cholinergic pathways in epileptic brains remains to be explored.

Acknowledgements. Supported by the Fund for Basic Research administered by the Israel Academy of Science and Humanities and an ETP grant to MSJ.

Ajmone Marsan, C. (1969) Acute effects of topical agents. In: H. H. Jasper, A. A. Ward, and A. Pope (eds), Basic Mechanisms of the Epilepsies. Little, Brown, Boston, pp. 229–328.

Baker, W. W., and Benedict, F. (1968) Analysis of local discharges induced by intrahippocampal microinjection of carbachol or diisopropylfluorophosphate (DFP). Int. J. Neuropharmac. 7: 135–147.

Benardo, L. S., and Prince, D. A. (1982) Cholinergic excitation of mammalian hippocampal pyramidal cells. Brain Res. 249: 315–331.

Christian, E. P., and Dudek, F. E. (1988) Electrophysiological evidence from glutamate microapplications for local excitatory circuits in the CA1 area of rat hippocampal slices. J. Neurophysiol. 59: 110–123.

Cole, A. E., and Nicoll, R. A. (1984) Characterization of a slow post-synaptic potential recorded *in vitro* from rat hippocampal pyramidal cells. J. Physiol. 352: 173–188.

Fitz, J. G., and McNamara, J. O. (1979) Muscarinic cholinergic regulation of epileptic spiking in kindling. Brain Res. 178: 117–127.

Gahwiler, B. H., and Brown, D. A. (1987) Muscarine effects calcium-currents in rat hippocampal pyramidal cells *in vitro*. Neurosci. Lett. 76: 301–306.

Green, J. D. (1964) The hippocampus. Physiol. Rev. 44: 561–608.

Green, J. D., and Maxwell, D. S. (1961) Hippocampal electrical activity. I. Morphological aspects. Electroenceph. clin. Neurophysiol. 13: 837–846.

Hounsgaard, J. (1978) Presynaptic inhibitory action of acetylcholine in area CA1 of the hippocampus. Exp. Neurol. 62: 787–797.

Jensen, M. S., and Yaari, Y. (1988) The relationship between interictal and ictal paroxysms in an *in vitro* model of focal hippocampal epilepsy. Ann. Neurol. 24: 591–598.

Kandel, E. R., and Spencer, W. A. (1961) Excitation and inhibition of single pyramidal cells during hippocampal seizure. Exp. Neurol. 4: 163–179.

Konnerth, A., Heinemann, U., and Yaari, Y. (1986) Nonsynaptic epileptogenesis in the mammalian hippocampus *in vitro*. I. Development of seizure-like activity in low extracellular calcium. J. Neurophysiol. 56: 409–423.

Korn, S. J., Giacchino, J. L., Chamberlin, N. L., and Dingledine, R. (1987) Epileptiform burst activity induced by potassium in the hippocampus and its regulation by GABA-mediated inhibition. J. Neurophysiol. 57: 325–340.

Lewis, P. R., Shute, C. C. D., and Silver, A. (1967) Confirmation from choline acetylase of a massive cholinergic innervation to the rat hippocampus. J. Physiol 191: 215–224.

Madison, D. V., and Nicoll, R. A. (1984) Control of the repetitive discharge of rat CA1 pyramidal neurones *in vitro*. J. Physiol. 354: 319–331.

McNamara, J. O., Carwile, S., Hope, V., Luther, J., and Miller, P. (1980) Effects of oral choline on human complex partial seizures. Neurology 30: 1334–1336.

Misgeld, U., Calabresi, P., and Dodt, H. U. (1986) Muscarinic modulation of calcium-dependent plateau potentials in rat neostriatal neurons. Pflügers Arch. 407: 482–487.

Murphy, S., Pearce, B., and Morrow, C. (1986) Astrocytes have both M_1 and M_2 muscarinic receptor sybtypes. Brain Res. 364: 177–180.

Rutecki, P. A., Lebeda, F. J., and Johnston, D. (1985) Epileptiform activity induced by changes in extracellular potassium in hippocampus. J. Neurophysiol. 54: 1363–1374.

Somjen, G. G. (1984) Interstitial ion concentration and the role of neuroglia in seizures. In: Wheal, H. V., and Schwartzkroin, P. A. (eds), Electrophysiology of Epilepsy. Academic Press, London, pp. 303–341.

Traynelis, S. F., and Dingledine, R. (1988) Potassium induced spontaneous electrographic seizures in the rat hippocampal slice. J. Neurophysiol. 59: 259–276.

Turski, W. A., Czuczwar, S. J., Kleinrok, Z., and Turski, L. (1983) Cholinomimetics produce seizures and brain damage in rats. Experientia 39: 1408–1411.

Valentino, R. J., and Dingledine, R. (1981) Presynaptic inhibitory effect of acetylcholine in the hippocampus. J. Neurosci. 1: 784–792.

Wong, R. K. S., and Prince, D. A. (1978) Participation of calcium spikes during intrinsic burst firing in hippocampal neurons. Brain Res. 159: 385–390.

Yaari, Y., and Jensen, M. S. (1988) Nonsynaptic mechanisms and interical-ictal transitions in the mammalian hippocampus. In: Dichter, M. A. (ed.), Mechanisms of Epileptogenesis: From Membranes to Man, the Transition from Interical to Ictal Activity. Plenum, New York, pp. 183–198.

Yaari, Y., and Jensen, M. S. (1989) Two types of epileptic foci generating brief and sustained paroxysms in the *in vitro* rat hippocampus. In: Fariello, R. G. Avanzini, G., and Heinemann, U. (eds), Workshop on Neurotransmitters in Epilepsy, Vol. IV. Raven Press, New York, in press.

Yaari, Y., Konnerth, A., and Heinemann, U. (1986) Nonsynaptic epileptogenesis in the mammalian hippocampus *in vitro*. II. Role of extracellular potassium. J. Neurophysiol. 56: 424–438.

Zuckermann, E. C., and Glaser G. H. (1968) Hippocampal epileptic activity induced by localized ventricular perfusion with high-potassium cerebrospinal fluid. Exp. Neurol. 20: 87–110.

The cholinergic nucleus basalis: A key structure in neocortical arousal

György Buzsáki and Fred H. Gage

*Department of Neurosciences M-024, University of California at San Diego,
La Jolla, CA 92093, USA*

Summary. Single unit studies indicate that increased activity in the cholinergic nucleus basalis (NB) correlates with behavioral activation and neocortical desynchronization. Lesions of the NB result in neocortical slow delta waves, similar to the action of antimuscarinic drugs, and the lesion releases the oscillation of GABAergic neurons in the reticular nucleus of the thalamus, resulting in high voltage neocortical spindles. Extensive damage of the thalamus does not produce slowing of neocortical activity but it abolishes neocortical spindles. We suggest that the NB plays a key role in neocortical activation by a) blocking the afterhyperpolarizations and accommodation in neocortical pyramidal neurons and b) suppressing the rhythm generation in the reticular nucleus-thalamocortical circuitry. We further suggest that the NB system may serve as a structural basis for the concept of the generalized activation described by Moruzzi and Magoun (1949).

Introduction

A basic question concerning the physiology of the cerebral cortex involves the activating afferents that keep the neocortex in the aroused or waking state. A cholinergic activating input has long been implicated because administration of antimuscarinic drugs, such as atropine and scopolamine, results in slow delta waves, and, under central cholinergic blockade, sensory stimuli no longer have a sufficient 'activating' or 'desynchronizing' effect on the neocortical EEG (Wikler, 1952). For the past four decades it was generally accepted that the reticulo-thalamic system served as the structural basis of neocortical activation. Cholinergic (Shute and Lewis, 1967) and other neurons in the brainstem reticular formation were suggested as exciting the neocortical neurons directly (Krnjevic et al., 1971; Phillis, 1980; Vanderwolf and Robinson, 1981) or indirectly, via the reticular nucleus (RT; Jasper, 1949; Papez, 1956) and later the intralaminar nuclei of the thalamus (Steriade, 1970). In turn, a diffuse projection from these thalamic nuclei to cerebral cortex was thought to represent the final common pathway of the ascending reticular activating system (Moruzzi and Magoun, 1949).

In this chapter we will summarize data indicating that the reticulo-thalamic neocortical arousal model is no longer tenable and suggest that the cholinergic nucleus basalis (NB) is the major anatomical substrate of neocortical activation. In addition, we will put forward a hypothesis for a physiological explanation of EEG and behavioral arousal.

Reevaluation of the reticulo-thalamic-neocortex model of arousal

There are several reasons why the reticulo-thalamic-neocortex model should be reevaluated. First, anatomical evidence indicates that direct cholinergic reticulo-cortical and RT-neocortical projections are nonexistent (Fibiger, 1982; Mesulam et al., 1983; Jones, 1985). The only reticulo-thalamo-cortical path with the required diffuse projection has remained the intralaminar nuclei, which have widespread terminations in the neocortex and are innervated by cholinergic cell groups located in the midbrain (Hallanger et al., 1987). However, it has recently been suggested that the intralaminar nuclei are responsible for the induction of neocortical slow activity and not activation-related desynchronization (Fox and Amstrong-James, 1986). Second, a major cholinergic system originating in the basal forebrain with widespread cortical targets has been discovered (Divac, 1975; Jones et al., 1976; Mesulam and Van Hoesen, 1976) without any assigned physiological function. Third, increased power of slow delta EEG activity was reported in Alzheimer patients with severe loss of basal forebrain cholinergic neurons (Coben et al., 1985; Pentilla et al., 1985). Fourth, the neurotransmitter of thalamo-cortical neurons, including those of the intralaminar nuclei, is probably glutamate (Jones, 1985; Fox and Amstrong-James, 1986). Recent studies on the biophysical and pharmacological properties of cortical cells (Nicoll, 1988) indicate that glutamate alone is not capable of maintaining an efficient responding state in pyramidal cells (see arguments below).

In the rat, qualitatively, three basic EEG patterns can be distinguished. In the alert animal the EEG consists of predominantly low voltage fast waves (desynchronized or arousal pattern). In the drowsy or sleeping rat, large-amplitude slow waves in the delta band (1–4 Hz) dominate (Longo, 1956; Vanderwolf, 1975, 1988). In the motionless but awake rat, an additional highly synchronized pattern, characterized by large rhythmic spikes and waves (high voltage spindle, HVS) has also been described (Klingberg and Pickenhain, 1968).

In an attempt to reveal the role of the NB in neocortical activation we set out to investigate the correlation between neocortical EEG and cellular activity in the NB, thalamus and neocortex. We also studied the consequences of circumscribed ibotenic acid lesions of the NB, RT and other thalamic nuclei on the neocortical electrical activity. Finally, we compared the power and spatial distribution of the EEG in both the young and aged rat.

Relationship between neocortical EEG and activity of NB neurons

In our studies, highest frequency activity of NB neurons was observed during running, followed by drinking and immobility. The decrease in the firing rate correlated with the increase in the power of slow activity in the neocortex (Buzsáki et al., 1988). At least some of these neurons projected to the neocortex, as demonstrated by antidromic activation in response to neocortical volleys. A similar inverse correlation between NB neuronal discharge frequency and neocortical slow waves has been reported in urethane-anesthetized rats (Détári and Vanderwolf, 1987). In awake, but diazepam-sedated rats, three quarters of NB neurons exhibited robust increases in firing rates during the presentation of a conditioned tone cue (Rigdon and Pirch, 1986). Based on spontaneous firing repertories, Détári et al. (1984) has distinguished 5 subgroups of neurons in the pallidal/peripallidal region of the cat. From 75 to 100% of cells in all groups had higher frequency discharge rates in the awake state and the paradoxical phase of sleep, than in slow wave sleep. The majority of the neurons had higher than 10 Hz firing frequency, and neurons in 3 subgroups fired in 'bursts' during slow wave sleep, and with single spikes during awake behaviors. A similar frequency increase with sleep-awake transitions was reported in another study (Szymusiak and McGinty, 1986).

From these studies it appears that the frequency changes of NB neurons correlate well with both behavioral and neocortical electrical arousal. Increased discharge frequency in NB neurons may therefore be considered a putative cause for neocortical desynchronization. The cholinergic nature of these NB neurons awaits further confirmation.

Effect of NB and thalamic lesions on neocortical activity

Lo Conte et al. (1982) were the first to examine the neocortical consequences of NB lesions. They used electrolytic lesions and reported that the neocortical EEG became depressed. Stewart et al. (1984), using extensive kainic acid destruction of the NB, found increased delta activity in the neocortex during immobility. Unfortunately, in this latter study the lesion also produced extensive damage to areas surrounding the NB, including the thalamus, and interpretation of the findings was not straightforward because of the distal cortical damage caused by kainic acid lesion (Arendash et al., 1987).

In our studies, circumscribed lesions in the peripallidal or magno-cellular preoptic regions of the NB gave rise to a delta focus in the neocortical maps of delta waves (Fig. 1). Subsequent histological evaluation of the brains revealed that the delta focus corresponded to neocortical areas where AChE-positive fiber density was mostly reduced. Furthermore, the asymmetric distribution of delta power in the lesioned rat could be reduced by scopolamine, indicating that the increased slow activity in the lesioned hemisphere was due mainly to the malfunctioning of the cholinergic NB projection (Buzsáki et al., 1988a).

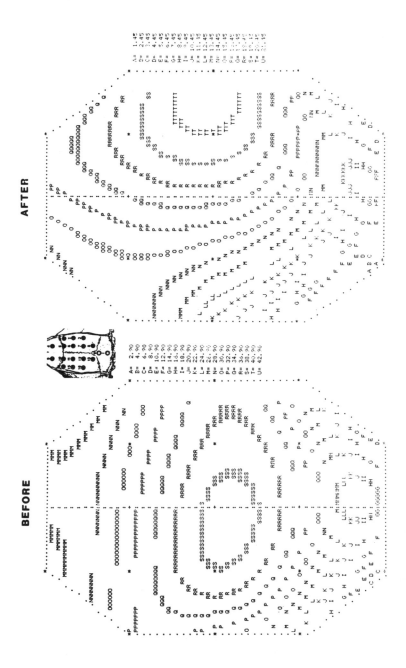

Figure 1. Spatial distribution of the power of delta (1–4 Hz) activity in the neocortex of the alert but immobile rat before and after ibotenic acid lesion of the right NB. The locations of the 16 recording electrodes are indicated in the inset. Power increases in alphabetical order. Note delta focus over the parietal area after NB lesion.

Large or small lesions of the thalamus, including the intralaminar nuclei, had no effect on the neocortical EEG correlates of behavior, except abolishing HVS (Buzsáki et al., 1988a; Vanderwolf and Stewart, 1988).

High-voltage spindles in the neocortex

HVS occurred occasionally, while the rat was sitting motionless, with highest amplitude over the sensorimotor area. Depth profile measurements indicated that the laminar profiles of HVS and delta waves were different in the neocortex. Delta waves displayed an amplitude minimum and abrupt reversal of phase in layer V, while HVS had no amplitude minima and a gradual phase-shift from layer I to layer VI, suggesting that different dipoles are involved in the generation of these respective EEG patterns. All neurons in the neocortex and thalamocortical nuclei fired exclusively during the spike component of the HVS. In contrast, neurons in the GABAergic RT discharged maximally during the wave component. Furthermore, rhythmic firing of RT cells preceded the neocortical appearance of the HVS. Ibotenic acid lesions of the RT reduced or completely abolished spontaneous or pentylene-tetrazol-induced HVS (Buzsáki et al., 1988a) and barbiturate spindles (Steriade et al., 1985) ipsilateral to the lesion.

Some neurons in the NB also fired rhythmically during the HVS. Importantly, an abrupt decrease in the firing of NB neurons often preceded or coincided with the onset of HVS (Fig. 2). Bilateral lesions of the NB increased the incidence of HVS. In aged rats with shrunken NB neurons, the incidence and duration of HVS was 4 to 8 times above the levels in young controls (Buzsáki et al., 1988b).

Generation of neocortical slow waves: An hypothesis

Why is the neocortex desynchronized during aroused states and dominated by slow delta waves during drowsy and sleep states? What is the neurophysiological mechanism behind this strict correlation?

Depth profile measurements in the neocortex of the cat (Ball et al., 1977; Calvet et al., 1973), rabbit (Rappelsberger et al., 1981), and rat (Buzsáki et al., 1988a) revealed that surface negative-deep positive delta waves during slow wave sleep correlate with the suppression or cessation of discharges of layer V pyramidal neurons. Our studies indicate further that this suppression of neuronal firing is present in all cortical layers (Fig. 3). The polarity reversal of delta waves occurred at the border of layers IV and V. In anesthetized animals the deep positive waves correlate with hyperpolarization of pyramidal cells (Creutzfeldt et al., 1966).

164

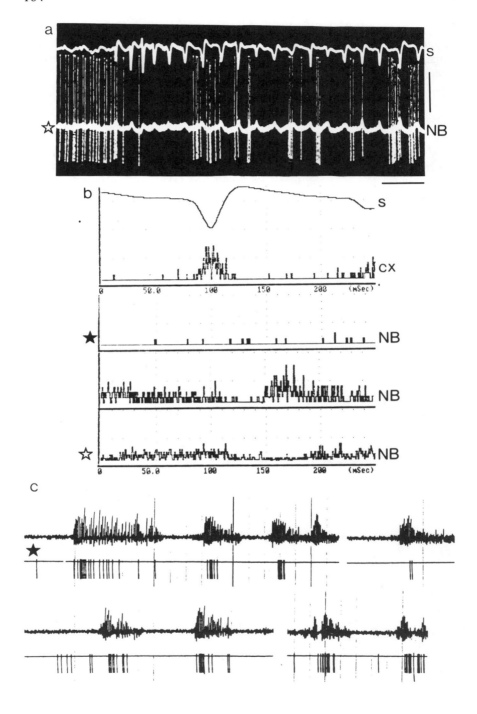

The depth profile of the slow delta waves and the observed field-unit relationship are compatible with the hypothesis that the extracellularly recorded delta waves reflect inhibition of pyramidal cells mediated by GABAergic interneurons (Ribak, 1978). GABA released at the somata of layer V pyramidal cells would open the Cl^- channels and produce an active outward current whose extracellular spatial summation corresponds to deep positivity. A simultaneously occurring passive inward current at the distal dendrites would set up extracellular (surface) negativity. Indeed, with their widespread action, GABAergic interneurons may play an important role in affecting large numbers of pyramidal cells. Assuming that subcortical inputs terminate also on GABAergic interneurons (Somogyi et al., 1983), the subcortical afferents may globally affect the whole neocortical mantle.

A major problem with this model is the lack of explanation of the desynchronized state. It is not clear how the synchronized-desynchronized states are switched by subcortical inputs in a cortical network operating with an excitatory transmitter, glutamate, and inhibitory substance, GABA. Such an isolated system (i.e. a subcortically deafferented cortical slab) will generate continuous, large, synchronized waves (our unpublished findings).

We hypothesize that the synchronized state of the neocortex cannot be altered (i.e. desynchronized) by even strong glutamate stimulation, such as mediated by activation of the thalamus. Strong excitation of pyramidal cells will result in a burst of action potentials followed by a long-lasting afterhyperpolarization (AHP; Connors et al., 1982). Since the burst is an intrinsic property of the pyramidal cell, the information transmitted by the neuron to its target will be rather independent of its input. In other words, the bursting mode of pyramidal cell activity is probably the least optimal state for information processing and transmission.

An implicit prediction of the inhibitory interneuron-pyramidal cell model of slow wave generation is that GABAergic cells should fire maximally during the deep-positive delta waves. In our studies (Buzsàki et al., 1988a), however, we failed to find such a correlation. All putative, physiologically identified, neocortical interneurons decreased their firing rates during the deep-positive slow waves. Based on this finding and on the similar depth profiles of several different types of slow waves, we suggested that extracellular waves are not, or not exclusively, a result of

Figure 2. NB control of thalamocortical HVS. *a* Neuronal activity in NB in relation to HVS. Note decreased firing of the cell at the onset of HVS and phase-locked firing before the spike component of HVS recorded from the cortical surface (s). *b* Perievent histograms (n = 50) of a cortical cell (cx) and 3 types of NB cells recorded in the same rat. *c* Polygraph recording of HVS and discriminated unit discharges. Open and black stars in *b*, neurons shown in *a* and *c*, respectively. Calibrations: 0.4 s (*a*); 2 s (*c*); 1 mV (s in *a* and *c*); 0.5 mV (NB in *a*). Reprinted from Journal of Neuroscience (1988) with permission.

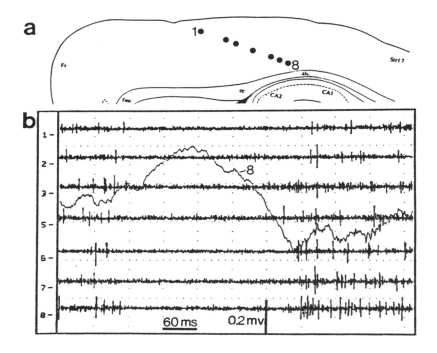

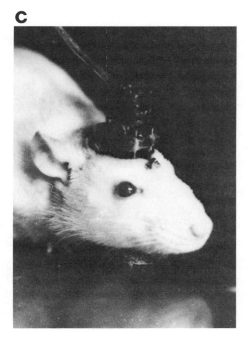

synaptic activity, but may reflect summation of long-lasting AHPs of dominantly layer V pyramidal neurons (Buzsáki et al., 1988a). A major part of the AHP is due to a Ca^{2+}-mediated K^+-conductance change. The K^+ current is outgoing at the level of the soma. The long-lasting nature of AHP favors the summation of outward currents of individual pyramidal cells resulting in a large local field positivity.

Physiological role of NB in neocortical activation

An attractive feature of the AHP hypothesis of slow wave generation is that it is capable of explaining the correlation between EEG activity and behavioral arousal. Neurotransmitters of the diffusely projecting basal forebrain and possibly brainstem nuclei (e.g. acetylcholine, noradrenaline, histamine) possess a common functional denominator: namely, all of them suppress the Ca^{2+}-mediated K^+ conductance in pyramidal cells and block the accommodation of spike discharges to depolarizing events (Krnjevic et al., 1971; Benardo and Prince, 1982; Cole and Nicoll, 1984). We assume that behavioral arousal and neocortical desynchronization are linked by the blockade of AHPs in pyramidal cells as a result of the increased activity of basal forebrain cholinergic and/or brainstem amine-containing neurons. Suppression of AHPs will abolish synchronous extracellular current flow and the neocortex will display low-amplitude, desynchronized EEG. At the same time, the lack of long-lasting AHP and blockade of accommodation will allow the cortical cells to respond briskly and without failure to the specific information transmitted from the thalamus and other cortical areas. We suggest that the cholinergic NB may be viewed as the major anatomical substrate of the parallel ascending activation system of Moruzzi and Magoun (1949).

The neocortex displays desynchronized activity not only during awake states but also during REM sleep. Several studies have examined the activity of locus coeruleus (LC) neurons in unanesthetized rat, cat, and monkey, and have reported that LC cells are most active in waking, discharge less in slow wave sleep, and are virtually quiescent in REM sleep (Foote et al., 1980; Aston-Jones and Bloom, 1981). Serotoninergic median raphe cells have a very low spontaneous firing rate and, like LC neurons, discharge most during waking, at reduced

Figure 3. Multiple electrode recording of unitary activity in the neocortex. *a* Reconstruction of the tips of the microelectrodes (out of 16) which yielded unitary activity. *b* Unitary activity in different cortical layers and simultaneously recorded delta wave recorded from layer V/VI (electrode 8). Note complete suppression of cellular discharges in all layers on the positive portion of the delta wave (positivity up). *c* Movable multielectrode headstage and compact 16-channel MOSFET preamplifier system.

rates during SWS, and at lowest levels in REM sleep (Trulson and Jacobs, 1979). On the other hand, the discharge frequency of NB neurons during REM sleep is as high as in the awake animal (Détári et al., 1987). Thus, it appears that the ascending cholinergic system alone is capable of keeping the neocortex in its operative mode. Brainstem aminergic afferents may have an important adjuvant effect in neocortical activation (Vanderwolf and Baker, 1986; Vanderwolf and Stewart, 1986; Vanderwolf, 1988), but the primary activating system appears to be the NB-neocortical projection (Buzsáki et al., 1988b).

A recently discovered, important target of the basal forebrain is the RT thalamus (Hallanger et al., 1987; Steriade et al., 1987). As summarized above, our findings with unit recordings, lesion techniques, and aged animals indicate that the NB has a tonic suppressive action on the 'pacemaker' cells of the RT. Indeed, early experiments indicate that acetylcholine has an inhibitory action on the activity of RT cells (Ben-Ari et al., 1976). When the activity of the cholinergic NB neurons falls below a critical level, highly synchronous population bursts appear in the GABAergic population of the RT. Rhythmic firing of these cells will impose hyperpolarization on the thalamo-cortical cells and release from hyperpolarization will de-inactivate the low-threshold Ca^{2+} channels present in these cells (Jahnsen and Llinas, 1984). As a result, a large population of thalamo-cortical neurons will fire rhythmically, and the rhythmicity, in turn, is transmitted to wide areas of the neocortex (Buzsáki et al., 1988a).

Conclusions

The cholinergic NB plays an essential role in activation of the cerebral cortex via a dual mechanism. First, the cortical cholinergic projection from NB directly influences neocortical pyramidal cells mainly by blocking AHP and accommodation in these cells. Second, the cholinergic input from NB to RT suppresses the rhythmic oscillations of the 'pacemaker' neurons of this thalamic nucleus. Thus, increased activity of NB neurons, e.g. during walking, will ensure neocortical activation by direct release of acetylcholine in the neocortex and by dampening the oscillatory influences of the thalamus.

Acknowledgements. The work reviewed here was supported by the J.D. French Foundation for Alzheimer's Disease, the Sandoz Foundation, NIA, and the California State Department of Health. We thank Sheryl Christenson for typing the manuscript.

Arendash, G., Millard, W. J., Dunn, A. J., and Meyer, E. M. (1987) Long-term neuropathological and neurochemical effects of nucleus basalis lesions in the rat. Science 238: 952–956.

Aston-Jones, G., and Bloom, F. E. (1981) Norepinephrine-containing locus coeruleus neurons in behaving rats exhibit pronounced responses to non-noxious stimuli. J. Neurosci. 1: 887–900.

Ball, C. J., Gloor, P., and Schaul, N. (1977) The cortical electromicrophysiology of cats. Electronencephal. clin. Neurophysiol. 43: 346–363.

Benardo, L. S., and Prince, D. A. (1982) Cholinergic excitation of mammalian hippocampal pyramidal cells. Brain Res. 249: 315–331.

Ben-Ari, Y., Dingledine, R., Kanazawa, I., and Kelly, J. S. (1976) Inhibitory effects of acetylcholine on neurons in the feline nucleus reticularis thalami. J. Physiol. (Lond.) 261: 647–671.

Buzsáki, G., Bickford, T. G., Ponomareff, G., Thal, L. J., Mandel, R. J., and Gage, F. H. (1988a) Nucleus basalis and thalamic control of neocortical activity in the freely moving rat. J. Neurosci. 26: 735–744.

Buzsáki, G., Bickford, R. G., Amstrong, D. M., Ponomareff, G., Chen, K., Ruiz, R., Thal, L. J., and Gage, F. H. (1988b) EEG activity in the neocortex of freely moving young and aged rats. Neuroscience 8: 4007–4026.

Calvet, J., Fourment, H., and Michel, T. (1973) Electrical activity in neocortical projection and association areas during slow wave sleep. Brain Res. 52: 173–187.

Coben, L. A., Danzinger, W. L., and Starandt, M. (1985) A longitudinal study of mild senile dementia of Azheimer type: changes at 1 year and 2.5 years. Electroencephal. clin. Neurophysiol. 61: 101–112.

Cole, A. E., and Nicoll, R. A. (1984) Characterization of a slow cholinergic postsynaptic potential recorded in vitro from rat hippocampoal pyramidal cells. J. Physiol. (Lond.) 352: 173–188.

Connors, B. W., Gutnick, M. J., and Prince, D. A. (1982) Electrophysiological properties of neocortical neurons in vitro. J. Neurophysiol. 48: 1302–1320.

Creutzfeldt, O., Watanabe, S., and Lux, H. D. (1966) Relations between EEG phenomena and potentials of single cortical cells. I. Evoked responses after thalamic and epicortical stimulation. Electronencephal. clin. Neurophysiol. 20: 1–18.

Détári, L., Juhász, G., and Kukorelli, T. (1984) Firing properties of cat basal forebrain neurons during sleep-wakefulness cycles. Electroenceph. clin. Neurophysiol. 58: 362–368.

Détári, L., and Vanderwolf, C. H. (1987) Activity of identified cortically projecting and other basal forebrain neurons during large slow waves and cortical activation in anaethetized rats. Brain Res. 437: 1–10.

Divac, I. (1975) Magnocellular nuclei of the basal forebrain project to neocortex, brain stem, and olfactory bulb. Review of some functional correlates. Brain Res. 93: 385–398.

Fibiger, H. C. (1982) The organization and some projections of cholinergic neurons of the mammalian forebrain. Brain Res. Rev. 4: 327–388.

Foote, S. L., Aston-Jones, G., and Bloom, F. E. (1980) Impulse activity of locus coeruleus neurons in awake rats and monkeys is a function of sensory stimulation and arousal. Proc. natl Acad. Sci. USA 77: 3033–3037.

Fox, K., and Armstrong-James, M. (1986) The role of the anterior intralaminar nuclei and N-methyl D-aspartate receptors in the generation of spontaneous bursts in rat neocortical neurones. Exp. Brain Res. 63: 505–518.

Hallanger, A. E., Levey, A. I., Lee, H. J., Rye, D. B., and Wainer, B. H. (1987) The origins of cholinergic and other subcortical afferents to the thalamus in the rat. J. comp. Neurol. 262: 105–124.

Jahnsen, H., and Llinas, R. (1984) Electrophysiological properties of guinea-pig thalamic neurones; an in vitro study. J. Physiol. (Lond.) 349: 205–226.

Jasper, H. H. (1949) Diffuse projection systems: the integrative action of the thalamic reticular system. Electroenceph. clin. Neurophysiol. 1: 405–420.

Jones, E. G. (1985) The Thalamus. Plenum, New York.

Jones, E. G., Burton, H., Saper, C. B., and Swanson, L. W. (1976) Midbrain diencephalic and cortical relationships of the basal nucleus of Meynert and associated structures in primates. J. comp. Neurol. 167: 385–420.

Klingberg, F., and Pickenhain, L. (1968) Das Aufreten von "Spindelentladungen" bei der Ratte in Beziehung zum Verhalten. Acta biol. med. germ. 20: 45–54.

Krnjevic, K., Pumain, R., and Renaud, L. (1971) The mechanism of excitation by acetylcholine in the cerebral cortex. J. Physiol. (Lond.) 215: 247–268.

Lo Conte, G., Casamenti, E., Bigl, V., Milaneschi, E., and Pepeu, G. (1982) Effect of magnocellular forebrain nuclei lesions on acetylcholine output from the cerebral cortex, electrocorticogram and behavior. Arch. Ital. Biol. 120: 176–188.

Longo, V. G. (1956) Effects of scopolamine and atropine on electroencephalographic and behavioral reactions due to hypothalmic stimulation. J. Pharmac. 116: 198–208.

Mesulam, M.-M., and Van Hoesen, G. W. (1976) Acetylcholinesterase-rich projections from the basal forebrain of the rhesus monkey to neocortex. Brain Res. 109: 152–157.

Mesulam, M.-M., Mufson, E. J., Wainer, B. H., and Levey, A. I. (1983) Central cholinergic pathways in the rat: an overview based on an alternative nomenclature (Ch1–Ch6). Neuroscience 4: 1185–1201.

Moruzzi, G., and Magoun, H. W. (1949) Brainstem reticular formation and activation of the EEG. Electroenceph. clin. Neurophysiol. 1: 455–473.

Nicoll, R. A. (1988) The coupling of neurotransmitter receptors to ion channels in the brain. Science 241: 545–551.

Papez, J. W. (1956) Central reticular path to intralaminar and reticular neurons of the thalamus for activating EEG related to consciousness. Electroenceph. clin. Neurophysiol. 8: 117–128.

Pentilla, M., Partanen, J. V., Soinen, H., and Reikkinen, P. J. (1985) Quantitative analysis of occipital EEG in different stages of Alzheimer's disease. Electroenceph. clin. Neurophysiol. 60: 1–6.

Phillis, J. W. (1980) Acetylcholine release from the cerebral cortex and its role in cortical arousal. Brain Res. 7: 378–389.

Rappelsberger, P., Pockberger, H., and Petsche, H. (1981) The contribution of the cortical layers to the generation of the the EEG: field potential and current source density analyses in the rabbit's visual cortex. Electroenceph. clin. Neurophysiol. 53: 254–269.

Ribak, C. E. (1978) Aspinous and sparsely-spinous stellate neurons in the visual cortex of rats containing glutamic acid decarboxylase. J. Neurocytol. 7: 461–478.

Rigdon, G. C., and Pirch, J. H. (1986) Nucleus basalis involvement in conditioned neuronal responses in the rat frontal cortex. J. Neurosci. 6: 2535–2542.

Shute, C. C. D., and Lewis, P. R. (1967) The ascending cholinergic reticular systems: neocortical, olfactory and subcortical projections. Brain 90: 497–520.

Somogyi, P., Kisvárday, Z. F., Martin, K. A. C., and Whitteridge, D. (1983) Synaptic connections of morphologically identified and physiologically characterized large basket cells in the striate cortex of cat. Neuroscience 10: 261–294.

Steriade, M. (1970) Ascending control of thalamic and cortical responsiveness. Int. Rev. Neurobiol. 12: 87–144.

Steriade, M., Deschenes, M., Domich, L., and Mulle, C. (1985) Abolition of spindle oscillations in thalamic neurons disconnected from nucleus reticularis thalami. J. Neurophysiol. 54: 1473–1497.

Steriade, M., Parent, A., Pare, D., and Smith, Y. (1987) Cholinergic and non-cholinergic neurons of cat basal forebrain project to reticular and medio-dorsal thalamic nuclei. Brain Res. 408: 372–376.

Stewart, D. J., MacFabe, D. F., and Vanderwolf, C. H. (1984) Cholinergic activation of the electrocorticogram: Role of substantia innominate and effects of atropine and quinuclidinyl benzylate. Brain Res. 322: 219–232.

Szymusiak, D., and McGinty, D. (1986) Sleep-related neuronal discharge in the basal forebrain of cats. Brain Res. 370: 82–92.

Trulson, M. E., and Jacobs, B. L. (1979) Raphe unit activity in freely moving cats: correlation with level of behavioral arousal. Brain Res. 163: 135–150.

Vanderwolf, C. H. (1975) Neocortical and hippocampal activation in relation to behavior: Effects of atropine, eserine, phenothiazines, and amphetamine. J. comp. Physiol. Psychol. 88: 300–323.

Vanderwolf, C. H. (1988) Cerebral activity and behavior: Control by central cholinergic and serotonergic systems. Int. Rev. Neurobiol. in press.

Vanderwolf, C. H., and Baker, G. B. (1986) Evidence that serotonin mediates non-cholinergic neocortical low voltage fast activity, non-cholinergic hippocampal rhythmical slow activity and contributes to intelligent behavior. Brain Res. 374: 342–356.

Vanderwolf, C. H., and Robinson, T. H. (1918) Reticulocortical activity and behavior: A critique of the arousal theory and a new synthesis. Behav. Brain Sci. 4: 459–514.

Vanderwolf, C. H., and Stewart, D. J. (1986) Joint cholinergic-serotonergic control of neocortical and hippocampal electrical activity in relation to behavior: Effects of scopolamine, ditran, trifluoperazine and amphetamine. Physiol. Behav. 38: 57–65.

Vanderwolf, C. H., and Stewart, D. J., (1988) Thalamic control of neocortical activation: A critical re-evaluation. Brain Res. Bull. 20: 529–553.

Wikler, A. (1952) Pharmacologic dissociation on behavior and EEG sleep patterns in dogs: Morphine, N-allylnormorphine and atropine. Proc. Soc. exp. Biol., N.Y. 79: 261–265.

Cholinergic mechanisms in the telencephalon of cat and chicken

Christian M. Müller

Max-Planck-Institut für Hirnforschung, Deutschordenstr. 45, 6000 Frankfurt/M. 71, Federal Republic of Germany

Introduction

Cholinergic afferents to the mammalian cortex have gained considerable interest because of their possible involvement in learning and memory (Everitt et al., 1987; Rigdon and Pirch, 1986), as well as their role in attentional processing (Phillis, 1968; Singer, 1979). These afferents originate predominantly from several nuclei in the basal forebrain (Mesulam et al., 1983). Similar cholinergic afferents are known from non-mammalian species, e.g. from reptiles and birds (Mufson et al., 1984; Davies and Horn, 1983). Receptor binding studies also reveal the presence of acetylcholine receptors in the bird telencephalon (Wächtler, 1985). Like in mammalian species, sensory processing is impaired in birds when cholinergic antagonists are given (Cleeves and Green, 1982). These data indicate an overall similarity of the organization and effects of the cholinergic system in mammals and non-mammalian vertebrates.

From single-unit studies in mammals it is known that microiontophoretic application of cholinergic agonists has three distinct effects on cortical neuronal responses. Namely facilitation of evoked responses, direct excitation, and response suppression (Krnjević et al., 1971; Sillito and Kemp, 1983). All three actions of acetylcholine are mediated via muscarinic receptors. To date, the interrelation of the different effects of ACh is only poorly understood. Little information is present on the action of ACh in non-mammalian telencephalon. In an attempt to gain insight into the role of cholinergic afferents to the sensory telencephalon I studied the effects of ACh on neuronal responses in the auditory forebrain nuclei of the chicken and the visual cortex of the cat. For both structures we have detailed information about the functional organization. One of the organizational principles that is shared by both structures is the layering. While we can subdivide the cortex into six laminae, four layers can be distinguished in the auditory forebrain center (three layers in the neostriatum, and the adjacent hyperstriatal portion; Müller and Scheich, 1988). One of these layers receives the thalamic input in

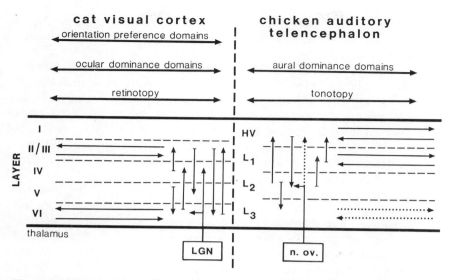

Figure 1. Thalamo-telencephalic and intratelencephalic connections in the caudal auditory telencephalon of birds (based on: Bonke et al., 1979; Müller, 1987a) and the visual cortex of the cat (based on: Gilbert, 1983). Notice the presence of reciprocal horizontal (intralaminar) and vertical (interlaminar) interconnectivities in both species. Functional topographies are arranged orthogonal to the layering of the structures.

both structures and relays it to the other laminae. In addition, there are vertical interconnectivities between the layers and horizontal connections within individual laminae (Fig. 1). A second organizational principle of both sensory structures is the presence of functional topographies extending orthogonal to the layering. In the visual cortex these are the retinotopic organization, the system of ocular dominance stripes, and the system of orientation preference stripes (for review see Hubel and Wiesel, 1977). In the auditory system the prominent functional organization is the tonotopy, i.e. a spatial representation of tone frequencies (Heil and Scheich, 1985).

The interpretation of the data will specifically focus on the integration of cholinergic mechanisms into the functional architecture of the studied sensory nuclei. Comparison of the functional topography of cholinergic actions in the avian auditory telencephalon and the mammalian visual cortex suggests a model of sensory processing in which the different actions of ACh can be accommodated in modulating sensory processing. Implications of the model for further research strategies will be discussed.

Methods

Two approaches were undertaken to study the action of ACh in the avian forebrain. First, I studied the effects of microiontophoretical application of ACh onto the responses of single-units in fully awake chickens. The animals underwent a surgical preparation under deep Halothane anesthesia at least one day prior to the experiments. This included the opening of the scull dorsal to the caudal auditory telencephalon to allow electrode access during the experiments. In addition, a metal anchor was fixed onto the scull which served as a fixation point during single-unit recordings. Multi-barrel electrodes with a tungsten wire inserted into the central barrel were used for recording neuronal activity and applying neuroactive substances by iontophoresis. A detailed description of the methods has been published previously (Müller and Scheich, 1988).

A further analysis of the pharmacological properties of the effects of ACh was done by intracellular recordings in an *in vitro* slice preparation of the chicken auditory telencephalon. Slices that contained all four substructures of the auditory telencephalon were cut on a vibratome and transferred to a superfusion recording chamber. Standard recording and stimulation techniques were employed (for details see Müller, 1987a). Cholinergic drugs were either added to the superfusion medium or pressure-applied via a micropipette.

Studies in the visual cortex of the cat were done in anesthetized and paralyzed preparations. For single-unit recording and microiontophoretic application of neuroactive substances we used piggyback electrodes consisting of a glass recording pipette glued alongside a seven-barrel iontophoresis electrode. Details of the experimental procedures may be found in recent publications (Francesconi et al., 1988; Müller and Singer, in print).

Results

Cholinergic effects in the chicken auditory forebrain—in vivo

Iontophoretic application of ACh in the chicken auditory forebrain has two major effects. The most common action of ACh is a facilitation of evoked responses, without a clear effect on maintained activity levels. Within several seconds after the start of drug application neuronal responses to auditory stimuli are enhanced. In addition, there usually occurs a prolongation of phasic responses. These changes in the neuronal responsiveness are blocked by the muscarinic receptor antagonist scopolamine. In a few cells I compared the effects of iontophoretic application of ACh with changes in neuronal responses induced by potentially arousing stimuli. Both actions of ACh, i.e. an increase in response amplitude and a prolongation of phasic responses to tone-stimulation, were also present during behavioral arousal (Müller, 1987a). Beside the changes in the response strength, also changes in the response selectivity are present during the application of ACh. About 20% of the cells in the auditory forebrain do not respond to simple stimuli, like pure tones, but only to complex sound structures, e.g. vocalizations. Iontophoresis of ACh often uncovers tone-responses in these cells. In addition to that, we observed a widening of auditory tuning curves ('auditory receptive fields') in tone-responsive cells. Interestingly, similar changes could also be seen following the blockage of GABAergic inhibition by iontophoretic application of bicuculline (Müller and Scheich, 1988).

The second effect observed during application of ACh, *in vivo*, was a rapid suppression of both maintained firing rates, as well as evoked responses. This effect has an onset time of less than a second, which is similar to the inhibition induced by iontophoresis of GABA (Müller, 1987a; Müller and Scheich, 1988).

Reconstruction of the location of the recorded units in histological sections revealed that facilitatory effects of ACh were present predominantly in cells located in the neostriatal layers of the auditory telencephalon, while the suppressive effects were usually observed in hyperstriatal units. Although the reconstruction of recording sites from multiple experimental sessions by small electrolytic lesions bears some inaccuracies, this was a first hint for a topographic distinction of the facilitatory and suppressive effects of ACh on telencephalic auditory units.

Cholinergic effects in the chicken auditory forebrain—in vitro

In order to obtain further information about the topography and the pharmacology of cholinergic effects auditory forebrain neurons of the chicken were intracellularly recorded in slice preparations of this structure. In this preparation the border between the hyperstriatal and the neostriatal part of the auditory structure is clearly visible with transillumination. Thus, either neostriatal or hyperstriatal neurons can be recorded by aiming the recording electrode under visual control at one specific site. Stable recordings were obtained from 62 neurons. In all neurons located in the neostriatal part (N = 37) application of ACh resulted in a slow depolarization, which was accompanied by an increase in the membrane resistance. This depolarization increased when the membrane potential was clamped to levels more positive to the resting potential. This voltage dependency is to be expected if one assumes that ACh blocks the M-current (Adams and Brown, 1982; Müller, 1987a). An additional effect of ACh was the reduction of spike-accommodation during depolarizing current pulses.

Cells located in the hyperstriatal parts of the auditory telencephalon showed a different response to ACh. Superfusion of the slice with ACh

(5 mM) reduced or even suppressed excitatory postsynaptic potentials evoked by afferent stimulation. At the same time most of the cells showed a reduction of the membrane resistance and a slight hyperpolarization of the membrane potential. When ACh was washed out of the recording chamber, a second effect occurred consisting of a depolarization and an increase of the membrane resistance above predrug levels. This effect was identical to the cholinergic effects seen in neostriatal cells. Both responses were eliminated when the muscarinic receptor blocker scopolamine (10 μM) was added to the superfusion medium. As the suppressive effect of ACh closely resembled the one observed after local application of GABA, the influence of bicuculline, a selective GABA$_A$ receptor antagonist, was also tested on ACh-induced hyperpolarizations. In all studied cells bicuculline (5 μM) selectively blocked the hyperpolarizing response to locally applied ACh.

Cholinergic effects on cells in the cat visual cortex

The effect of iontophoretic application of ACh was studied in 119 units from the visual cortex of the cat. In the majority of the cells (57%) ACh-application resulted in an enhanced spike frequency in response to visual stimulation with only minor changes in the maintained activity levels. This effect typically built up over a period of several tens of seconds and was antagonized by concurrent iontophoresis of scopolamine. In addition to the increase in response amplitude, we also observed a reduced stimulus selectivity in a subpopulation of cells. These units responded to only one direction of stimulus movement before ACh-application, whereas additional clear responses to the other stimulus direction were evident during the application of ACh. Comparable changes in the response specificity were seen during the blockage of GABAergic transmission by iontophoretically administered bicuculline.

A second effect of ACh iontophoresis was a suppression of maintained and evoked activity. This action was observed in 26% of the cells studied. Like the facilitatory action, also the suppressive effect was antagonized by the muscarinic ACh-receptor antagonist scopolamine. In addition, the effect of concurrent application of bicuculline was tested. In all cells studied, blockage of GABAergic inhibition antagonized the suppressive effect of ACh. In some of these neurons ACh even induced a facilitation during bicuculline iontophoresis. The time course of the suppressant action of ACh was considerably faster than the facilitatory action. Usually this effect occurred within a few hundred milliseconds after starting the iontophoresis.

In a few cells (7%) ACh-application resulted in a fast excitation having an onset-time similar to the suppressive effect. This excitation was not parallelled by an increase in the response amplitude, but the visually evoked activity was sometimes even obscured by the increased maintained firing level.

Discussion

General effects of ACh on telencephalic activity

Three effects of ACh on telencephalic neurons can be differentiated: 1) facilitation, 2) suppression, and 3) excitation. Facilitation of neuronal responses is the most common action. It is characterized by a slow onset and takes at least 500 ms and up to several seconds until the effect is maximal. The major effects are an increase in the response amplitude and a prolongation of responses most probably due to a change in the accommodation properties of the cells. These effects can be explained by an ACh-induced blockage of voltage and/or calcium-dependent potassium channels, both in mammals (Krnjevic et al., 1971; Cole and Nicoll, 1983) and in the chicken (Müller, 1987a). Facilitation mediated by ACh is found in all layers of the cortex (Sillito and Kemp, 1983; Sato et al., 1987) and also throughout the auditory telencephalon of the chicken. However, in the latter it is usually obscured by the suppression in the hyperstriatal part and is only visible during bicuculline application or following the fast decay of the suppressive effect.

The suppressive effect of ACh has a much faster onset than the facilitatory action. Suppression typically occurs within a hundred milliseconds and recovers with a similarly short time constant. Experiments on *in vitro* slice preparations of the mammalian cortex (McCormick and Prince, 1986), as well as those from the chicken auditory forebrain (Müller, 1987a) suggest that the suppressive effect of ACh is mediated by a rapid activation of GABAergic interneurons. The fact that cholinergic suppression is antagonized by bicuculline in the cat visual cortex further supports this interpretation. In contrast to the facilitatory response, cholinergic suppression is restricted to cortical layers II–IV (Sillito and Kemp, 1983; Sato et al., 1987) and to the hyperstriatal part of the auditory telencephalon of the chicken (Müller, 1987a). The absence of cholinergic suppression in the chicken neostriatum cannot be due to the fact that ACh does not reach the appropriate GABAergic interneurons, as even bath application of ACh does not reveal this effect, despite the presence of synaptically evoked IPSP's (Müller, 1987a). Thus, it has to be assumed that cholinergic suppression (i.e. activation of GABAergic interneurons) occurs only in restricted regions of both the cortex and the chicken auditory telencephalon.

A small subpopulation of visual cortical cells in the cat have also been shown to be rapidly excited by ACh. These cells are good candidates for

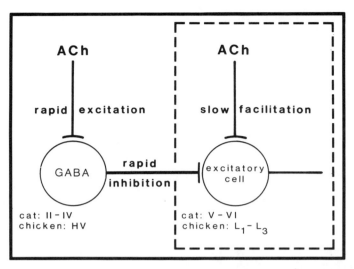

Figure 2. Characteristics and topography of cholinergic effects on telencephalic neurons. Slow facilitation affects excitatory cells in all layers of the sensory nuclei. This action of ACh is often obscured by the indirect suppression mediated by a rapid activation of GABAergic neurons. The latter effect is observed only in the superficial layers of the cat cortex and the hyperstriatal lamina of the chicken auditory forebrain.

GABAergic inhibitory interneurons, as the time course of the excitation closely resembles the time course of the (indirect) cholinergic suppression. The failure of recording such neurons in the chicken auditory forebrain might be due to the fact that GABAergic interneurons are both smaller than presumed excitatory cells and are usually located in the direct vicinity of non-GABAergic cells (Müller, 1987b). This arrangement probably reduces the chance to obtain single-cell recordings from inhibitory cells.

In summary, it can be concluded that cholinergic actions differ topographically. As schematically shown in Figure 2, at least a subpopulation of GABAergic interneurons as well as excitatory cells are influenced by ACh in the superficial cortical layers and the hyperstriatal parts of the chicken auditory telencephalon; while the GABAergic cells are excited by ACh, excitatory cells show cholinergic facilitation and indirect suppression. In the deep cortical layers, as well as in the neostriatal parts of the chicken auditory telencephalon, only the slow facilitation is observed.

A model of cholinergic modulation of telencephalic circuitry

In order to elaborate a model of cholinergic influences on telencephalic circuitry I want to point out two characteristics of cholinergic effects. First, the two time courses of cholinergic facilitation versus the excitation of GABAergic interneurons mediating the rapid indirect suppression. While the excitation and the subsequent suppression occur within a very short time span, facilitatory responses build up over a longer time period. The second point is the distinct presence of suppressive effects only in the upper layers of the cat cortex and in the hyperstriatal portion of the chicken auditory telencephalon. These areas share one feature in regard to the interconnectivities in the telencephalon. Both, the cortical layers II–IV, as well as the hyperstriatum are relays of multiple intratelencephalic excitatory loops. This is exemplified for the cat cortex in Figure 3A which shows the excitatory loop systems. If we assume that the cholinergic afferents to the telencephalon are phasically activated, then a characteristic sequence of effects can be predicted. First, there is a rapid activation of GABAergic neurons which suppresses on-going activity in those neuronal elements that participate in multiple intratelencephalic excitatory loop systems. This can be expected to lead to a rapid elimination of stabilized activity patterns in

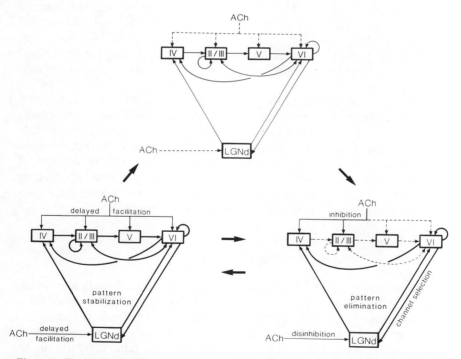

Figure 3. Functional model of the action of cholinergic afferents to the visual cortex. Phasic activation of the cholinergic system leads to a sequence of effects: 1) Elimination of activity patterns in the excitatory loop systems by rapid inhibition in the superficial layers. 2) Formation of new activity patterns after the cholinergic suppression has faded is supported by delayed facilitation (for details see text).

most of the reverberating networks. This inhibition is transient and followed by the delayed facilitation. The latter is likely to support the elaboration of new activity patterns by increasing the efficacy of synaptic transmission.

The proposed model is supported by preliminary experiments in the visual cortex of the cat. Electrical stimulation of the reticular formation triggers a similar sequence of events: a short-term inhibition that is followed by facilitation. Such stimulation has been shown to lead to an increased ACh-release in the cortex (Phillis, 1968). Thus, one function of the corticopetal cholinergic afferents might be to first eliminate on-going activity, and to subsequently support the formation of new activity patterns.

The absence of cholinergic suppression in the deep layers of the cortex leaves the recurrent excitatory loop system interconnecting the visual thalamus and cortical layer VI unaffected. Layer VI cells have been shown to be able to respond independently from the other cortical layers (Schwark et al., 1986) and the cortico-thalamic loop has been implicated to participate in the selection of thalamo-cortical activity (Koch, 1987; Varela and Singer, 1987). Thus, it can be assumed that the deep cortical layers exert a preselection of activated thalamo-cortical afferent, while activity in the excitatory loops is reset by cholinergic suppression. Once cholinergic suppression has faded, this preselected set of afferents subsequently can induce new activity patterns in the cortical loop systems.

Appropriately timed phasic activation of cholinergic projections would thus be well suited to facilitate transitions between telencephalic activity patterns in response to changing sensory input. Interestingly orienting responses and saccadic eye movements which predictably lead to a change in visual scenes are associated with corollary activity of central origin (for review see Singer, 1977). The proposed cholinergic mechanism might be a computational mechanism that allows the analysis of sensory input unbiased by prior sensory stimulation.

Questions arising from the proposed model

The proposed model implicates that cholinergic afferents are phasically activated during arousal. To date, only very limited information is available about response patterns of corticopetal cholinergic cells. No attempts have been made to study responses of cholinergic cells in non-mammalian species. Though some studies report neuronal responses from cells located in the basal forebrain of monkeys during attentive behavior (Mitchell et al., 1987), the cholinergic nature of these cells cannot be inferred from the topography of the recording sites. Recent data give further evidence that corticopetal cholinergic neurons are intermingled with non-cholinergic cells projecting to the cortex (Fisher et al., 1988). Thus, one question concerns the timing of responses of cholinergic cells projecting to the cortex in behaviorally relevant situations.

A second question relates to the uniformity of cholinergic afferents to the cortex. It is yet unknown whether the facilitatory and the inhibitory effects shown by local ACh application are mediated by the same or by different corticopetal afferents. Thus, the topography of nuclei of origin of cholinergic afferents has to be elaborated and to be related to postsynaptic targets and/or postsynaptic actions. Although the proposed model is partially speculative, it sets a framework of expectations and, thus, might guide future research on the mechanisms of cholinergic modulation of the telencephalon.

Adams, P. R., and Brown, D. A. (1982) Synaptic inhibition of the M-current slow excitatory post-synaptic potential mechanism in bullfrog sympathetic neurones. J. Physiol. 332: 263–272.

Bonke, B. A., Bonke, D., and Scheich, H. (1979) Connectivity of the auditory forebrain nuclei in the guinea fowl (Numida meleagris). Cell Tiss. Res. 200: 101–121.

Cleeves, L., and Green, S. E. (1982) Effects of scopolamine on visual difference thresholds in the pigeon. Physiol. Psychol. 10: 306–312.

Cole, A. E., and Nicoll, R. A. (1983) Acetylcholine mediates a slow synaptic potential in hippocampal pyramidal cells. Science 221: 1299–3101.

Davies, D. C., and Horn, G. (1983) Putative cholinergic afferents of the chick hyperstriatum ventrale: a combined acetylcholinesterase and retrograde fluorescence labeling study. Neurosci. Lett. 38: 103–107.

Everitt, B. J., Robbins, T. W., Evenden, J. L., Marston, H. M., Jones, G. H., and Sirkia, T. E. (1987) The effects of excitotoxic lesions of the substantia innominata, ventral and dorsal globus pallidus on the acquisition and retention of a conditional visual discrimination: implications for cholinergic hypotheses of learning and memory. Neuroscience 22: 441–469.

Fisher, R. S., Buchwald, N. A., Hull, C. D., and Levine, M. S. (1988) GABAergic basal forebrain neurons project to the neocortex: the localization of glutamic acid decarboxylase and choline acetyltransferase in feline corticopetal neurons. J. comp. Neurol. 272: 489–502.

Francesconi, W., Müller, C. M., and Singer, W. (1988) Cholinergic mechanisms in the reticular control of transmission in the cat lateral geniculate nucleus. J. Neurophysiol. 59: 1690–1718.

Gilbert, C. D. (1983) Microcircuitry of the visual cortex. A. Rev. Neurosci. 6: 217–247.

Heil, P., and Scheich, H. (1985) Quantitative analysis and two-dimensional reconstruction of the tonotopic organization of the auditory field L in the chick from 2-deoxyglucose data. Exp. Brain Res. 58: 532–543.

Hubel, D. H., and Wiesel, T. N. (1977) Functional architecture of macaque monkey visual cortex. Proc. R. Soc. Lond. B 198: 1–59.

Koch, C. (1987) The action of the corticofugal pathway on sensory thalamic nuclei: a hypothesis. Neuroscience 23: 399–406.

Krnjevic, K., Pumain, R., and Renaud, L. (1971) The mechanism of excitation by acetylcholine in the cerebral cortex. J. Physiol. 215: 247–268.

McCormick, D. A., and Prince, D. A. (1986) Mechanism of action of acetylcholine in the guinea pig cerebral cortex, in vitro. J. Physiol. 357: 169–194.

Mesulam, M. M., Mufson, E. J., Levey, A. L., and Wainer, B. H. (1983) Cholinergic innervation of cortex by basal forebrain: cytochemistry and cortical connections of the septal area, diagonal band nuclei, nucleus basalis (substantia innominata), and hypothalamus in the rhesus monkey. J. comp. Neurol. 214: 170–197.

Mitchell, S. J., Richardson, R. T., Baker, F. H., and DeLong, M. R. (1987) The primate nucleus basalis of Meynert: neuronal activity related to a visuomotor tracking task. Exp. Brain Res. 68: 506–515.

Müller, C. M. (1987a) Differential effects of acetylcholine in the chicken auditory neostriatum and hyperstriatum ventrale—studies in vivo and in vitro. J. comp. Physiol. A 161: 857–866.

Müller, C. M. (1987b) Distribution of GABAergic perikarya and terminals in the centers of the higher auditory pathway of the chicken. Cell Tiss. Res. 252: 99–106.

Müller, C. M., and Scheich, H. (1988) Contribution of GABAergic inhibition to the response characteristics of auditory units in the avian forebrain. J. Neurophysiol. 59: 1673–1689.

Müller, C. M., and Singer, W. (1989) Acetylcholine induced inhibition in the cat visual cortex is mediated by a GABAergic mechanism. Brain Res., in press.

Mufson, E. J., Desan, P. H., Mesulam, M. M., Wainer, B. H., and Levey, A. L. (1984) Choline acetyltransferase-like immunoreactivity in the forebrain of the red-eared turtle (Pseudemys scripta elegans). Brain Res. 323: 103–108.

Phillis, J. W. (1968) Acetylcholine release from the cerebral cortex: its role in cortical arousal. Brain Res. 7: 378–389.

Rigdon, G. C., and Pirch, J. H. (1986) Nucleus basalis involvement in conditioned neuronal responses in the rat frontal cortex. J. Neurosci. 6: 2535–2542.

Sato, H., Hata, Y., Masui H., and Tsumoto, T. (1987) A functional role of cholinergic innervation to neurons in the cat visual cortex. J. Neurophysiol. 58: 765–780.

Schwark, H. D., Malpeli, J. G., Weyand, T. G., and Lee, C. (1986) Cat area 17. II. Response properties of infragranular layer neurons in the absence of supragranular layer activity. J. Neurophysiol. 56: 1074–1087.

Sillito, A. M., and Kemp, J. A. (1983) Cholinergic modulation of the functional organization of the cat visual cortex. Brain Res. 289: 143–155.

Singer, W. (1977) Control of thalamic transmission by corticofugal and ascending reticular pathways in the visual system. Physiol. Rev. 57: 386–420.

Singer, W. (1979) Central-core control of visual cortex functions, In: Schmitt, F. O. and Worden, F. G. (eds), The Neurosciences Fourth Study Program, MIT Press, Cambridge, pp. 1093–1110.

Varela, F. J., and Singer, W. (1987) Neuronal dynamics in the visual corticothalamic pathway revealed through binocular rivalry. Exp. Brain Res. 66: 10–20.

Wächtler, K. (1985) Regional distribution of muscarinic acetylcholine receptors in the telencephalon of the pigeon (Columba livia f. domestica). J. Hirnforsch. 26: 85–89.

Central nicotinic acetylcholine receptors in the chicken and *Drosophila* CNS: Biochemical and molecular biology approaches

Heinrich Betz, Eckart D. Gundelfinger, Irm Hermans-Borgmeyer, Erich Sawruk, Patrick Schloß and Bertram Schmitt

Zentrum für Molekulare Biologie der Universität Heidelberg, Im Neuenheimer Feld 282, 6900 Heidelberg, Federal Republic of Germany

Summary. Putative neuronal nicotinic acetylcholine receptors (nAChRs) were investigated using biochemical and molecular biology approaches. In the chick visual system, a nicotinic cholinergic binding site was localized on a polypeptide of Mr 57 000 using the potent antagonist, α-bungarotoxin (α-Btx). Ion flux experiments indicate that this membrane protein is different from the toxin-insensitive nAChR present in vertebrate neurons.

Crosshybridization with a *Torpedo* nAChR cDNA probe allowed isolation of a cholinergic receptor cDNA (ARD) from the *Drosophila* CNS. Analysis of the corresponding gene, its transcripts and the ARD protein are presented here.

Introduction

Electrophysiological and pharmacological studies indicate considerable differences between the nicotinic acetylcholine receptors (nAChRs) found in skeletal muscle and their counterparts present in the brain. So far, investigation of neuronal nAChR structure has relied on the use of probes derived from studies on cholinergic receptors in skeletal muscle and *Torpedo* electric organ. Here we describe experiments in which the potent antagonist of neuromuscular nAChR, α-bungarotoxin (α-Btx), and a *Torpedo* AChR cDNA probe were used to characterize putative neuronal nicotinic receptors in the chicken and *Drosophila* central nervous system (CNS).

The α-Btx receptor of the chick visual system

Different investigators have used α-Btx to identify putative nAChRs in the invertebrate and vertebrate CNS. This approach presumes that the toxin recognizes receptor proteins homologous to muscle nAChR also in other tissues ('pharmacological homology screening'). The avian visual system appears to be ideally suited for such an attempt since it contains high concentrations of high-affinity binding sites for ^{125}I-α-Btx which display a typical nicotinic cholinergic pharmacology (Vogel and

Nirenberg, 1976; Wang et al., 1978; Betz, 1981). In support of a potential AChR function, the α-Btx binding site in chick retina has been localized to synapses by electron microscope histochemistry (Vogel and Nirenberg, 1976). Previous crosslinking and affinity purification experiments have identified a high-affinity α-Btx binding site of chick retina and optic lobe on a polypeptide of Mr 57 000 (57 K) (Betz et al., 1982; Norman et al., 1982). Also, a polypeptide of 49 K has been described in affinity-purified preparations of the optic lobe α-Btx receptor and shown to exhibit partial sequence homology with muscle nAChR subunits (Conti-Tronconi et al., 1985). Indirect evidence suggests that the latter polypeptide may be a proteolytic fragment of the 57 K protein.

We have investigated the polypeptide composition and putative nAChR function of the avian α-Btx receptor by analyzing highly purified receptor preparations and raising monoclonal antibodies (mAbs) which precipitate the detergent-solubilized toxin receptor. An improved affinity purification employing sequential chromatography on α-Btx and wheat germ agglutinin columns resulted in isolation of two major polypeptide species of ≈ 57 K and 65–69 K (Hermans-Borgmeyer et al., 1988). The 57 K band probably corresponds to the toxin-binding polypeptide, as it can be detected upon ligand blotting of crude optic lobe extracts with ^{125}I-α-Btx. Furthermore, mAbs against the chick α-Btx receptor (mAbs OAR) immunoprecipitate a polypeptide of the same molecular weight from ^{125}I-labelled receptor preparations and [^{35}S]methionine-labelled detergent extracts of the rat phaeochromocytoma cell line PC12 (Betz and Pfeiffer, 1984). This cell line contains both functional nAChR and high-affinity α-Btx binding sites. The role of the 65–69 K material is presently not clear; it may correspond to additional subunits of the native high-molecular weight (about 300 K) α-Btx receptor.

To investigate the functional relationship between the α-Btx receptor and neuronal nAChRs, we have exploited the immunological crossreactivity between the avian and rat α-Btx binding proteins revealed by our monoclonal antibodies (Betz and Pfeiffer, 1984). AChR function was monitored using a highly sensitive Li$^+$ influx assay which allows quantitative determinations on entire culture dishes (Hermans-Borgmeyer et al., 1988). Addition of the agonist carbamylcholine produced a ca. 5-fold increase in intracellular Li$^+$ with maximal responses seen around 0.1–1 mM agonist concentration. This response was blocked by d-tubocurarine, but not by α-Btx and the mAbs OAR against the α-Btx receptor of chick optic lobe tissue.

In analogy to the antigenic modulation of nAChR in muscle, cultivation of PC 12 cells in the presence of crossreacting mAb produces a down-regulation of externally exposed α-Btx binding sites. Here, we attempted to correlate the number of cell surface α-Btx sites with functional nicotinic receptors by using the Li$^+$ influx assay. Although some of our mAbs produced a highly significant reduction of α-Btx binding sites after overnight incubation, no changes in Li^+ uptake could be observed (Table 1). Thus, the α-Btx receptor and nAChR in PC 12 cells must be separate membrane proteins, a conclusion consistent with previous data of Patrick and Stallcup (1977a, b).

A similar dissociation of α-Btx binding sites and nAChR probably also exists in the avian visual system. Using [^{3}H]acetylcholine in the presence of the muscarinic antagonist atropine, high-affinity acetylcholine binding sites of nicotinic pharmacology were identified in chick optic lobe tissue which resemble the α-Btx receptor with respect to subcellular localization, hydrodynamic properties, lectin binding and agonist affinity rank order (Schneider et al., 1985). They differ, however, from the toxin receptor in nicotinic antagonist affinity, regional distribution and thermal stability. Interestingly, the Triton X-100 solubilized agonist receptor could be separated from the α-Btx binding component by either affinity chromatography on a α-Btx matrix or immunoprecipitation with mAb OAR (Schneider et al., 1985). This high-affinity 'agonist receptor' probably corresponds to a non-α-Btx binding nAChR of the chicken CNS.

What may be the function of the α-Btx receptor in the avian and mammalian brain? At present, we have no definitive answer to this question. An intriguing speculation is that the avian α-Btx binding protein may correspond to another receptor of yet unknown function. In view of the recently discovered structural homology between different ligand-gated ion channels (Grenningloh et al., 1987; Schofield et al., 1987), the α-Btx receptor might indeed constitute a neurotransmitter-gated ion channel without a known ligand. Alternatively, the snake toxin might recognize a receptor for a polypeptide hormone, a possibility which deserves attention because of the presence of growth and differentiation factors and other surface ligands in snake venoms.

Table 1. ^{125}I-α-Btx binding and Li^+ influx in PC 12 cells after cultivation in the presence of OAR monoclonal antibodies[a].

mAb in culture	^{125}I-α-Btx bound (% control)	Carb-induced Li^+ influx (% control)
OAR 1a	73 ± 12*	98 ± 4
OAR 4b	116 ± 21	110 ± 16
OAR 5a	70 ± 6*	83 ± 16
OAR 8a	88 ± 9	92 ± 13
OAR 11b	43 ± 12*	93 ± 2
$\beta 1$	95 ± 8	82 ± 8

[a] PC 12 cells were cultivated on polylysine-coated culture dishes in the presence of the indicated mAb for 17 h, washed to remove bound mAb, and specific ^{125}I-α-Btx binding and carbamylcholine-induced Li^+ influx were determined. *indicates values significantly different from controls. $\beta 1$ is an mAb against the snake venom protein β-bungarotoxin.

Indeed, a report demonstrating inhibition of α-Btx binding to neuronal membranes by nerve growth factor has been published (Schmidt, 1977).

The ARD protein of *Drosophila*

An alternative approach to central nAChRs is based on the use of cDNA probes isolated from *Torpedo* electric organ or vertebrate skeletal muscle ('DNA homology screening'). Using this strategy, we and others have isolated cDNAs and genomic sequences of *Drosophila melanogaster* which encode proteins highly homologous to vertebrate nAChR subunits (Hermans-Borgmeyer et al., 1986; Gundelfinger et al., 1986; Bossy et al., 1988). *Drosophila* appears to be a particularly suitable organism to analyze the molecular genetics of neuronal nicotinic receptors as in insects neuromuscular transmission is mediated by amino acids, whereas acetylcholine is an abundant excitatory transmitter in the CNS. Furthermore, nicotine is a potent insecticide, indicating that nAChRs are indeed important for neuronal communication in these organisms.

Using a *Torpedo* nAChR gamma subunit cDNA as a probe, clones were isolated from a cDNA library of *Drosophila* heads which encode a protein (ARD protein) possessing 33–47% homology with vertebrate muscle and neuronal nAChR polypeptides and exhibiting structural features of a non-ligand binding subunit (Hermans-Borgmeyer et al., 1986). The corresponding *ard* gene has seven exons whose borders correspond to the exon-intron transitions found in vertebrate nAChR genes (Sawruk et al., 1988). Interestingly, a rather similar organization has also been reported for genes encoding putative neuronal nAChR in the chicken, suggesting that neuronal and muscular receptors diverged in evolution already before the separation of invertebrate and vertebrate animal kingdoms.

The time course of *ard* mRNA expression follows the major periods of neural differentiation in the fly, suggesting that the *ard* cDNA indeed encodes a neuronal receptor subunit (Hermans-Borgmeyer et al., 1986). We recently analyzed the localization of *ard* mRNA *in situ* hybridization. Transcripts were only found in neuronal tissue of late embryos, larvae, pupae, and newly enclosed flies (Hermans-Borgmeyer et al., 1989). Furthermore, the use of intron-specific probes in Northern and *in situ* hybridization studies revealed partially spliced transcripts to be abundant in the adult fly CNS. Regulation of *ard* RNA processing thus may provide a post-transcriptional mechanism for the control of acetylcholine sensitivity.

The association of the ARD protein with a neuronal nAChR of *Drosophila* is also supported by experiments in which we showed it to be part of an α-Btx binding protein present in fly head membranes. To produce ARD-specific antibodies, fusion constructs were designed using two different vector systems in order to express both extracellular (amino acids 65–212) and intracellular (amino acids 305–404, and 295–486) regions of the polypeptide (Schloß et al., 1988). After affinity purification on immobilized fusion proteins, the antibodies were employed in immunoprecipitation experiments.

Several groups have shown that the insect CNS including that of *Drosophila* contains high-affinity α-Btx binding sites of nicotinic cholinergic specificity (Dudai, 1978; Schmidt-Nielsen et al., 1977; Rudloff, 1978). Furthermore, an α-Btx binding protein has been purified from locust ganglionic tissue and, after reconstitution in planar bilayers, shown to form acetylcholine-gated channels of characteristics similar to those of vertebrate muscle nAChR (Hanke and Breer, 1986). Thus, α-Btx in insects seems to bind to neuronal nAChR. In our laboratory, binding experiments using ^{125}I-α-Btx uncovered two different high-affinity sites. From both of these sites the toxin can be displaced by nicotine and d-tubocurarine, and thus they are of nicotinic cholinergic specificity (Schloß et al., 1988). Interestingly, only one of these sites was precipitated by two of our ARD-fusion protein antisera (Fig. 1). Obviously, heterogeneity of nAChRs exists in the insect CNS. Whatever the nature of the second toxin binding protein may be, these data clearly establish the ARD protein as a component of membrane protein capable of binding nicotinic cholinergic ligands.

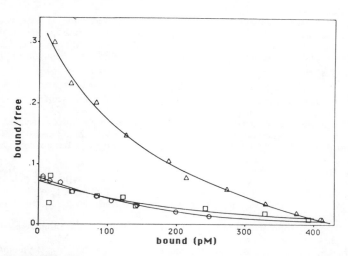

Figure 1. Scatchard plots of ^{125}I-α-Btx binding to detergent solubilized *Drosophila* head membranes after incubation with ARD antisera. Detergent extracts were incubated with an excess of antisera (AS) against ARD fusion proteins for 2 h; immune complexes then were precipitated using fixed *S. aureus* cells (see Schloß et al., 1988). Control extracts (△) display two binding sites with Kd values of 3.7×10^{-10} M and 3.9×10^{-9} M. After immunoprecipitation with AS 21 (□) or AS 6 (○), the high affinity site 1 is eliminated. Remaining sites bind the toxin with Kd values of 2.7×10^{-9} M and 7×10^{-9} M, respectively.

Conclusions

Here, pharmacological and molecular biological approaches aimed at the identification of putative nAChRs in the vertebrate and invertebrate CNS have been described. While use of the classical nicotinic antagonist α-Btx appears to have failed in identifying members of this class of membrane proteins in vertebrates, the present data from both biochemistry and molecular genetics on *Drosophila* are consistent with α-Btx binding to a membrane protein which is identical or closely related to vertebrate neuronal nAChR. Although final proof for AChR function of the ARD protein complex (e.g. oocyte expression) is still missing, the *ard* gene can definitively be said to encode a subunit of a ligand-gated ion channel. Future experiments should unravel whether its ability to bind nicotinic cholinergic ligands is functionally significant or just a remnant of common pharmacological properties not yet eliminated by evolution from a common receptor/ion channel subunit ancestor.

Comparison of the 'pharmacological' versus the 'cloning' approach clearly demonstrates the superiority of the latter. In particular, DNA sequencing rapidly reveals features of structural homology in polypeptide sequences predicted from cDNA or genomic clones. Furthermore, cloned sequences can be of immediate use in further immunological and functional studies employing both fusion proteins and oocyte or cell expression systems.

Acknowledgements. We thank C. Schröder, H. Krischke and C. Udri for technical assistance and I. Baro and Jutta Rami for help with the preparation of the manuscript. This work was supported by the Deutsche Forschungsgemeinschaft (Gu 230/1-1 and SFB 317), Bundesministerium für Forschung und Technologie (BCT 365/1) and the Fonds der Chemischen Industrie. E.S. holds a fellowship of the Boehringer-Ingelheim Fonds.

Betz, H. (1981) Characterization of the α-bungarotoxin receptor in chick-embryo retina. Eur. J. Biochem. 117: 131–139.

Betz, H., Graham, D., and Rehm, H. (1982) Identification of polypeptides associated with a putative neuronal nicotinic acetylcholine receptor. J. biol. Chem. 257: 11390–11394.

Betz, H., and Pfeiffer, F. (1984) Monoclonal antibodies against the α-bungarotoxin-binding protein of chick optic lobe. J. Neurosci. 4: 2095–2105.

Bossy, B., Ballivet, M., and Spierer, P. (1988) Conservation of neural nicotinic acetylcholine receptors from *Drosophila* to vertebrate central nervous system. EMBO J. 7: 611–618.

Conti-Tronconi, B. M., Dunn, S. M. J., Barnard, E. A., Dolly, J. O., Lai, F. A., Ray, N., and Raftery, M. A. (1985) Brain and muscle acetylcholine receptors are different but homologous proteins. Proc. natl Acad. Sci. USA 82: 5208–5212.

Dudai, Y. (1978) Properties of an α-bungarotoxin-binding cholinergic nicotinic receptor from *Drosophila melanogaster*. Biochim. biophys. Acta 593: 505–517.

Grenningloh, G., Rienitz, A., Schmitt, B., Methfessel, C., Zensen, M., Beyreuther, K., Gundelfinger, E. D., and Betz, H. (1987) The strychnine-binding subunit of the glycine receptor shows homology with nicotinic acetylcholine receptors. Nature 328: 215–220.

Gundelfinger, E. D., Hermans-Borgmeyer, I., Zopf, D., Sawruk, E., and Betz, H. (1986) Characterization of the mRNA and the gene of a putative neuronal nicotinic acetylcholine receptor protein from *Drosophila*. NATO ASI Series 3: 437–446.

Hanke, W., and Breer, H. (1986) Channel properties of an insect neuronal acetylcholine receptor protein reconstituted in planar lipid bilayers. Nature 321: 171–174.

Hermans-Borgmeyer, I., Zopf, D., Rysek, R. P., Hovemann, B., Betz, H., and Gundelfinger, E. D. (1986) Primary structure of a developmentally regulated nicotinic acetylcholine receptor protein from *Drosophila*. EMBO J. 5: 1503–1508.

Hermans-Borgmeyer, I., Sawruk, E., Schloß, P., Gundelfinger, E. D., and Betz, H. (1988) Biochemical and molecular biology approaches to central nicotinic acetylcholine receptors, in: Clementi, F. (ed.), Brain Nicotinic Acetylcholine Receptors. NATO ASI Series 25: 77–88.

Hermans-Borgmeyer, I., Hoffmeister, S., Sawruk, E., Betz, H., Schmitt, B., and Gundelfinger, E. D. (1989) Neuronal nicotinic acetylcholine receptors in *Drosophila*: mature and immature transcripts of the *ard* gene in the developing central nervous system. Neuron 2: 1147–1156.

Norman, R. I., Mehraban, F., Barnard, E. A., and Dolly, J. O. (1982) Nicotinic acetylcholine receptor from chick optic lobe. Proc. natl Acad. Sci. USA 79: 1321–1325.

Patrick, J., and Stallcup, W. B. (1977a) Immunological distinction between acetylcholine receptor and the α-bungarotoxin-binding component on sympathetic neurons. Proc. natl Acad. Sci. USA 74: 4689–4692.

Patrick, J., and Stallcup, W. B. (1977b) α-Bungarotoxin binding and cholinergic receptor function on a rat sympathetic nerve line. J. biol. Chem. 252: 8629–8633.

Rudloff, E. (1978) Acetylcholine receptors in the central nervous system of *Drosophila melanogaster*. Exp. Cell Res. 111: 185–190.

Sawruk, E., Hermans-Borgmeyer, I., Betz, H., and Gundelfinger, E. D. (1988) Characterization of an invertebrate nicotinic acetylcholine receptor gene: The *ard* gene of *Drosophila melanogaster*. FEBS Lett. 235: 40–46.

Schloß, P., Hermans-Borgmeyer, I., Betz, H., and Gundelfinger, E. D. (1988) Neuronal acetylcholine receptors in *Drosophila*: The ARD protein is a component of a high-affinity α-bungarotoxin binding complex. EMBO J. 7: 2889–2894.

Schmidt, J. (1977) Drug binding properties of an α-bungarotoxin binding component from rat brain. Molec. Pharmac. 13: 283–290.

Schmidt-Nielsen, B. K., Gepner, J. I., Teng, N. N. H., and Hall, L. M. (1977) Characterization of an α-bungarotoxin binding component from *Drosophila melanogaster*. J. Neurochem. 29: 1013–1029.

Schneider, M., Adee, C., Betz, H., and Schmidt, J. (1985) Biochemical characterization of two nicotinic receptors from the optic lobe of the chick. J. biol. Chem. 27: 14505–14512.

Schofield, P. R., Darlison, M. G., Fujita, N., Burt, D. R., Stephenson, F. A., Rodriguez, H., Rhee, L. M., Ramachandran, J., Reale, V., Glencorse, T. A., Seeburg, P. H., and Barnard, E. A. (1987) Sequence and functional expression of the GABA$_A$ receptor shows a ligand-gated receptor super-family. Nature 328: 221–227.

Vogel, Z., and Nirenberg, M. (1976) Localization of acetylcholine receptors during synaptogenesis in retina. Proc. natl Acad. Sci. USA 73: 1806–1810.

Wang, G.-K., Molinaro, S., and Schmidt, J. (1978) Ligand responses of α-bungarotoxin binding sites from skeletal muscle and optic lobe of the chick. J. biol. Chem. 253: 8507–8512.

Modulation of the sensitivity of nicotinic receptors in autonomic ganglia

Takashi Akasu and Takayuki Tokimasa

Department of Physiology, Kurume University School of Medicine, 67 Asahi-machi, Kurume 830, Japan

Summary. This article reviews some of the evidence suggesting that a variety of endogenous substances either facilitates or inhibits the sensitivity of nicotinic acetylcholine (ACh) receptors at the subsynaptic membrane of cholinergic synapses. It is noteworthy that 5-hydroxytryptamine and histamine act as competitive antagonists, like curare, presumably changing the affinity of ACh for the specific binding site on the nicotinic receptor. Catecholamine, neuropeptides, prostaglandin and glucocorticoids act as non-competitive antagonists on an allosteric site on the receptor-ionic channel complex. ATP and LH-RH (in a subpopulation of sympathetic neurons) caused a facilitation of the sensitivity of nicotinic receptors. The mode of actions of endogenous substances which modulate the nicotinic receptor-sensitivity is similar to those of pharmacological agents. Therefore, these neurotransmitters and neurohormones have been termed endogenous 'antagonists' or 'sensitizers'.

Introduction

Acetylcholine (ACh) released from preganglionic nerve terminals mediates cholinergic nicotinic transmission by initiating the fast excitatory postsynaptic potential (e.p.s.p.) at postganglionic neurons in sympathetic ganglia. When the amplitude of the fast e.p.s.p. is large enough to generate the action potential, neuronal information received from preganglionic nerves propagates along the postganglionic nerve axon. The efficiency of the nicotinic transmission is represented by the size (amplitude and duration) of the fast e.p.s.p., and it can be modulated pre- and postsynaptically by various neurotransmitters other than ACh.

With regard to the postsynaptic modulation of cholinergic transmission, ample evidence has accumulated suggesting that endogenous substances can modify the sensitivity to nicotinic action of ACh, regulating the generation of ACh-induced postsynaptic current at the subsynaptic membrane of the ganglion cell (Koketsu and Akasu, 1986). This is a conceptually distinct type of postsynaptic modification of nicotinic transmission which occurs at the receptor level.

In this article, we will deal with our experimental evidence or that obtained through literature which suggests interaction of the nicotinic response with multiple transmitters at the receptor level in vertebrate cholinergic synapses. Endogenous substances which modulate nicotinic responses are summarized in Table 1.

Inhibition of the ACh-receptor sensitivity

5-Hydroxytryptamine. Colomo et al. (1968) reported that 5-hydroxy-tryptamine (5-HT) decreased the amplitude of depolarization (ACh potential) produced by ionophoretic application of ACh to the end-plate membrane. Subsequently, Magazanik et al. (1976) also reported that 5-HT reduced the postsynaptic response by acting on the channel associated with nicotinic receptors. It is well known that amplitude and time course of the ACh potential are sensitive to the resting membrane potential and conductance. Therefore, ACh-induced postsynaptic current (ACh current) recorded by the voltage-clamp method was used as the indicator of the receptor-sensitivity. Akasu et al. (1981a) demonstrated that 5-HT directly depressed the ACh current at the frog end-plate. Recently, the mode of inhibition of the nicotinic response was analyzed at the bullfrog sympathetic ganglion cell (Akasu and Koketsu, 1986). 5-HT decreased the amplitude of the ACh current, while it did not alter the decay time constants of both the fast excitatory postsynaptic current (fast e.p.s.c.) and the miniature e.p.s.c. The reversal potential of the fast e.p.s.c. was not changed by 5-HT. These results suggest that the depression of the ACh current produced by 5-HT is not due to a change in the kinetics of opening and closing of the ionic channels coupled with nicotinic receptors.

Dose-response curve of the ACh current can be obtained by plotting the amplitude of the ACh current against the logarithm of relative ACh quantities applied by ionophoresis. 5-HT caused a parallel shift to the right of the dose-response curve for ACh, suggesting a competitive antagonism (Akasu and Koketsu, 1986). Analysis using a double reciprocal plot (Lineweaver-Burk plot) revealed that 5-HT increased the apparent dissociation constant (K_m) of ACh for the receptor, while it did not change the maximum ACh current (V_{max}). Erabutoxin-b (ETX-b), a sea snake venom irreversibly blocks nicotinic receptors acting on the specific binding site for ACh (Kato et al., 1978). If 5-HT acts on the ACh binding site, sufficient quantities of 5-HT can protect the end-plate from ETX-b binding. In the presence of 5-HT (1 mM), ETX-b could not produce the irreversible block of neuromuscular transmission (Koketsu et al., 1982a). Histamine also blocked the nicotinic receptor in a competitive manner (Ariyoshi et al., 1985).

5-HT has been found to modulate the sensitivity of nicotinic receptors in parasympathetic ganglia. Application of 5-HT to the vesical parasympathetic ganglia isolated from the rabbit urinary bladder wall caused a depression of the ACh potential. In these neurons, an intracellular metabolic process might be involved in the mechanism of the inhibition of the ACh response. Forskolin, known to be a potent activator of the adenylate cyclase strongly reduced the amplitude of the ACh potential. Forskolin shifted the dose-response curve of ACh potential downward, suggesting a non-competitive antagonism. 3-Isobutyl-l-methylxanthine, a phosphodiesterase inhibitor also produced an inhibition of the ACh potential.

Catecholamine. In 1955, Hutter and Loewenstein reported that noradrenaline increased the depolarization induced by application of ACh to frog skeletal muscles. Koketsu et al. (1982b) examined the effect of adrenaline on the responses to ACh at the frog end-plate as well as bullfrog sympathetic ganglia. It was demonstrated that adrenaline depressed the amplitudes of ACh potential and current. Adrenaline shifted the dose-response relationship of the ACh current downward. Lineweaver-Burk plot suggested that adrenaline depressed the maximum response to ACh (V_{max}), while it did not change the apparent dissociation constant (K_m). These results indicate that adrenaline blocks the nicotinic receptor in a non-competitive manner. Catecholamine did not prevent the irreversible blocking of end-plate receptors by ETX-b (Koketsu et al., 1982a). It is suggested that catecholamine may act on an allosteric site rather than the specific binding site on the receptor-ionic channel complex.

Facilitation of the ACh-receptor sensitivity

Adenosine triphosphate. Ewald (1976) has proposed a possibility that the nicotinic ACh response is directly facilitated by endogenous substances. Adenosine triphosphate (ATP) and its related compounds augmented the depolarization produced by ACh at the rat diaphragm muscle. This concept is supported by the result that adenine nucleotides increase the sensitivity of nicotinic receptors (Akasu et al., 1981b; Akasu and Koketsu, 1985). ATP (0.05–2 mM) augmented the amplitude of both potential and current responses induced by ACh. ATP did not change the time courses of the ACh current and the fast e.p.s.p. ATP also increased the amplitude of the current induced by carbachol which is known to be not hydrolyzed by cholinesterase. These results suggested that the augmentation of ACh responses cannot be due to inhibition of cholinesterase activity.

ATP shifted the dose-response curve upward. A double reciprocal plot revealed that ATP increased the V_{max} of ACh currents without producing any effect on the affinity (K_m) of ACh for the nicotinic receptor (Akasu and Koketsu, 1985). ATP probably acts on an allosteric site of the receptor-ionophore complex to increase the sensitivity of nicotinic receptors. Theophylline (2 mM), an antagonist of a P_1-subtype adenosine receptor (Burnstock 1981), did not block the ability of ATP in producing the facilitation of the sensitivity of nicotinic receptors (Akasu and Koketsu, 1985). Thus, it seems that the receptor responsible for the potentiating effect of ATP is a P_2-purinoceptor subtype.

Late slow e.p.s.p. and LH-RH. Schulman and Weight (1976) reported that the amplitude of the fast e.p.s.p. was increased during the late slow e.p.s.p. at the sympathetic ganglia of the bullfrog. They suggested that the facilitation of the fast e.p.s.p. is due to an increased membrane resistance. Luteinizing hormone-releasing hormone (LH-RH), an endogenous neuropeptide initially isolated from porcine hypothalamus, is known to be the putative transmitter for the late slow e.p.s.p. (Jan et al., 1979). Previously, it was shown that exogenously applied mammalian LH-RH depressed the ACh responses in sympathetic neurons (Hasuo and Akasu, 1986; Akasu and Hasuo, 1987) and in frog skeletal muscle end-plate (Akasu et al., 1983a). Dose-response analysis showed that LH-RH depressed the sensitivity of the nicotinic receptor in a non-competitive manner.

However, recently our results have provided a different type of LH-RH effect on the receptor function in sympathetic ganglia. LH-RH produced a facilitation of the nicotinic receptor sensitivity, although this phenomenon was seen only in a small population of sympathetic neurons. We examined an interaction between ACh potential and peptidergic postsynaptic response, the late slow e.p.s.p., at the same neurons in bullfrog sympathetic ganglia. In 55–65% neurons tested, no change in the amplitude of ACh responses was seen during the late slow e.p.s.p. In other neurons (35–45%), the amplitude of the ACh potential was increased by $25 \pm 2\%$ of the control. In this population of sympathetic neurons, bath-application of LH-RH (4 μM) caused a $33 \pm 8\%$ increase in the ACh current. LH-RH also increased the current induced by carbachol (CCh), suggesting that LH-RH increased the sensitivity of nicotinic receptors (Akasu and Hasuo, 1987; Hasuo and Akasu, 1988). Interestingly, LH-RH shifted in parallel the dose-response curve of CCh current to the left. The Lineweaver-Burk plot showed that LH-RH

Table 1. Modulation of nicotinic receptors

Type of modulation	Substances	Experimental techniques	Kinetics
Inhibition	5-HT	a, b	competitive
	catecholamine	b	non-competitive
	histamine	a, b	competitive
	LH-RH†,††	b	non-competitive
	substance P	a, b	non-competitive
	prostaglandin E_1	a	competitive
	glucocorticoid	a, b	non-competitive
Facilitation	ATP	b	$V_{max}\uparrow$
	LH-RH†	a, b	$K_m\downarrow$

a, membrane potential recordings; b, voltage-clamp; †, sympathetic neurons; ††, skeletal muscle and end-plate.

decreased the K_m, while it did not change the V_{max} (Hasuo and Akasu, 1988). LH-RH may increase the affinity of ACh to the recognition site on the nicotinic receptor.

Possible mechanisms of receptor modulation

Binding of ACh to the nicotinic receptor located on the subsynaptic membrane results in a large increase in the non-selective cation conductance, which in turn generates the fast e.p.s.p. In amphibian cholinergic synapses, 5-HT depresses the fast e.p.s.p. by decreasing the sensitivity of nicotinic receptors in a competitive manner. 5-HT did not change the properties of ion channels coupled with nicotinic receptors. Therefore, shortening of the life time of ionic channels is probably not the mechanism for 5-HT induced inhibition of the ACh current. Interaction of 5-HT with ETX-b suggests that 5-HT acts on the specific binding site for ACh and reduces the binding affinity with the nicotinic receptor molecule (Koketsu et al., 1982a). A simple representation of the reaction of 5-HT which competitively antagonizes the nicotinic receptor is shown below (Akasu and Koketsu, 1986).

$$A + R \underset{}{\overset{K_m}{\rightleftharpoons}} AR \underset{\alpha}{\overset{\beta}{\rightleftharpoons}} AR^*$$
$$+$$
$$5\text{-HT}$$
$$K_i \updownarrow K_i$$
$$5\text{-HT} \cdot R$$

where A and R are the amount of ACh and the number of nicotinic receptors, respectively. AR and AR* represent the closed (inactive) and open (activated) states of the receptor-ionic channel complex. K_m, α and β are the rate constants of reactions. K_i represents the dissociation constant for 5-HT · R complex, obtained from Dixon plot for the 5-HT action (Akasu and Koketsu, 1986). The combination of 5-HT with the receptor-channel complex increases the amount of non-conducting complex, 5-HT · R, which has no ability to open the ionic channels.

Adrenaline also depresses the receptor sensitivity in a non-competitive manner, acting on an allosteric site on the receptor. However, adrenaline did not change the properties of ion channels. Probably, adrenaline reduces the conductance of single ion channels associated with the nicotinic receptor without producing changes in opening and closing channel kinetics. Analysis with patch-clamp or noise analysis of ACh-current may provide a direct evidence for adrenaline action on the single ion-channel properties.

Neuropeptides have also been found to cause a modulation to ACh responses in various tissues (Akasu et al., 1983a, b; Clapham and

Neher, 1984; Hasuo and Akasu, 1986, 1988; Akasu and Hasuo, 1987; Ryall and Belcher, 1977). Clapham and Neher (1984) have recently demonstrated that substance P reduces the ACh current by blocking open ionic channels in bovine adrenal chromaffin cells. Substance P and LH-RH facilitate the rate of desensitization of nicotinic receptors at the frog skeletal muscle end-plate (Akasu et al., 1984). Cull-Candy et al. (1988) also reported that clonidine, an α_2 adrenoceptor agonist, depressed the ACh current of bovine chromaffin cell by increasing the rate of desensitization. Some endogenous antagonists may act on a certain step between receptor and ionic channels to depress the ACh responses.

Acknowledgements. The study was supported by a Grant-in-Aid for Scientific Research from the Ministry of Education, Science and Culture of Japan.

Akasu, T., and Hasuo, H. (1987) The facilitation of nicotinic receptor sensitivity during the late slow excitatory postsynaptic potential. Neurosci. Res. Suppl. 5: 142.

Akasu, T. Hirai, K., and Koketsu, K. (1981a) 5-Hydroxytryptamine controls ACh-receptor sensitivity of bullfrog sympathetic ganglion cells. Brain Res. 221: 217–220.

Akasu, T., Hirai, K., and Koketsu, K. (1981b) Increase of acetylcholine-receptor sensitivity by adenosine triphosphate: a novel action of ATP on ACh-sensitivity. Br. J. Pharmac. 74: 505–507.

Akasu, T., Kojima, M., and Koketsu, K. (1983a) Luteinizing hormone-releasing hormone modulates nicotinic ACh-receptor sensitivity in amphibian cholinergic transmission. Brain Res. 279: 347–351.

Akasu, T., Kojima, M., and Koketsu, K. (1983b) Substance P modulates the sensitivity of the nicotinic receptor in amphibian cholinergic transmission. Br. J. Pharmac. 80: 123–131.

Akasu, T., and Koketsu, K. (1985) Effect of adenosine triphosphate on the sensitivity of the nicotinic acetylcholine-receptor in the bullfrog sympathetic ganglion cell. Br. J. Pharmac. 84: 525–531.

Akasu, T., and Koketsu, K. (1986) 5-Hydroxytryptamine decreases the sensitivity of nicotinic acetylcholine receptor in bull-frog sympathetic ganglion cells. J. Physiol. (Lond.) 380: 93–109.

Akasu, T., Ohta, Y., and Koketsu, K. (1984) Neuropeptides facilitate the desensitization of nicotinic acetylcholine-receptor in frog skeletal muscle endplate. Brain Res. 290: 342–347.

Ariyoshi, M., Hasuo, H., Koketsu, K., Ohta, Y., and Tokimasa, T. (1985) Histamine is an antagonist of the acetylcholine receptor at the frog endplate. Br. J. Pharmac. 85: 65–73.

Burnstock, G. (1981) Neurotransmitters and trophic factors in the autonomic nervous system. J. Physiol. (Lond.) 313: 1–35.

Clapham, D. E., and Neher, E. (1984) Substance P reduces acetylcholine-induced currents in isolated bovine chromaffin cells. J. Physiol. (Lond.) 347: 255–277.

Colomo, F., Rahamimoff, R., and Stefani, E. (1968) An action of 5-hydroxytryptamine on the frog motor end-plate. Eur. J. Pharmac. 3: 272–274.

Cull-Candy, S. G., Mathie, A., and Powis, D. A. (1988) Acetylcholine receptor channels and their block by clonidine in cultured bovine chromaffin cells. J. Physiol. (Lond.) 402: 255–278.

Ewald, D. A. (1976) Potentiation of postjunctional cholinergic sensitivity of rat diaphragm muscle by high-energy-phosphate adenine nucleotides. J. Membr. Biol. 29: 47–65.

Hasuo, H., and Akasu, T. (1986) Luteinizing hormone-releasing hormone inhibits nicotinic transmission in bullfrog sympathetic ganglia. Neurosci. Res. 3: 444–450.

Hasuo, H., and Akasu, T. (1988) Facilitation of nicotinic acetylcholine responses during the late slow EPSP in a subpopulation of bullfrog sympathetic neurons. Kurume med. J. 35: 43–48.

Hutter, O. F., and Loewenstein, W. R. (1955) Nature of neuromuscular facilitation by sympathetic stimulation in the frog. J. Physiol. (Lond.) 130: 559–571.

Jan, Y. N., Jan, L. Y., and Kuffler, S. W. (1979) A peptide as a possible transmitter in sympathetic ganglia of the frog. Proc. natl Acad. Sci. USA 76: 1501–1505.

Kato, E., Kuba, K., and Koketsu, K. (1978) Effects of erabutoxins on neuromuscular transmission in frog skeletal muscles. J. Pharmac. exp. Ther. 204: 446–453.

Koketsu, K., and Akasu, T. (1986) Postsynaptic modulation. in: Karczmar, A. G., Koketsu, K., and Nishi, S. (eds), Autonomic and Enteric Ganglia. Plenum Press, New York, pp. 273–295.

Koketsu, K., Akasu, T., Miyagawa, M., and Hirai, K. (1982a) Biogenic antagonists of the nicotinic receptor: their interactions with erabutoxin. Brain Res. 250: 391–393.

Koketsu, K., Miyagawa, M., and Akasu, T. (1982b) Catecholamine modulates nicotinic ACh-receptor sensitivity. Brain Res. 236: 487–491.

Magazanik, L. G., Illes, P., and Snetkov, V. A. (1976) The action of 5-HT on the potential dependence of the muscle end-plate current. Dokl. Acad. Nauk. USSR 227: 1968.

Ryall, R. W. and Belcher, G. (1977) Substance P selectively blocks nicotinic receptors on Renshaw cells: a possible synaptic inhibitory mechanism. Brain Res. 137: 376–380.

Schulman, J. A., and Weight, F. F. (1976) Synaptic transmission: long-lasting potentiation by a postsynaptic mechanism. Science 194: 1437–1439.

Muscarinic modulation of acetylcholine release: Receptor subtypes and possible mechanisms

H. Kilbinger, H. Schwörer and K. D. Süß

Pharmakologisches Institut der Universität Mainz, Obere Zahlbacher Str. 67, D-6500 Mainz, Federal Republic of Germany

Introduction

The release of acetylcholine from central and peripheral neurones can be inhibited and facilitated by muscarine autoreceptors, i.e. receptors located on the cholinergic neurone. In the last few years evidence has accumulated that muscarine receptors are heterogeneous. This chapter describes attempts that have been made to classify the muscarine autoreceptors. In addition, some possible mechanisms behind the neuronal muscarine receptors are examined.

Pharmacological characterization of muscarine autoreceptors

Functional and radioligand binding studies have shown that different subtypes of muscarine receptors exist. Receptors with a high affinity to the antagonists pirenzepine (pA_2 value of about 8.5) or dicyclomine (pA_2 of about 9) have been called M1 receptors, and those with low affinities to pirenzepine ($pA_2 < 7$) or dicyclomine (pA_2 7.0–7.5) the M2 receptor (see Birdsall and Hulme, 1985; Kilbinger and Stein, 1988). M2 receptors are again heterogeneous. Those which have a high affinity to the antagonists methoctramine (Melchiorre et al., 1987) or AF-DX 116 (Giachetti et al., 1986) have been named $M2_\alpha$ (cardiac type), and those with high affinities for 4-DAMP (Barlow et al., 1976) or hexahydro-siladifenidol (Mutschler and Lambrecht, 1984) represent the $M2_\beta$ subtype (ileal smooth muscle type).

Inhibitory M2 receptors

Inhibitory muscarine autoreceptors have been characterized in release experiments by determining the pA_2 values for the presynaptic effects of muscarine receptor antagonists. Inhibition of the depolarization-evoked outflow of acetylcholine by an exogenous agonist was used as a presynaptic parameter, and the pA_2 values were evaluated by constructing concentration-response curves for the effects of the agonist in the absence and presence of different concentrations of the respective antagonist. Table 1 summarizes the presynaptic pA_2 values which have been

Table 1. pA_2 values of antagonists at inhibitory presynaptic muscarine autoreceptors

Antagonist	Tissue	pA_2 value	References
Pirenzepine	Rabbit striatum	7.1[a]	James and Cubbedu, 1987
	Rabbit hippocampus	6.7	Schwarzwälder et al., 1988
	Rabbit caudate nucleus	6.9	Schwarzwälder et al., 1988
	Rat cerebral cortex	6.9	Roberts and Tutty, 1986
	Guinea-pig ileum	6.9[a]	Kilbinger et al., 1984
	Guinea-pig ileum	7.0[a]	North et al., 1985
Dicyclomine	Guinea-pig ileum	7.5	Kilbinger and Stein, 1988
Methoctramine	Rabbit hippocampus	7.0	Schwarzwälder et al., 1988
	Rabbit caudate nucleus	6.6	Schwarzwälder et al., 1988
AF-DX 116	Rabbit hippocampus	7.3	Schwarzwälder et al., 1988
	Rabbit caudate nucleus	7.2	Schwarzwälder et al., 1988
	Guinea-pig ileum	6.7[a]	Kilbinger, 1987
4-DAMP	Rabbit hippocampus	8.2	Schwarzwälder et al., 1988
	Rabbit caudate nucleus	8.3	Schwarzwälder et al., 1988
	Guinea-pig ileum	8.4[a]	Kilbinger et al., 1984
	Guinea-pig ileum	8.7[a]	North et al., 1985
Hexahydrosi-ladifenidol	Rabbit hippocampus	7.4	Schwarzwälder et al., 1988
	Rabbit caudate nucleus	7.3	Schwarzwälder et al., 1988
	Guinea-pig ileum	8.1[a]	Fuder et al., 1985
Atropine	Rabbit striatum	8.9[a]	James and Cubbeddu, 1987
	Rabbit hippocampus	9.5	Schwarzwälder et al., 1988
	Rabbit caudate nucleus	9.8	Schwarzwälder et al., 1988
	Rat cerebral cortex	8.4	Marchi and Raiteri, 1985
	Rat cerebral cortex	8.9	Roberts and Tutty, 1986
	Guinea-pig ileum	8.8	Kilbinger, 1977
	Electric organ, Torpedo	9.1[a]	Dunant and Walker, 1982
Scopolamine	Guinea-pig ileum	9.0[a]	Kilbinger, et al., 1984
	Guinea-pig ileum	8.9[a]	North et al., 1985
Methylatropine	Guinea-pig ileum	8.9[a]	Kilbinger, et al., 1984
Trihexyphenidyl	Guinea-pig ileum	8.2[a]	Kilbinger et al., 1984
Clozapine	Guinea-pig ileum	6.9[a]	Kilbinger et al., 1984
QNB	Rat cerebral cortex	8.5	Marchi and Raiteri, 1985
Secoverine	Rat cerebral cortex	7.5	Marchi and Raiteri, 1985
Gallamine	Rat cerebral cortex	4	Roberts and Tutty, 1986

The table is based on studies on isolated tissues in which depolarization-evoked release of acetylcholine was measured. In two studies (Ref. North et al., 1985; Dunant and Walker, 1982) the nicotinic postsynaptic potentials were measured. [a]Determined from Schild plots; slopes were close to unity.

published so far. The values for pirenzepine were all of about 7 which indicates that the inhibitory muscarine autoreceptors in the tissues listed in Table 1 are M2 receptors. Recent investigations suggest that the vagus nerve in the chicken heart (Jeck et al., 1988) and the cholinergic neurones in the guinea-pig ileum (Kawashima et al., 1988) are also endowed with release-inhibitory M1 receptors, but pA_2 values were not determined in these studies.

According to the above-mentioned classification, the inhibitory muscarine receptors of guinea-pig myenteric neurones belong to the $M2_\beta$ subtype since hexahydrosiladifenidol has a high, and AF-DX 116 a low affinity to this receptor (Table 1). On the other hand, preliminary studies on rabbit hippocampus and caudate nucleus (Schwarzwälder et al., 1988; Table 1) suggest that the acetylcholine release from these tissues is inhibited by both $M2_\alpha$ and $M2_\beta$ muscarine receptors.

Facilitatory M1 receptors

Muscarinic agonists like muscarine, pilocarpine or methacholine not only cause an inhibition of the electrically evoked acetylcholine release from guinea-pig myenteric neurones, but also enhance spontaneous acetylcholine release (Kilbinger, 1978; Kilbinger and Nafziger, 1985). The receptors mediating this effect were found to have high affinities to pirenzepine (pA_2 8.5; Kilbinger and Nafziger, 1985) and dicyclomine (pA_2 9.3; Kilbinger and Stein, 1988) and are thus M1 receptors.

Possible mechanisms

Muscarinic agonists cause inhibition of adenylate cyclase activity in neural tissues (McKinney and Richelson, 1984) and cyclic AMP might therefore be involved in the mechanism of autoinhibition of acetylcholine release. One should then expect that any increase in intra-axonal cyclic AMP reduces the degree of inhibition. The muscarinic inhibition of adenylate cyclase may be mediated by a GTP-binding protein, such as Gi or Go which can be inactivated by pertussis toxin (Dolphin, 1987).

No evidence for the involvement of a G protein or of adenylate cyclase in the muscarinic modulation of acetylcholine release has been provided so far. We have therefore studied the effects of the adenylate cyclase activator, forskolin, and of pertussis toxin on the release of acetylcholine from the longitudinal muscle-myenteric plexus preparation of the guinea-pig.

Forskolin and the muscarinic inhibition of acetylcholine release

The preparations were incubated with [3H]choline and after a washout period of 60 min the strips were stimulated twice (S1, S2) at 0.1 Hz (10 min). Forskolin was added to the medium 20 min before S1. The electrically evoked release of [3H]acetylcholine during S1 was significantly larger in the presence of forskolin ($4.59 \pm 0.23\%$ of the tritium tissue content, $N = 20$) than in control experiments without forskolin ($3.03 \pm 0.19\%$, $N = 25$). The ratio S2/S1 was similar in the presence (1.01 ± 0.05, $N = 6$) and absence (0.87 ± 0.05, $N = 9$) of forskolin. Muscarine ($0.01-1\ \mu M$) added to the medium 24 min before S2, caused the same degree of inhibition of the evoked release of [3H]acetylcholine in the absence and presence of forskolin, and the EC 50 values for muscarine were similar (63 and 46 nmol/l, respectively).

Pertussis toxin and the M2 receptor-mediated inhibition of acetylcholine release

Guinea-pigs were injected with pertussis toxin (List) ($60\ \mu g/kg$) or vehicle through the jugular vein. After 4 days the guinea-pigs were killed and the small intestine removed.

In order to test the effectiveness of the *in vivo* pertussis toxin pretreatment, longitudinal muscle strips of the small intestine were suspended in Tyrode solution and the effect of adenosine on the twitch response to 0.1 Hz stimulation was studied. The inhibition of the smooth muscle contraction by adenosine was significantly attenuated on strips from guinea-pigs pretreated with pertussis toxin (EC 50 $7\ \mu M$), as compared to strips from vehicle-treated animals (EC 50 $2\ \mu M$). A similar effect of pertussis toxin pretreatment on the adenosine-mediated inhibition of the twitch response of the guinea-pig ileum has been described by Lux and Schulz (1986).

Table 2. Effect of oxotremorine on electrically evoked release of [3H]acetylcholine from myenteric plexus preparations of guinea-pigs pretreated with pertussis toxin or vehicle

Pretreatment	Oxotremorine before S2(μM)	S1 (%)	S2/S1	N
Vehicle	—	5.12 ± 0.42	0.86 ± 0.03	4
Vehicle	0.1	$5.02 + 0.35$	0.29 ± 0.02	4
		$5.07 \pm 0.25*$		8
PTx	—	6.23 ± 0.21	0.78 ± 0.04	10
PTx	0.1	$6.10 + 0.28$	0.28 ± 0.02	8
		$6.18 \pm 0.19*$		18

*p < 0.01.

Strips were preincubated with [3H]choline and after a washout period of 60 min stimulated electrically (S1, S2; 1 Hz 3 min). The muscarinic agonist oxotremorine was added before S2. The results are shown in Table 2. The release of [3H]acetylcholine during S1 was significantly increased after pertussis toxin pretreatment. However, the inhibition by oxotremorine of the electrically evoked [3H]acetylcholine release was similar in the control group and after pertussis toxin pretreatment.

Pertussis toxin and the M1 receptor-mediated facilitation of acetylcholine release

Strips from guinea-pigs were preincubated with [3H]choline and subsequently superfused with pilocarpine and muscarine. The muscarinic agonists evoked an increased release of [3H]acetylcholine, but the effects of pilocarpine and muscarine were significantly attenuated on preparations from guinea-pigs pretreated with pertussis toxin (Table 3). For comparison, the effects of serotonin and eledoisin are included. Both drugs facilitate the release of [3H]acetylcholine through stimulation of specific receptors (5-HT$_3$ and SP-E receptors, respectively). Pretreatment of the guinea-pigs with pertussis toxin did not affect the facilitatory actions of serotonin and eledoisin on acetylcholine release.

Conclusions

Muscarinic agonists inhibit the electrically evoked release of acetylcholine from the guinea-pig myenteric plexus via stimulation of receptors which belong to the M2$_\beta$ subtype. Interaction experiments with

Table 3. Stimulation of [3H]acetylcholine release by agonists at M1, 5-HT$_3$ and substance P receptors in myenteric plexus preparations of guinea-pigs pretreated with pertussis toxin

Pretreatment	Agonist (μM)	[3H]ACh-release (%)	N
Vehicle	Pilocarpine 30	3.1 ± 0.2	4
PTx	Pilocarpine 30	$1.6 \pm 0.3**$	5
Vehicle	Muscarine 10	2.6 ± 0.3	5
PTx	Muscarine 10	$1.7 \pm 0.1*$	3
Vehicle	Serotonin 10	1.8 ± 0.1	11
PTx	Serotonin 10	2.4 ± 0.4 ns	4
Vehicle	Eledoisin 0.1	2.4 ± 0.3	7
PTx	Eledoisin 0.1	2.9 ± 0.3 ns	4

**$p < 0.01$; *$p < 0.05$. [3H]acetylcholine (ACh) release is given in per cent of the tissue tritium content.

202

forskolin indicate that this decrease in release is not mediated through an inhibition of an intra-axonal adenylate cyclase. Moreover, there is no evidence for the involvement of a pertussis toxin sensitive G protein in the M2 autoreceptor-mediated inhibition of acetylcholine release.

Muscarinic agonists also initiate a release of acetylcholine from guinea-pig myenteric neurones via stimulation of M1 receptors. Pertussis toxin pretreatment significantly attenuates this effect, which suggests that a G protein may be involved in the M1 receptor-mediated effect.

Acknowledgement. This work was supported by the Deutsche Forschungsgemeinschaft (Ki 210/6-3).

Barlow, R. B., Berry, K. J., Glenton, P. A. M., Nikolaou, N. M., and Soh, K. S. (1976) A comparison of affinity constants for muscarine-sensitive acetylcholine receptors in guinea-pig atrial pacemaker cells at 29°C and in ileum at 29°C and 37°C. Br. J. Pharmac. 58: 613–620.

Birdsall, N. J. M., and Hulme, E. C. (1985) Multiple muscarinic receptors: further problems in receptor classification. In: Kalsner, S. (ed.), Trends in autonomic pharmacology, vol. 3, pp. 17–34. Taylor and Francis, London.

Dolphin, A. C. (1987) Nucleotide binding proteins in signal transduction and disease. TINS 10: 53–57.

Dunant, Y., and Walker, A. I. (1982) Cholinergic inhibition of acetylcholine release in the electric organ of Torpedo. Eur. J. Pharmac. 78: 201–212.

Fuder, H., Kilbinger, H., and Müller, H. (1985) Organ selectivity of hexahydrosiladifenidol in blocking pre- and postjunctional muscarinic receptors studied in guinea-pig ileum and rat heart. Eur. J. Pharmac. 113: 125–127.

Giachetti, A., Micheletti, R., and Montagna, E. (1986) Cardioselective profile of AF-DX 116, a muscarine M₂ receptor antagonist. Life Sci. 38; 1663–1672.

James, M. K., and Cubeddu, L. X. (1987) Pharmacologic characterization and functional role of muscarinic autoreceptors in the rabbit striatum. J. Pharmac. Ther. 240: 203–215.

Jeck, D., Lindmar, R., Löffelholz, K., and Wanke, M. (1988) Subtypes of muscarinic receptor on cholinergic nerves and atrial cells of chicken and guinea-pig hearts. Br. J. Pharmac. 93: 357–366.

Kawashima, K., Fujimoto, K., Suzuki, T., and Oohata, H. (1988) Direct determination of acetylcholine release by radioimmunoassay and presence of presynaptic M₁ muscarinic receptors in guinea pig ileum. J. Pharmac. exp. Ther. 244: 1036–1039.

Kilbinger, H. (1977) Modulation by oxotremorine and atropine of acetylcholine relase evoked by electrical stimulation of the myenteric plexus of the guinea-pig ileum. Nauyn-Schmiedeberg's Arch. Pharmak. 300: 145–151.

Kilbinger, H. (1978) Muscarinic modulation of acetylcholine release from the myenteric plexus of the guinea pig small intestine. In: Jenden, D. J. (ed.), Cholinergic Mechanism and Psychopharmacology, pp. 401–410. Plenum Press, New York.

Kilbinger, H. (1987) Control of acetylcholine release by muscarinic autoreceptors. In: Cohen, S., and Sokolovsky, M. (eds), International Symposium on Muscarinic Cholinergic Mechanisms, pp. 219–228. Freund Publ House, London.

Kilbinger, H., Halim, S., Lambrecht, G., Weiler, W., and Wessler, I. (1984) Comparison of affinities of muscarinic antagonists to pre- and postjunctional receptors in the guinea-pig ileum. Eur. J. Pharmac. 103: 313–320.

Kilbinger, H., and Nafziger, M. (1985) Two types of neuronal muscarine receptors modulating acetylcholine release from guinea-pig myenteric plexus. Naunyn-Schmiedeberg's Arch. Pharmak. 328: 304–309.

Kilbinger, H., and Stein, A. (1988) Dicyclomine discriminates between M1 and M2 muscarinic receptors in the guinea-pig ileum. Br. J. Pharmac. 94: 1270–1274.

Lux, B., and Schulz, R. (1986) Effect of cholera toxin and pertussis toxin on opioid tolerance and dependence in the guinea-pig myenteric plexus. J. Pharmac. exp. Ther. 237; 995–1000.

Marchi, M., and Raiteri, M. (1985) On the presence in the cerebral cortex of muscarinic receptor subtypes which differ in neuronal localization, function and pharmacological properties. J. Pharmac. exp. Ther. 235: 230–233.

McKinney, M., and Richelson, E. (1984) The coupling of the neuronal muscarinic receptor to responses. A. Rev. Pharmac. Toxic. 24: 121–146.

Melchiorre, C., Cassinelli, A., and Quaglia, W. (1987) Differential blockade of muscarinic receptor subtypes by polymethylene tetraamines. Novel class of selective antagonists of cardiac M2 muscarinic receptors. J. med. Chem. 30: 201–204.

Mutschler, E., and Lambrecht, G. (1984) Selective muscarinic agonists and antagonists in functional tests. Trends pharmac. Sci. 5: Suppl, 39–44.

North, R., Slack, B. E., and Surprenant, A. (1985) Muscarinic M1 and M2 receptors mediate depolarization and presynaptic inhibition in guinea-pig enteric nervous system. J. Physiol. 368: 435–452.

Roberts, E., and Tutty, C. (1986) Quantification of the effects of muscarinic antagonists on the release of [3H]acetylcholine from rat cerebral cortex slices. Br. J. Pharmac. 88: 358 P.

Schwarzwälder, U., Johann, S., and Melchiorre, C. (1988) Subtypes of muscarinic autoreceptors in rabbit hippocampus and caudate nucleus. Naunyn-Schmiedeberg's Arch Pharmak. 337: R 89.

Characterization of muscarinic receptors modulating acetylcholine release in the rat neostriatum

Molly H. Weiler

University of Wisconsin, School of Pharmacy, 425 N. Charter St., Madison, WI 53706, USA

Summary. Two classical and two nonclassical muscarinic agents were tested for their effects on neostriatal acetylcholine (ACh) release. The classical muscarinic antagonist, atropine (0.1–2 μM), increased ACh release in a dose-dependent fashion, whereas the 'nonclassical' antagonist, pirenzepine (2–200 μM), had no effect on ACh release. The muscarinic agonist, oxotremorine (1–100 μM), and its analog, oxotremorine-M (0.1–50 μM), both decreased ACh release in a dose-dependent fashion. The inhibitory potency of oxotremorine-M was approximately 20 times that of oxotremorine, and the inhibitory effects of both agonists were blocked by atropine. The results obtained with the muscarinic antagonists indicate that 'M-2 type' muscarinic receptors mediate ACh release in the neostriatum, but further pharmacological analyses are required to assign with certainty the receptor subtype involved in neostriatal ACh release.

Introduction

The release of acetylcholine (ACh) from neurons in both the central and peripheral nervous system can be modulated by muscarinic agents. In the neostriatum this muscarinic modulation of ACh release has been examined in neurochemical studies (James and Cubeddu, 1984; Weiler et al., 1984) and in electrophysiological studies (Dodt and Misgeld, 1986; Misgeld et al., 1982). Muscarinic agonists that decreased ACh release from neostriatal slices also decreased evoked extra- and intracellular potentials in neostriatal neurons, whereas muscarinic antagonists facilitated ACh release and increased neostriatal-evoked potentials (Misgeld et al., 1982; Weiler et al., 1984). Evidence that this muscarinic modulation of ACh release has an important role in the synaptic transmission of neostriatal neurons has also been documented (Dodt and Misgeld, 1986).

It is generally agreed that there are at least two subtypes of muscarinic receptors that can be classified by radioligand binding tests or by pharmacological tests. Given that muscarinic receptors in the neostriatum play an important role in modulating ACh release in the neostriatum, it would be important to know whether these muscarinic receptors are of a particular subtype. *In vivo* pharmacological evidence indicates that the M-2 receptor is involved in the regulation of ACh release in the neostriatum (Ladinsky et al., 1987), but by nature of the *in vivo* experiments, ACh release was not measured. In the present studies, therefore, the muscarinic receptor subtype involved in neostriatal release regulation was further investigated by monitoring ACh release from neostriatal slices *in vitro*. To pharmacologically characterize these receptors, the classical muscarinic agents, atropine and oxotremorine, the M-1 selective antagonist, pirenzepine, and the oxotremorine analog, oxotremorine-M, were tested for their effects on the potassium-evoked release of ACh from neostriatal slices.

Methods

Slice preparation and incubations. Neostriatal slices from Fischer 344 rats (3 months) were prepared and incubated in Krebs Ringer (KR) bicarbonate buffer as previously described (Weiler et al., 1984). Following a 3-h preincubation, individual slices were transferred to tissue holders, incubated 10 min in fresh KR buffer, then tested for spontaneous release (SP1) and potassium-stimulated (K1) release during a 5-min incubation in aliquots (1 ml) of regular KR buffer and 25 mM K^+ KR buffer, respectively. The slices were then transferred to regular KR buffer containing the muscarinic agent to be tested (saline was used for the controls) and incubated for 20 min. After this 20-min period the slices were again tested for spontaneous (SP2) release and potassium-stimulated (K2) release in the presence of a muscarinic agent. The ACh released from each slice in the presence of a muscarinic agent during the second release periods (K2 and SP2) was compared to the ACh released from that same slice in the absence of the test drug during the first release periods (K1 and SP1), and the data expressed as K2/K1 or SP2/SP1 ratios.

Assays. ACh and choline (Ch) were assayed by gas chromatography mass spectrometry according to the methods described by Jenden et al. (1973) and Freeman et al. (1975). Selected ion monitoring was used to determine the quantity of ACh and Ch in each sample relative to deuterated internal standards for ACh and Ch.

Tissue protein content was determined by the method of Lowry et al. (1951). Bovine serum albumin was used as the protein standard.

Source of chemicals. Muscarinic agents: atropine sulfate and oxotremorine sesquifumarate were obtained from Sigma Chemical Co. (St. Louis, MO, USA), pirenzepine dihydrochloride from Karl Thomae, GMBH (Biberach, West Germany). Oxotremorine-M was provided by Dr Bjorn Ringdahl (Department of Pharmacology, University of California Los Angeles).

Chemicals for ACh/Ch assay: tris(hydroxymethyl)-methylamino-propane sulfonic acid (TAPS) was obtained from Sigma Chem. Co. (St. Louis, MO, USA), dipicrylamine (2,2',4,4',6,6'-hexadinitropheny-lamine) from Pfaltz and Bauer, Inc. (Waterbury, CT, USA), dichloromethane (99 + %) from Aldrich (Milwaukee, WI, USA). The deuterated internal standards and all other reagents were provided by Dr Donald Jenden (Department of Pharmacology, Los Angeles).

All other chemicals were reagent grade.

Statistics. All results are expressed as the mean $\pm$ SEM, and the sample size, n, represents the number of independent observations. In all cases, each observation for a given concentration of a given muscarinic agent was obtained from a different animal. Multigroup data were compared by a one-way analysis of variance, and the difference between the mean of a control group and the mean of an experimental group was determined by the two-tailed Student's t-test.

Results

In previous studies it was shown that muscarinic agents modulated potassium-evoked ACh release from neostriatal slices in a manner that paralleled their effects on intrinsically evoked synaptic population spikes (Weiler et al., 1984). Also, the relative effects of muscarinic agents were apparently influenced by ACh levels in the synapse since AChE inhibition tended to enhance the facilitation of ACh release by muscarinic antagonists or attenuate the inhibition of ACh release by muscarinic agonists. These relative changes with AChE inhibition have also been observed in other cholinergic systems (Kilbinger et al., 1984; Szerb, 1979).

In the present study, the release of endogenous ACh evoked by 25 mM potassium was determined in the absence of an acetylcholinesterase (AChE) inhibitor, and a significant percentage of the choline (Ch) released was assumed to be the hydrolysis product of released ACh. This was confirmed by incubating slices in a high potassium KR buffer containing 8 mM magnesium and 0 mM calcium. Under these conditions, the potassium-evoked release of Ch decreased by 78–81% (n = 5 slices; p < 0.001).

Effects of the muscarinic antagonists, atropine and pirenzepine, on neostriatal ACh release

Atropine (0.1–2 μM) increased the potassium-evoked release of ACh from neostriatal slices when AChE was not inhibited (Fig. 1). The

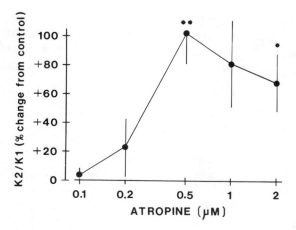

Figure 1. Increase in ACh release ratio (K2/K1) by atropine. Atropine-induced changes in K2/K1 are expressed as the % change from control and plotted versus the log concentration (μM) of atropine. Each point represents the mean $\pm$ SE of the mean, n = 4–6 separate observations. ** significantly different from control K2/K1 ratio at $p < 0.05$. *$p < 0.1$.

increase occurred in a dose-response manner and was apparently maximal in the presence of 0.5 μM atropine. Only at this concentration was the release significantly greater than control ($p < 0.01$). In the presence of 2 μM atropine the increase in ACh release was significant at $p < 0.1$.

In marked contrast, pirenzepine induced no significant changes in ACh release from the neostriatal slices (Fig. 2). Concentrations ranging from 2–200 μM were tested, and, if anything, the higher concentrations of pirenzepine induced a decrease in the release of ACh from the neostriatal slices. As mentioned above, the stimulatory effects of muscarinic antagonists are more pronounced when AChE is inhibited, so in a separate experiment the effects of pirenzepine on ACh release were

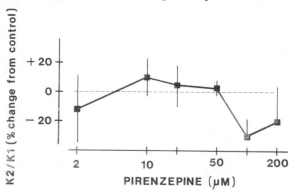

Figure 2. Effect of pirenzepine on ACh release ratio (K2/K1). Data for this and all subsequent log dose-response graphs are expressed in a manner similar to that described for figure 1. n = 3 – 6 separate observations.

208

examined in the presence of the AChE inhibitor, paraoxon (40 μM). Under these conditions 200 μM pirenzepine induced an 80% (n = 2) increase in the K2/K1 ACh release ratio, and 2 μM pirenzepine induced less than a 5% change (n = 3) in the K2/K1 ACh release ratio.

Effects of the muscarinic agonists, oxotremorine and oxotremorine-M, on neostriatal ACh release

The classical muscarinic agonist, oxotremorine, and an oxotremorine analog, oxotremorine-M, had an overall inhibitory effect on neostriatal ACh release. The maximal inhibition induced by oxotremorine (Fig. 3) was about 30% at 20 μM (p < 0.05) and by oxotremorine-M (Fig. 4) was about 40% at 5 μM (p < 0.01). Several equieffective concentrations of these two agonists were compared (Tallarida and Murray, 1979), and the relative potency of oxotremorine-M to oxotremorine was 0.044 ± 0.007 (n = 7). That is, at equieffective concentrations, oxo-tremorine-M was approximately 25 times more potent than oxo-tremorine.

In a separate experiment the effects of these two muscarinic agents were tested in the presence of atropine. AChE was inhibited for these experiments. Atropine (0.1 μM) alone induced a 66% increase in the release of ACh, and when it was added with oxotremorine the K2/K1 ratio was only 7% greater than control. Whereas in the previous experiments oxotremorine-M produced its maximal inhibitory effect on ACh release within the 5–50 μM range, the concomitant incubation of oxotremorine-M (10 μM) with atropine resulted in a 25% increase in ACh release, which was not significantly different from control (n = 3 observations). Thus, the inhibitory effects on ACh release of oxo-tremorine and oxotremorine-M were blocked by atropine.

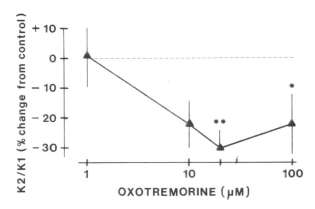

Figure 3. Decrease in ACh release ratio (K2/K1) by oxotremorine. Data is presented as described in legend for Figure 1. n = 6–10 separate observations. **p < 0.05. *p < 0.1.

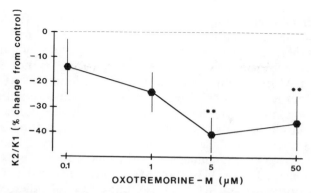

Figure 4. Decrease in ACh release ratio (K2/K1) by oxotremorine-M. Data is presented as described in legend for Figure 1. n = 3–9 separate observations. **p < 0.01 for 5 μM, p < 0.05 for 50 μM.

Effects of muscarinic agents on the spontaneous efflux of choline

The highest concentrations of the muscarinic agents were also tested for their effects on the spontaneous efflux of Ch from the neostriatal slices. Atropine (2 μM), pirenzepine (200 μM), oxotremorine (100 μM) and oxotremorine-M (50 μM) induced no significant changes in Ch efflux during the second spontaneous release period (SP2).

Discussion

In contrast to the effects of the classical muscarinic receptor antagonist, atropine, the effects on neostriatal release of the M-1 selective antagonist, pirenzepine, were minimal over a broad concentration range. These results indicate that in the neostriatum the muscarinic modulation of ACh release is mediated by an M-2 receptor subtype and are supportive of the concept that a major function of the M-2 receptor is to regulate transmitter release in the brain (Potter et al., 1984; Mash and Potter, 1986). In the myenteric cholinergic neurons the inhibitory modulation of evoked ACh release also is mediated by M-2 receptors (Kilbinger, 1984; Surprenant, 1986), since micromolar as opposed to nanomolar concentrations of pirenzepine blocked the inhibition of ACh release by oxotremorine. In the present study, the antagonists were tested for their effects on endogenous ACh rather than on an exogenously added agonist, and even micromolar concentrations of pirenzepine had no effect on the potassium-evoked release of endogenous ACh whereas atropine did. Whether nanomolar concentrations of pirenzepine alter ACh release was not examined in this study, but it seems unlikely that lower concentrations would have an effect on ACh release if the higher concentrations tested in this study were ineffective. Conversely, since concentrations of pirenzepine above 200 μM were not tested, it is not certain whether higher concentrations of pirenzepine would affect neostriatal ACh release when AChE is not inhibited. Inhibition of AChE resulted in a substantial facilitation in ACh release by 200 μM but not by 2 μM pirenzepine, which indicates that the putative M-1 antagonist can influence ACh release at higher concentrations.

The muscarinic agonists, oxotremorine and oxotremorine-M are not receptor selective, and their inhibitory effects on ACh release were blocked by atropine. Preliminary experiments with the putative M-1 selective agonist, McN-A-343 (100 μM), indicate that it has no significant effects on neostriatal ACh release, but effects of other concentrations of McN-A-343 need to be tested. Moreover, whether pirenzepine, even though itself noneffective, blocks the effects of these muscarinic agonists remains to be examined. The data of this study indicate, however, that quite different responses of ACh release can be elicited by different muscarinic agents, and some of the effects of the different agents or even the same muscarinic agent depend upon elevated levels of ACh, i.e., inhibition of AChE, as has been observed in other cholinergic systems (Kilbinger, 1985) and with other cholinergic agents (Nordstrom et al., 1983).

In the present study, the potency of oxotremorine-M in inhibiting ACh release was approximately 20 times that of oxotremorine. The maximal inhibitory response elicited by oxotremorine-M was also about 25% greater (40% vs 30% inhibition), but due to the variation of the responses this was not statistically significant. In other systems these two agents do produce different pharmacological (Ringdahl, 1985), biochemical (Fisher and Bartus, 1985), and electrophysiological effects (McCormick and Prince, 1985). These pharmacological and biochemical differences are apparently due to differences in the receptor reserve of different tissues (Ringdahl, 1987), in receptor coupling of muscarinic events in different brain areas (Fisher and Bartus, 1985), or even to interspecies differences (Fisher, 1986). Therefore, further pharmacological analyses, e.g. pA_2 estimations and relative affinity and efficacy determinations, are required to characterize the muscarinic receptor involved in ACh release from the neostriatum (Kenakin, 1984).

Acknowledgments. The author would like to thank Joleen Stolp for her technical assistance. The arrangement by Dr Donald J. Jenden and his staff for the use of the GCMS facilities at the University of California Los Angeles is greatly appreciated. These studies were supported by USPHS Grant AG05953.

Dodt, H. U., and Misgeld, U. (1986) Muscarinic slow excitation and muscarinic inhibition of synaptic transmission in the rat neostriatum. J. Physiol. 380: 593–608.
Fisher, S. K. (1986) Inositol lipids and signal transduction at CNS muscarinic receptors. Trends pharmac. Sci. Suppl: 61–65.

Fisher, S. K., and Bartus, R. T. (1985) Regional differences in the coupling of muscarinic receptors to inositol phospholipid hydrolysis in guinea pig brain. J. Neurochem. 45: 1085–1095.

Freeman, J. J., Choi, R. L., and Jenden, D. J. (1975) Plasma choline, its turnover and exchange with brain choline. J. Neurochem. 24: 729–734.

James, M. K., and Cubeddu, L. X. (1984) Frequency-dependent muscarinic receptor modulation of acetylcholine and dopamine release from rabbit striatum. J. Pharmac. exp. Ther. 229: 98–104.

Jenden, D. J., Roch, M, and Booth, R. (1973) Simultaneous measurement of endogenous and deuterium labeled tracer variants of choline and acetylcholine in subpicomole quantities by gas chromatography mass spectrometry. Analyt. Biochem. 55: 438–448.

Kenakin, T. P. (1984) The classification of drugs and drug receptors in isolated tissues. Pharmac. Rev. 36: 165–222.

Kilbinger, H. (1985) Subtypes of muscarinic receptors modulating acetylcholine release from myenteric nerves. In: Lux, G., and Daniel, E. E. (eds), Muscarinic Receptor Subtypes in the GI Tract. Springer-Verlag, Berlin, pp. 37–42.

Kilbinger, H., Halim, S., Lambreact, G., Weiler, W., and Wessler, I. (1984) Comparison of affinities of muscarinic antagonists to pre- and postjunctional receptors in the guinea-pig ileum. Eur. J. Pharmac. 103: 313–320.

Ladinsky, H., Vinci, R., Wang, J., Palazzi, E., Cicioni, P., and Consolo, S. (1987) On the functional role of muscarinic receptor subtypes in the regulation of cholinergic neurotransmission in striatum and hippocampus of the rat. In: Dowdall, Hawthorne (eds), Cellular and Molecular Basis of Cholinergic Function. Ellis Horwood Ltd, Chichester England, pp. 193–199.

Lowry, O. H., Rosebrough, N. J., Farr, A. L., and Randall, R. J. (1951) Protein measurement with the Folin phenol reagent. J. biol. Chem. 193: 265–275.

Mash, D. C., and Potter, L. T. (1986) Autoradiographic localization of M1 and M2 muscarinic receptors in the rat brain. Neuroscience 19: 551–564.

McCormick, D. A., and Prince, D. A. (1985) Two types of muscarinic response to acetylcholine in mammalian cortical neurons. Proc. natl Acad. Sci. 82: 6344–6348.

Misgeld, U., Weiler, M. H., and Cheong, D. K. (1982) Atropine enhances nicotinic cholinergic EPSPs in rat neostriatal slices. Brain Res. 253: 317–320.

Nordstrom, O., Alberts, P., Westlind, A., Unden, A., and Bartfai, T. (1983) Presynaptic antagonist-postsynaptic agonist at muscarinic cholinergic synapses. N-methyl-N-(1-methyl-4-pyrrolidino-2-butynyl)acetamide. Molec. Pharmac. 24: 1–5.

Potter, L. T., Flynn, D. D., Hanchett, H. E., Kalinoski, D. L., Luber-Narod, J., and Mash, D. C. (1984) Independent M1 and M2 receptors: ligands, autoradiography and functions. Trends Pharmac. Sci. Suppl: 22–31.

Ringdahl, B. (1985) Structural requirements for muscarinic receptor occupation and receptor activation by oxotremorine analogs in the guinea-pig ileum. J. Pharmac. exp. Ther. 232: 67–73.

Ringdahl, B. (1987) Selectivity of partial agonists related to oxotremorine based on differences in muscarinic receptor reserve between the guinea pig ileum and urinary bladder. Molec. Pharmac. 31: 351–356.

Suprenant, A. (1986) Muscarinic receptors in the submucous plexus and their roles in mucosal ion transport. Trends pharmac. Sci. Suppl: 23–27.

Szerb, J. C. (1979) Autoregulation of acetylcholine release. In: Langer, S. Z., Starke, K., and Dubocovich, M. L. (eds) Autoregulation of acetylcholine release. Pergamon Press, Oxford, pp. 263–286.

Tallarida, R. J., and Murray, R. B. (1987) Manual of pharmacologic calculations with computer programs, 2nd edn. Springer-Verlag, New York.

Weiler, M. H., Misgeld, U., and Cheong, D. K. (1984) Presynaptic muscarinic modulation of nicotinic excitation in the rat neostriatum. Brain Res. 296: 111–120.

Distribution of cholinergic receptors in the rat and human neocortex

K. Zilles[a], H. Schröder[a], U. Schröder[a], E. Horvath[b], L. Werner[c],
P. G. M. Luiten[d], A. Maelicke[e] and A. D. Strosberg[f]

[a]Anatomical Institute, University of Köln, Joseph-Stelzmann-Str. 9, D-5000 Köln 41, FRG;
[b]Troponwerke, Neurobiological Department, Köln, FRG; [c]Paul-Flechsig-Hirnforschungs-institut, Leipzig, GDR; [d]Department of Animal Physiology, University of Groningen, The Netherlands; [e]Max-Planck-Institut für Ernährungsphysiologie, Dortmund, FRG; [f]Institut Pasteur, Paris, France

Summary. Autoradiographic labelling of muscarinic (M1, M2, NMS binding sites) and nicotinic receptors shows an inhomogeneous distribution over architectonically identified cortical areas of the rat brain with highest concentrations in the medial prefrontal and frontal areas. Beside this general trend the areal patterns of different receptors are slightly varying. The laminar distribution of these receptors in the rat and human neocortex is characterized by two different patterns, one with highest receptor densities in the supragranular layers (M1 receptors, NMS binding sites), the other with a preferential labelling of layer IV and (with a lower intensity) layer V (M2 and nicotinic receptors). M1 receptors and NMS binding sites are codistributed at the laminar level with each other and with $GABA_A$, D1, 5-HT_1 and glutamate receptors; M2 receptors are codistributed only with nicotinic receptors. Immuno-histochemical studies with antibodies against muscarinic and nicotinic receptors demonstrate that these structures occur mainly in pyramidal and spiny stellate cells and to a lesser extent (13%) in a variety of interneurons. The immunoreactivity is visible in the perikaryon, dendrites and postsynaptic membranes. Neurons are found in the human neocortex, which react exclusively with one of the two antibodies, but a fraction of the neurons (about 30%) contains antigenic sites reacting with both antibodies. This is interpreted as colocalization of nicotinic and muscarinic receptors in some cortical neurons.

Introduction

In vitro receptor autoradiography (e.g. Young and Kuhar, 1980) and immunohistochemistry with antibodies raised against receptors (Deutch et al., 1987; Fels et al., 1986; Matsuyama et al., 1988; Swanson et al., 1987) are presently the most powerful methods to study the topographical distribution of transmitter receptors in the CNS. The combination of image analysis and quantitative autoradiography (e.g. Zilles et al., 1986) offers the advantage of obtaining rapidly quantitative data about areal and laminar receptor distributions. However, local resolution of this method is not sufficient for the analysis of receptor distributions at the cellular or subcellular level. This can be achieved by immunohistochemistry, especially by a combination of light and electronmicroscopical methods. Immunohistochemistry, on the other hand, has the drawback of restricted quantifiability. We used, therefore, both methodical approaches to study the distribution of cholinergic receptors in the neocortex of human and rat brains. The present observation will focus on the following issues:

- areal pattern of cholinergic receptor distribution in the neocortex of the rat,
- laminar pattern within defined neocortical areas of the human and rat brain,
- codistribution at the laminar level between acetylcholine, glutamate, GABA, serotonin and dopamine receptors,
- cellular and subcellular location in morphologically defined neuronal types.

Material and methods

Whole brains of Wistar rats or cortical pieces of human postmortem or neurosurgical material were frozen for cryostate sectioning (section thickness 20 μm) and mounted in alternating series for autoradiography and for Nissl staining. Muscarinic receptors were labelled with the antagonist (^{3}H)-N-methylscopolamine (0.2 nM), the M1 subtype with the antagonist (^{3}H)pirenzepine (2–3 nM) and the M2 subtype with the agonist (^{3}H)oxotremorine-M (1–2 nM). Nicotinic receptors were labelled with 4 nM (^{3}H)nicotine. Glutamate binding sites were labelled with 100–150 nM (^{3}H)glutamate, the GABA-receptors with 5–8 nM (^{3}H)muscimol, the 5-HT$_1$ receptors with 0.5 nM (^{3}H)serotonin and the D1 receptors with 1 nM (^{3}H)-SCH 23390. The ligand concentration roughly corresponded to the half K_D-values determined in separate saturation analyses in cryostate sections. The labelling procedure followed standard methods the details of which were published by Zilles (1988, 1989) and Zilles et al. (1988). Quantitative evaluation of receptor densities was performed with an image analyzer (IBAS 2; Kontron, West Germany) and own computer programs (for details, see Zilles et al., 1986). Receptor localization was mapped by superimposing the adjacent Nissl-stained sections on the digitized receptor images.

Immunohistochemistry was performed on material from perfusion-fixed rat brains or immersion-fixed human cortical pieces from neurosurgical or autopsy material carefully checked to be free from tumorous infiltration or other pathological changes. The fixation fluid was Zamboni's solution. Floating vibratome sections (50 μm) were incubated with the monoclonal antibodies M35 (André et al., 1984; Luiten et al., 1988; Matsuyama et al., 1988) against muscarinic receptors and WF6 (Fels et al., 1986) against nicotinic receptors. The antibodies were used in working dilutions between 1:10 to 1:2000 (Schröder et al., submitted a, b). The immunoreactive sites were visualized by a biotin-streptavidin-peroxidase protocol. Controls were performed by the omission of the primary and/or secondary antisera, incubation with the peroxidase-streptavidin complex along or preabsorption with membrane fragments. All controls were negative.

Additionally, double labelling with both antibodies was performed in human cortical samples for detection of colocalization of nicotinic and muscarinic receptors in the same cell. In the case of WF6, a fluorescein-labelled anti-mouse IgG was used as a secondary antibody, whereas in the case of M35 a biotinylated anti-mouse IgM was used. The biotinylated IgM was subsequently immersed in a phycoerythrin stabilizer and was detected by red fluorescence. The fluorescein labelled IgG was visualized by green fluorescence. Double incubation with both primary antibodies in both possible sequences were performed with the same section. Controls were as described above and, additionally, the possibility of detection systems reacting with the 'false' primary antibodies (IgM with WF6, or IgG with M35) was tested. All controls were negative.

Results

Areal pattern

The areal pattern of the mean receptor densities of M1, M2 and nicotinic receptors in the rat brain is shown in Figure 1. The highest densities of M1 receptors were found in the classical motor cortex (Fr1–2), the medial prefrontal areas (Cg1, Cg3) and the orbital and insular cortex (MO, VO, VLO, LO, AID, AIV, AIP). The highest relative densities of M2 receptors were found in the major part of the motor cortex (Fr1–2) and medial prefrontal cortex (Cg1, Cg3). The sensory areas again had a lower receptor density and the allocortical regions had a very low density of M2 receptors, in contrast to the highest M1 density in these regions. The only exception was the septal complex, which showed a high M2, but low M1 density. The areal pattern of NMS binding sites differed slightly from both M1 and M2 distributions by having a receptor density in the allocortex, which was between the M1 and M2 values, and by showing a higher density in lateral frontal (Fr3), medial parietal (FL, HL) and medial occipital (Oc2M) areas than found in both subtypes. The areas with a lower mean M1 density are characterized architectonically by much higher contents in small granular nerve cells. Contrastingly, the neocortical areas with predominating larger pyramidal cells contain the higher receptor density. This may be due to a preferential location of muscarinic receptors in pyramidal cells and will be further analyzed in the last part of these observations.

The areal pattern of nicotinic receptors (analyzed with (^{3}H)nicotine) was distinct from the distribution of both muscarinic subtypes (Fig. 1).

The relatively highest densities of nicotinic receptors were found in the primary and medial secondary visual cortex and the medial prefrontal and granular retrosplenial cortex. The lowest densities occurred in the piriform cortex and the medial and lateral septum.

Laminar pattern

There was a clear difference in the laminar distribution of muscarinic receptor subtypes M1 and M2. M1 receptors were found to be preferentially localized in the supragranular layers, whereas M2 receptors were predominating in layer IV and to a lower degree in the deeper part of layer V (Fig. 2). The nicotinic receptors were preferentially labelled in layer IV, followed by layer V.

The highest densities of M1 receptors and NMS binding sites in the human striate area were found in layers II–IVa with a peak in layer III. The pattern in the agranular frontal cortex differed slightly, because layers II–III contained the highest density with a peak value in layer II. The laminar distributions of M1 and NMS binding sites were in correlation in the human striate area (r = 0.97) and frontal agranular cortex (r = 0.91) (Zilles et al., 1988), because M1 receptors constitute the major part of NMS binding sites in the human neocortex. This has been described also in displacement studies in the rat brain (Cortés and Palacios, 1986; Wamsley et al., 1980; Yamamura et al., 1985). The laminar distribution of M2 receptors in the human primary visual cortex showed maximal densities in layers IVa and IVc followed by layer V.

Codistribution

The similarity of laminar distribution patterns of receptors of different neurotransmitters ('codistribution') was analyzed by correlating the proportionate laminar densities. Significant codistributions were found in the primary cortical areas of the rat (Table 1) between NMS binding sites (or M1 receptor) and GABA$_A$ or glutamate binding sites, but not between M2 and M1 receptors, GABA$_A$ or glutamate binding sites. Moreover, a highly significant codistribution was found between M2 and nicotinic receptors. Codistributions in the human striate area (Table 2) or agranular frontal cortex showed the same assembly of significant local associations between receptors, which indicates an impressive similarity in neurochemical structure of the cortex of two different mammals, in which homologous areas are architectonically different. Additionally, dopamine D1 and serotonin 5-HT$_1$ receptor distributions were tested in the human cortex. D1 receptors showed no or only a low degree (with NMS and glutamate binding sites in the

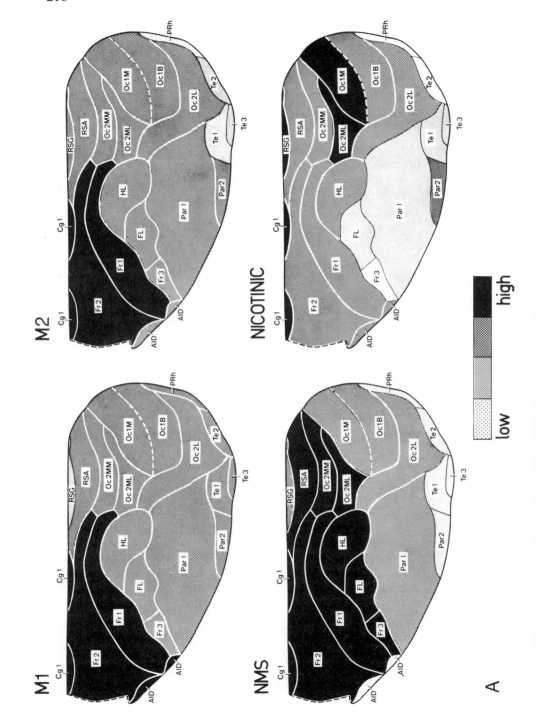

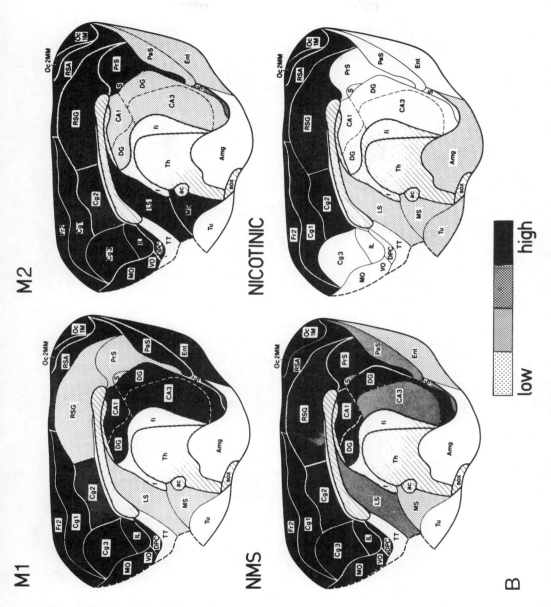

Figure 1. Areal pattern of the mean receptor densities of M1, M2, NMS and nicotinic binding sites in the rat forebrain. The whole range of the mean areal receptor densities for each receptor type is subdivided into four classes of increasing densities. The cytoarchitectonical delineation follows the subdivisions given in Zilles (1985). The dorsal (A) and medial (B) aspects of a hemisphere are shown.

218

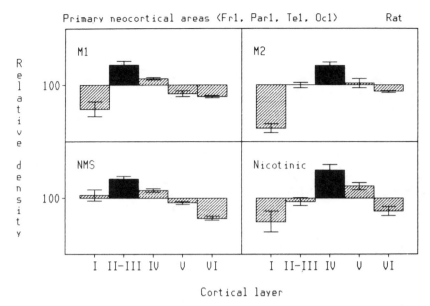

Figure 2. Laminar pattern of cholinergic receptors in the primary neocortical areas (Fr1, Par1, Te1, Oc1) of the rat. The mean density over all layers of each area was defined as 100% and the proportionate values for the different layers were calculated. The columns in this figure represent the mean values $\pm$ SEM of the four primary areas.

striate area) of codistribution with all the other receptors, whereas 5-HT$_1$ receptors showed a high degree of codistribution with glutamate, NMS and M1 receptors. These results suggest a close topographical relationship between M1, but not M2 receptors, and glutamate binding sites, and—to a lower degree—between M1 receptors and 5-HT$_1$

Table 1. Codistribution of different receptors in the primary cortical areas (Fr1, Par1, Te1, Oc1) of the rat. The codistribution is tested with the Spearman rank correlation test. First line, correlation coefficient R$_s$; second line, error probability P; ns, not significant.

	M1	M2	NMS	Nicotinic	Glutamate	GABA$_A$	5-HT$_1$
M1		ns	ns	ns	0.9 0.042	0.9 0.042	ns
M2			ns	1.0 0.008	ns	ns	ns
NMS				ns	0.9 0.042	0.9 0.042	ns
Nicotinic					ns	ns	ns
Glutamate						1.0 0.008	ns
GABA$_A$							ns

Table 2. Codistribution of different receptors in the human primary visual cortex (area 17). The codistribution is tested with the Spearman rank correlation test. First line, correlation coefficient R_s; second line, error probability P; ns, not significant.

	M1	M2	NMS	Glutamate	GABA$_A$	5-HT$_1$	D1
M1		ns	0.81 0.011	0.76 0.018	ns	0.79 0.014	ns
M2			ns	ns	0.64 0.048	ns	ns
NMS				0.90 0.002	0.93 0.001	0.79 0.014	0.69 0.035
Glutamate					0.74 0.023	0.90 0.002	0.76 0.018
GABA$_A$						ns	ns
5-HT$_1$							ns

receptors. The M1 receptors showed no significant codistribution with M2 and D1 receptors.

Cellular pattern

The cellular localization of muscarinic and nicotinic receptors was studied with immunohistochemical methods at the light and electro-microscopical levels both in the rat and human neocortex. The nicotinic and muscarinic antibodies had identical cellular localizations. Numerous neuronal cell bodies (Fig. 3A) and apical dendrites of pyramidal cells were immunopositive. Immunoprecipitate was found attached to neurotubular-like structures and to postsynaptic membranes (Fig. 3B). Perikaryal morphology in histological sections of the rat primary visual cortex counterstained with cresyl violet enabled a preliminary cell typing of immunopositive cells. The criteria were based on the results obtained from Golgi-stained and deimpregnated as well as Nissl-stained sections by Werner et al. (1982, 1985). The comparison of the results of these studies with electron microscopical observations and cell typing by Peters and Kara (1985a, b) demonstrates the reliability of the present approach. About 29% of the cortical neurons proved to be immunopositive when stained with the muscarinic receptor antibody M35. Eighty-one percent of the immunoreactive neurons in the rat primary visual cortex were pyramidal cells, 6% were interneurons (Fig. 4). The muscarinic receptors of interneurons were preferably localized in basket cells and in a group of cells which comprises bipolar and chandelier cells. Muscarinic receptors were also found in layer I interneurons, neuroglioform cells and Martinotti cells. The laminar distribution of cell bodies

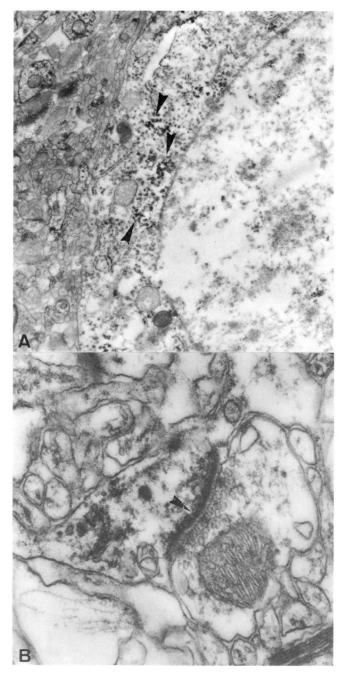

Figure 3. Immunoprecipitate (arrowheads) in a neuronal perikaryon (*A*) and a cortical synapse of the human brain (*B*) after immunocytochemistry with the nicotinic antibody WF6. Magnification: $11,000 \times$ (*A*), $61,000 \times$ (*B*).

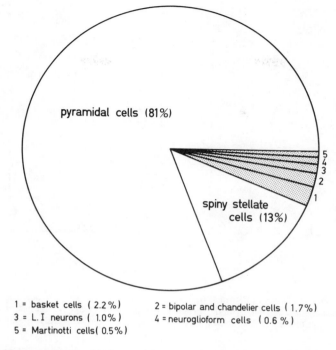

Figure 4. Neuronal types with muscarinic receptors demonstrated with the monoclonal antibody M35 in the primary visual cortex (Oc1) of the rat.

with M35 immunoreactivity showed the largest absolute number in layer VI, but a calculation of the number per volume unit revealed layers II and IV as places with the highest densities in neuronal cell bodies with muscarinic receptors (Fig. 5). Although a similarly detailed analysis of nicotinic receptors is still in progress, preliminary results appear to show the same pattern. This raises the question, whether nicotinic and muscarinic receptors are localized in the same or in different cells. A double labelling study showed that approximately 60% of the immunoreactive cells contain muscarinic receptors, 10% contain nicotinic receptors and 30% are double-labelled both with M35 and WF6 (Fig. 6). This means that most of the nicotinic receptors are colocalized with muscarinic receptors.

Discussion

Areal pattern

We have shown that muscarinic receptors and their subtypes are unevenly distributed throughout the rat neocortex. Areas with the

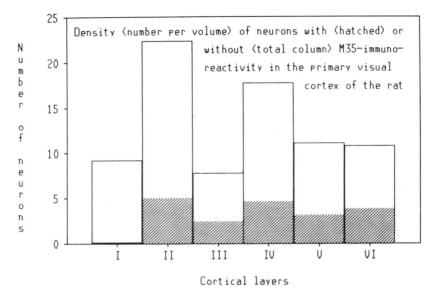

Figure 5. Laminar pattern of neuronal perikarya with muscarinic receptors demonstrated with the monoclonal antibody M35 in the primary visual cortex (Oc1) of the rat. The density (number per volume unit) of all neuronal cell bodies and neurons with M35-immunoreactivity is shown.

highest receptor densities are found in the medial prefrontal and motor cortex. This demonstrates that the observation of a single and often anatomically undefined neocortical area cannot be representative for the whole neocortex.

Spencer et al. (1986) described in a semiquantitative way a higher density of M1 receptors in the piriform cortex compared with the M2 subtype; contrastingly, they found a higher density of M2 receptors in the septum compared with the M1 subtype. Both findings are corroborated quantitatively by the present observations, but in contrast to Spencer et al. (1986) the highest densities of M1 receptors are found in the cingulate and frontal cortex, which are classified by these authors as areas with a lower density. This difference may be explained by the fact that Spencer et al. (1986) have made an estimate without quantification.

A poor correlation is found between the ChAT content (Brownstein and Palkovits, 1984) of various forebrain areas and their M1 receptor densities. The areal distribution of ChAT (Brownstein and Palkovits, 1984) and M2 receptors, however, shows a good correlation, because the forebrain areas with relatively high ChAT contents (medial prefrontal areas (Cg1, Cg3), caudate-putamen, accumbens nucleus, granular retrosplenial cortex, superficial part of the olfactory tubercle, medial septum with the vertical limb of the diagonal band) have the relatively highest

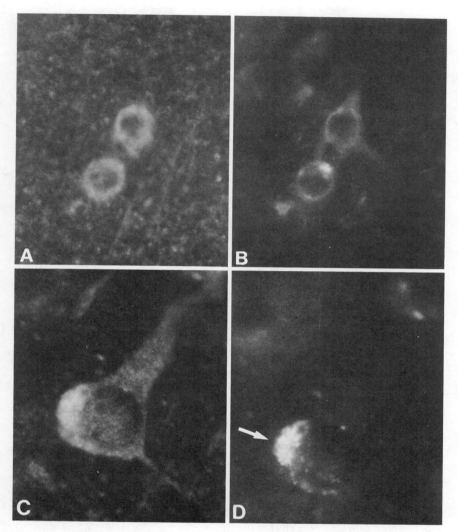

Figure 6. Localization of muscarinic (*A*, *C*) and nicotinic (*B*, *D*) receptors in the same neurons (*A* and *B*; *C* and *D*) of the human cortex. The two neurons in the upper part are immunoreactive with both antibodies indicating colocalization, whereas the pyramidal cell in the lower part of figure 6 shows only a reaction with the muscarinic antibody. This neuron contains lipofuscin granules (arrow), which exhibit autofluorescence. Magnification: 450 ×.

M2 receptor densities. Another coincidence occurs between the distribution patterns of ChAT-positive terminals and the groups of neocortical areas with higher or lower muscarinic receptor densities. Eckenstein et al. (1988) describe a distinct laminar distribution of ChAT-containing terminals in all of our areas with a low M1 density, whereas the high density areas comprise cortical regions with no distinct laminar pattern of ChAT-terminals.

The nicotinic receptors also show an uneven areal distribution with the highest density in the medial prefrontal, primary visual, medial secondary visual and granular retrosplenial cortex. The lowest densities are found in the septal region, olfactory tubercle and piriform cortex. Contrastingly, the septal region has high ChAT contents (Brownstein and Palkovits, 1984). This, however, does not imply a mismatch between an important cholinergic marker and nicotinic receptors, because the septum contains a high proportion of ACh-synthetizing cells projecting to the cortex. The enzyme is, therefore, accumulated mainly in the perikarya and only a minor proportion of this marker may be located in cholinergic terminals within this region.

Laminar pattern

NMS binding sites and M1 receptors reach densities in layers II–III highest of all neocortical areas in the rat. A striking similarity can be ascertained between the laminar patterns of M1 receptors and NMS binding sites with QNB (Kuhar and Yamamura, 1976; Wamsley et al., 1981) binding sites. M2 and nicotinic receptors are predominating in layers IV and V. This laminar pattern is in agreement with studies of Clarke et al. (1984), London et al. (1985) also using (^{3}H)nicotine. The use of (^{3}H)acetylcholine plus unlabelled atropine (Clarke et al., 1985; Schwartz, 1986) leads to comparable results.

The distribution of α-bungarotoxin binding sites (Clarke et al., 1985), which do not represent the acetylcholine binding sites in the brain, differs greatly from the pattern revealed by studies with (^{3}H)nicotine or (^{3}H)acetylcholine (Clarke et al., 1984, 1985; London et al., 1985; Schwartz, 1986; present results). No single muscarinic and nicotinic receptor distribution was found to match completely with the distribution pattern of axons from the basal forebrain, the major source of cortical cholinergic innervation (Eckenstein et al., 1988; Luiten et al., 1987; Rieck and Carey, 1984), with the biochemically determined laminar ChAT contents (Johnson et al., 1981; Parnavelas and McDonald, 1983) or the immunohistochemically revealed distribution of ChAT-containing axons (Eckenstein et al., 1988; Parnavelas et al., 1986; Wainer, 1984). Only in the case of the primary visual cortex could a nearly perfect match be observed between the laminar pattern of ChAT-containing terminals, AChE histochemistry, nicotinic and M2 receptor distributions.

Codistribution

Our codistribution studies revealed very similar laminar patterns between M1 or NMS, but not M2 receptors, and GABA$_A$, D1, 5-HT$_1$

receptors or glutamate binding sites. Furthermore, M2 receptors are not codistributed with any of the analyzed receptor types except nicotinic receptors. Nicotinic receptors have also no additional codistribution. Although most of the receptor mapping studies are focussed exclusively on one receptor type, a few publications, mostly concerned with the control of transmitter release, provide data, which are of some interest for the local associations between different receptors presented in this work. Cortical muscarinic receptors are reported to be localized on dopaminergic terminals (Andree et al., 1983; Marchi et al., 1987); muscarinic and $GABA_A$ receptors show correlated regulatory changes in the abstinence after barbital treatment (Nordberg et al., 1986). M1 receptors were found on 5-HT containing cortical terminals (Härfstrand et al., 1987; Marchi et al., 1986). These examples of possible interactions in the neocortex are supported by reports about comparable relations in the hippocampus and subcortical areas.

Cellular pattern

The localization of immunopositive material within single neurons after incubation with antibodies against nicotinic and muscarinic receptors is comparable with observations of other groups (Jacob et al., 1986; Matsuyama et al., 1988). The occurrence of immunostraining on ribosomes and endoplasmic reticulum in the perinuclear space, Golgi-apparatus, neurotubular-like structures and postsynaptic membranes may represent the anatomical equivalents of synthesis, packaging, transport and deposition in membranes (Matsuyama et al., 1988; Swanson et al., 1987).

A comparison of the laminar patterns of cholinergic receptors revealed by autoradiography or immunohistochemistry showed that autoradiography does not primarily reflect the localization of receptor containing perikarya, which were demonstrated in all layers, but most densely concentrated in layers II and IV. The layers with the most intense autoradiographic label are those with the highest concentration of immunopositive dendrites and postsynaptic densities. We think that this is caused by a much higher amount of receptors in the dendritic as compared to the perikaryal compartment. Functional interpretations of receptor mapping studies must take into consideration that immunohistochemistry demonstrates also a significant amount of receptors in extrasynaptic positions. Additionally, muscarinic (Hösli and Hösli, 1988; Murphy et al., 1986) and nicotinic (Hösli and Hösli, 1988) receptors have been localized to astrocytes and myelin (Larocca et al., 1987).

The WF6 antibody was raised against the α-subunit of *Torpedo* membranes (Fels et al., 1986). This implies the possibility that the α-bungarotoxin binding site is demonstrated. Autoradiographic labelling with α-bungarotoxin (Clarke et al., 1985), however, showed a maximal density in layer VI, whereas immunohistochemistry with the WF6 antibody presents a laminar pattern best comparable with the autoradiographic observations after application of nicotine ligands (Clarke et al., 1985; present observations), or after immunohistochemical detection of nicotinic receptors that do not bind α-bungarotoxin (Swanson et al., 1987). Moreover, the WF6 antibody stains neurons in regions in which a signal is visible after *in situ* hybridization with a cDNA probe to α_3 gene of the neuronal nicotinic receptor (Boulter et al., 1986; Goldman et al., 1987). Finally, Deutch et al. (1987) have shown that an antibody generated against receptors from *Torpedo* electric organ can crossreact with central nicotinic receptors of rats.

The localization of muscarinic receptors within pyramidal cells and interneurons of the rat primary visual cortex is in agreement with electrophysiological studies of McCormick and Prince (1985), who proposed cholinergic actions both on pyramidal cells and GABAergic interneurons projecting onto pyramidal cells. The colocalization of muscarinic and nicotinic receptors in cortical neurons may be the structural basis for muscarinic and nicotinic responses of one cell shown in a variety of neurons in the brain and spinal cord by electrophysiological observations (for review Kelly and Rogawski, 1987).

Acknowledgements. This work was supported by grants from the DFG (Zi 192/8-1; Schr 283/4-1). We thank P. Ahrens, H. Beck, M. Henschel, M. Rath and Dr A. Schleicher for technical assistance, Ch. Opfermann-Rüngeler for the graphical and I. Koch for the photographic work.

Andre, C., Guillet, J. G., DeBacher, J. P., Vanderheyden, P., Hoebeke, J., and Strosberg, A. D. (1984) Monoclonal antibodies against the native or denatured forms of muscarinic acetylcholine receptor. EMBO J. 2: 17–21.

Andree, T. H., Gottesfeld, Z., DeFrance, J. F., Sikes, R. W., and Enna, S. J. (1983) Evidence for cholinergic muscarinic receptors on mediodorsal thalamic projections to the anterior cingulate cortex. Neurosci. Lett. 40: 99–103.

Boulter, J., Evans, K., Goldman, D., Martin, G., Treco, D., Heinemann, S., and Patrick, J. (1986) Isolation of a cDNA clone coding for a possible neural nicotinic acetylcholine receptor α-subunit. Nature 319: 368–374.

Brownstein, M. J., and Palkovits, M. (1984) Catecholamines, serotonin, acetylcholine, and γ-aminobutyric acid in the rat brain: biochemical studies. In: Björklund, A., and Hökfelt, T. (eds), Handbook of Chemical Neuroanatomy, vol. 2. Elsevier, Amsterdam, pp. 23–54.

Clarke, P. S., Pert, C. B., and Pert, A. (1984) Autoradiographic distribution of nicotine receptors in rat brain. Brain Res. 232: 390–395.

Clarke, P. B. S., Schwartz, R. D., Paul, S. M., Pert, C. B., and Pert, A. (1985) Nicotinic binding in rat brain: autoradiographic comparison of [3H]acetylcholine, [^{3}H]nicotine, and [^{125}I]-α-bungarotoxin. J. Neurosci. 5: 1307–1315.

Cortés, R., and Palacios, J. M. (1986) Muscarinic cholinergic receptor subtypes in the rat brain. I. Quantitative autoradiographic studies. Brain Res. 362: 227–238.

Deutch, A. Y., Holliday, J., Roth, R. H., Chun, L. L. Y., and Hawrot, E. (1987) Immunohistochemical localization of a neuronal nicotinic acetylcholine receptor in mammalian brain. Proc. natl Acad. Sci. USA 84: 8697–8701.

Eckenstein, F. P., Baughman, R. W., and Quinn, J. (1988) An anatomical study of cholinergic innervation in rat cerebral cortex. Neuroscience 25: 457–474.

227

Fels, G., Plümer-Wilk, R., Schrieber, M., and Maelicke, A. (1986) A monoclonal antibody interfering with binding and response of the acetylcholine receptor. Biol. Chem. 261: 1546–1554.

Goldman, D., Deneris, E., Luyten, W., Kochhar, Patrick, J., and Heinemann, S. (1987) Members of a nicotine acetylcholine receptor gene family are expressed in different regions of the mammalian central nervous system. Cell 48: 965–973.

Härfstrand, A., Fuxe, K., Andersson, K., Agnati, L., Janson, A. M., and Nordberg, A. (1987) Partial di-mesencephalic hemitransections produce disappearance of [³H]nicotine binding in discrete regions of rat brain. Acta physiol. scand. 130: 161–163.

Hösli, E., and Hösli, L. (1988) Autoradiographic localization of binding sites for muscarinic and nicotinic agonists and antagonists on cultured astrocytes. Exp. Brain Res. 71: 450–454.

Jacob, M. H., Lindstrom, J. M., and Berg, D. K. (1986) Surface and intracellular distribution of a putative neuronal nicotinic acetylcholine receptor. J. Cell Biol. 103: 205–214.

Johnson, M. V., Young, A. C., and Coyle, J. T. (1981) Laminar distribution of cholinergic markers in neocortex: effects of lesions. J. Neurosci. Res. 6: 597–607.

Kelly, J. S., and Rogawski, M. A. (1987) Acetylcholine. In: Rogawski, M. A., and Barker, J. L. (eds), Neurotransmitter Actions in the Vertebrate Nervous System. Plenum Press, New York/London, pp. 143–197.

Kuhar, M. J., and Yamamura, H. I. (1976) Localization of cholinergic muscarinic receptors in rat brain by light microscopic radioautography. Brain Res. 110: 229–243.

Larocca, J. N., Ledeen, R. W., Dvorkin, B., and Makman, M. H. (1987) Muscarinic receptor binding and muscarinic receptor-mediated inhibition of adenylate cyclase in rat brain myelin. J. Neurosci. 7: 3869–3876.

London, E. D., Waller, S. B., and Wamsley, J. K. (1985) Autoradiographic localiztion of [³H]nicotine binding sites in the rat brain. Neurosci. Lett. 53: 179–184.

Luiten, P. G. M., Gaykema, R. P. A., Traber, J., and Spencer, D. G. (1987) Cortical projection patterns of magnocellular basal nucleus subdivisions as revealed by anterogradely transported Phaseolus vulgaris leucoagglutinin. Brain Res. 413: 229–250.

Luiten, P. G. M., Matsuyama, T., Strosberg, A. D., and Traber, J. (1988) Immunocytochemical localization of muscarinic acetylcholine receptor proteins in rat cerebral cortex and magnocellular basal forebrain nuclei. Neurochem. Int. 13 Suppl 1: 52.

Marchi, M., Paudice, P., Bella, M., and Raiteri, M. (1986) Dicyclomine- and pirenzepine-sensitive muscarinic receptors mediate inhibition of (³H)serotonin release in different rat brain areas. Eur. J. Pharmac. 129: 353–357.

Marchi, M., Paudice, P., Gemignani, A., and Raiteri, M. (1987) Is the muscarinic receptor that mediates potentiation of dopamine release negatively coupled to the cyclic system? J. Neurosci. Res. 17: 142–145.

Matsuyama, T., Luiten, P. G. M., Spencer, D. G., and Strosberg, A. D. (1988) Ultrastructural localization of immunoreactive sites for muscarinic acetylcholine receptor proteins in the rat cerebral cortex. Neurosci. Res. Commun. 2: 69–76.

McCormick, D. A., and Prince, D. A. (1985) Two types of muscarine response to acetylcholine in mammalian cortical neurons. Proc. natl Acad. Sci. USA 82: 6344–6348.

Murphy, S., Pearce, B., and Morrow, C. (1986) Astrocytes have both M_1 and M_2 muscarinic receptor subtypes. Brain Res. 364: 177–180.

Nordberg, A., Wahlström, G., and Eriksson, B. (1986) [³H]muscimol and [³H]quinuclidinyl benzilate binding in rat cortex in the abstinence after long-term barbital treatment. Acta physiol. scand. 126: 153–156.

Parnavelas, J. G., and McDonald, J. K. (1983) The cerebral cortex. In: Emson, P. C. (ed.), Chemical Neuroanatomy. Raven Press, New York, pp. 505–549.

Parnavelas, J. G., Kelly, W., Franke, E., and Eckenstein, F. (1986) Cholinergic neurons and fibres in the rat visual cortex. J. Neurocytol. 15: 329–336.

Peters, A., and Kara, D. A. (1985a) The neuronal composition of area 17 of rat visual cortex. I. The pyramidal cells. J. comp. Neurol. 234: 218–241.

Peters, A., and Kara, D. A. (1985b) The neuronal composition of area 17 of rat visual cortex. II. The nonpyramidal cells. J. comp. Neurol. 234: 242–263.

Rieck, R., and Carey, R. G. (1984) Evidence for a laminar organization of basal forebrain afferents to the visual cortex. Brain Res. 297: 374–380.

Schröder, H., Zilles, K., Luiten, P. G. M., and Strosberg, A. D. (submitted a) Immunocytochemical visualization of muscarinic cholinoceptors in the human cerebral cortex.

Schröder, H., Zilles, K., Maelicke, A., and Hajos, F. (submitted b) Immunohistochemical localization of cortical nicotinic cholinoceptors in rat and man.

Schwartz, R. D. (1986) Autoradiographic distribution of high affinity muscarinic and nicotinic cholinergic receptors labeled with [³H]acetylcholine in rat brain. Life Sci. 38: 2111–2119.

Spencer, D. G., Horváth, E., and Traber, J. (1986) Direct autoradiographic determination of M1 and M2 muscarinic acetylcholine receptor distribution in the rat brain: relation to cholinergic nuclei and projections. Brain Res. 380: 59–68.

Swanson, L. W., Simmons, D. M., Whiting, P. J., and Lindstrom, J. (1987) Immunohistochemical localization of neuronal nicotine receptors in the rodent central nervous system. J. Neurosci. 7: 3334–3342.

Wainer, B. H., Bolam, J. P., Freund, T. F., Henderson, Z., Totterdell, S., and Smith, A. D. (1984) Cholinergic synapses in the rat brain: a correlated light and electron microscope immunohistochemical study employing a monoclonal antibody against choline acetyltransferase. Brain Res. 308: 69–76.

Wamsley, J. K., Zarbin, M. A., Birdsall, N. J. M., and Kuhar, M. J. (1980) Muscarinic cholinergic receptors: autoradiographic localization of high and low affinity agonist binding sites. Brain Res. 200: 1–12.

Wamsley, J. K., Palacios, J. M., Young, W. S., and Kuhar, M. J. (1981) Autoradiographic determination of neurotransmitter receptor distribution in the cerebral and cerebellar cortices. J. Histochem. Cytochem. 29: 125–135.

Werner, L., Wilke, A., Blödner, R., Winkelmann, E., and Brauer, K. (1982) Topographical distribution of neuronal types in the albino rat's area 17. A qualitative study. Z. mikroskanat. Forsch. 96: 433–453.

Werner, L., Hedlich, A., and Winkelmann, E. (1985) Neuronentypen im visuellen Kortex der Ratte, identifiziert in Nissl- und deimprägnierten Golgi-Präparaten. J. Hirnforsch. 26: 173–186.

Yamamura, H. I., Vickroy, T. W., Gehlert, D. R., Wamsley, J. K., and Roeske, W. R. (1985) Autoradiographic localization of muscarinic agonist binding sites in the rat central nervous system with (+)-cis-[³H]methyldioxolane. Brain Res. 325: 340–344.

Young, W. S., and Kuhar, M. J. (1980) Serotonin receptor localization in rat brain by light microscopic autoradiography. Eur. J. Pharmac. 62: 237–239.

Zilles, K. (1985) The cortex of the rat. A stereotaxic atlas. Springer Verlag, Berlin/Heidelberg/New York/Tokyo.

Zilles, K. (1988) Receptor autoradiography in the hippocampus of man and rat. In: Frotscher, A., Kugler, P., Misgeld, U., and Zilles, K. (eds), Neurotransmission in the Hippocampus. Adv. Anat. Embryol. Cell Biol. 11: 61–80.

Zilles, K. (1989) Codistribution of transmitter receptors in the human and rat hippocampus. In: Chan-Palay, V., and Köhler, C. (eds), The Hippocampus—New Vistas. A. R. Liss, New York, pp. 171–187.

Zilles, K., Schleicher, A., Rath, M., Glaser, T., and Traber, J. (1986) Quantitative autoradiography of transmitter binding sites with an image analyzer. J. neurosci. Meth. 18: 207–220.

Zilles, K., Schleicher, A., Rath, M., and Bauer, A. (1988) Quantitative receptor autoradiography in the human brain. Meth. Asp. Histochem. 90: 129–137.

Effects of chronic in vivo replacement of choline with a false cholinergic precursor

Donald J. Jenden, Roger W. Russell, Ruth A. Booth, Beat J. Knusel, Sharlene D. Lauretz, Kathleen M. Rice and Margareth Roch

Department of Pharmacology, School of Medicine and Brain Research Institute, University of California, Los Angeles, CA 90024-1735, USA

It is generally agreed that degeneration of cholinergic projections from the basal forebrain to the neocortex and hippocampus is an early and uniform feature of Alzheimer's disease (Whitehouse et al., 1982; Coyle et al., 1983; Sims et al., 1983), although other transmitter systems may also be affected (Davies and Terry, 1981; Rossor, 1982). The basis for the selective vulnerability of certain cholinergic neurons is not clear, but it may result from a competition for free choline between biochemical pathways leading to the synthesis of phospholipids and acetylcholine (Wurtman et al., 1985; Jenden, 1986). This competition can only occur in cholinergic cells, and could be precipitated by a variety of inherited nutritional and environmental factors. It is likely to be age-related and might lead to a failure of synaptic transmission, of mechanisms for membrane renewal or both.

To evaluate this hypothesis we have sought an analog of choline which can replace it in all its known metabolic functions but is utilized less efficiently, so that an artificial competition would be established. Chronic administration of such a compound to experimental animals might be expected to lead to the development of a model syndrome resembling Alzheimer's disease. N-aminodeanol (NADe) (Fig. 1) has been shown to meet the basic requirements (Newton and Jenden, 1985; Newton et al., 1985a; Newton et al., 1985b). This compound is an isostere of choline in which one of the N-methyl groups is replaced by $-NH_2$. It shares most of the physicochemical and biochemical characteristics of choline, but is handled less efficiently; this is reflected in a higher K_M or K_T and an equal or lower V_{max} (Newton and Jenden, 1985). It is taken up in competition with choline by the high and low affinity choline transport systems, acetylated by choline acetyltransferase and stored in vesicles as a classical false transmitter, O-acetyl-N-aminodeanol (AcNADe). Acetylcholine is depleted as it is replaced by AcNADe, but the total stored ester remains approximately the same (Newton and Jenden, 1985). Both are released together by stimulation, provided that Ca^{++} is present, but both storage and release favor choline.

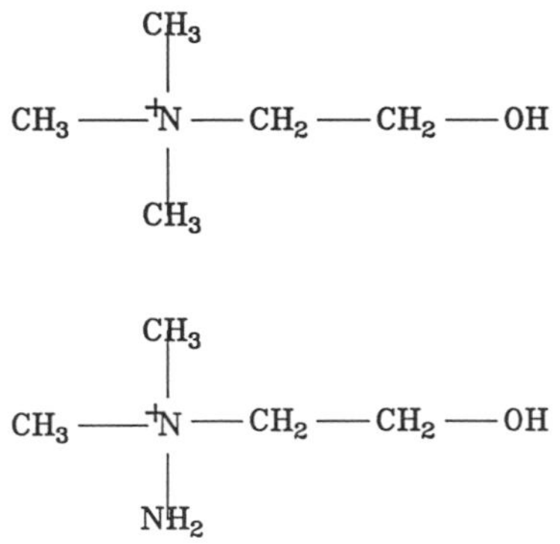

$$CH_3 \overset{\displaystyle CH_3}{\underset{\displaystyle CH_3}{\overset{|}{\underset{|}{\,^+N}}}} - CH_2 - CH_2 - OH \qquad \text{CHOLINE}$$

$$CH_3 \overset{\displaystyle CH_3}{\underset{\displaystyle NH_2}{\overset{|}{\underset{|}{\,^+N}}}} - CH_2 - CH_2 - OH \qquad \text{N-AMINODEANOL}$$

Figure 1. Structures of choline and the false cholinergic precursor, N-aminodeanol.

AcNADe interacts with both muscarinic and nicotinic receptors in competition with acetylcholine, but its potency is only 4% and 17% respectively of the potency of acetylcholine. This results in a profound interference with cholinergic transmission, particularly at muscarinic sites. Like acetylcholine, it is rapidly hydrolyzed by acetylcholinesterase.

Any attempt to replace choline with NADe in vivo is complicated by two factors. Firstly, phospholipid-bound choline represents a large pool which in brain is approximately $1000 \times$ larger than that of free choline, and which turns over very slowly. Secondly, choline can be synthesized as phosphatidylcholine in the liver by sequential methylation of phosphatidylethanolamine, which tends to obfuscate attempts to replace dietary choline with NADe. For these reasons chronic experiments have been undertaken in which weanling rats are placed on an artificial diet devoid of free or phospholipid bound choline and partially depleted of methionine, the primary metabolic source of methyl groups, but containing 5 g/kg of NADe. A control group was fed a diet containing an equivalent amount of choline in place of NADe. Under these conditions phospholipid-bound choline and free choline are progressively replaced by the corresponding species of NADe in plasma, liver, brain and all other tissues studied. Acetylcholine is replaced by AcNADe in brain, myenteric plexus and diaphragm. The rate of replacement is faster for free than for lipid-bound choline, and is faster in liver than in brain. After about 120–180 days, replacement reaches an asymptote of 75–80% for lipid-bound choline and 85–95% for free choline. Total (choline + NADe) levels are equal to or lower than control for phospholipid and equal to or higher than control for free compounds, presumably reflecting the lower efficiency with which NADe is utilized and transported.

Some physiological and behavioral effects of this regimen are immediately apparent while others develop progressively over a period of several months. After an initial period of slower growth, growth acceler- ates and in several experiments body weight was 0–15% less than controls. Total brain weight was also significantly $(20.9 \pm 1.8\%)$ less after 270 days (Table 1, groups I and II). Food and water intake did not differ significantly. No obvious differences were seen initially in general behavior, but striking differences appeared during handling and in- creased progressively with time. Hypertonia and hyperreactivity were seen uniformly after 15 days, and the animals were not infrequently aggressive toward the handler. These effects are similar to those seen after intraventricular administration of hemicholinium-3 (Russell and Macri, 1978) and after septal lesions (Olton and Gage, 1976). Hyper- reactivity, as measured by a startle response to an acoustic stimulus (Russell and Macri, 1978; Silverman, Chang and Russell, in press), is an early sign and in many animals became so pronounced that a brief stimulus (0.15 s, 10 kHz, 100 dB) elicited a clonic seizure. Nociceptive threshold (Crocker and Russell, 1984) was significantly lower in NADe- treated animals at all time points, from 15 days to 420 days. This is perhaps to be expected in a hypocholinergic state, since muscarinic agonists that penetrate into the central nervous system consistently produce the opposite effect (for review, see Karczmar, 1976).

Evidence of impaired learning and memory was derived from three experimental paradigms: inhibited (passive) avoidance, conditioned avoidance and learned alternation in a swimming maze. In the inhibited avoidance test, the ratio of retention latency to training latency was consistently reduced in four separate experiments involving treatment times from 15 to 120 days, and the difference was highly significant. In one experiment lasting 360 days retention trials were repeated at monthly intervals beginning with initial training at 210 days and contin- uing for 150 days. Latency was significantly shorter, indicating poorer memory, in NADe-treated animals at every trial (Fig. 2).

Comparisons of learning a conditioned avoidance response were made after treatment with NADe for various times. There was no significant difference between ChCl and NADe groups when learning occurred early (15 days). However, NADe animals that had their first practice at 120 days were significantly poorer in learning $(p = 0.0014)$ at that time and 30 days later $(p = 0.031)$. This difference disappeared

Table 1. Brain weight after 270 days on the respective diets

Group	Brain weight (g)
I	$1.78 \pm .027$ (9)
II	2.25 ± 0.38 (10)
III	1.95 ± 0.41 (8)

Group I was maintained on the NADe diet and group II on Ch diet throughout the 270 days. Group III was fed the NADe diet for 210 days and then changed to the Ch diet.

Aminodeanol Drug Study
Inhibited Avoidance

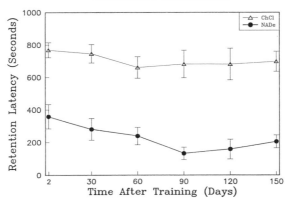

Figure 2. Measures of retention latencies (memory) of NADe and control groups on their respective diets for 360 days. Initial training occurred at 210 days and continued for 150 days.

when the interval between learning sessions was 60 days, when the memories of the control animals had also declined.

Rats which had been on the NADe diet for 405 days and corresponding controls were trained to alternate in a 2-arm water maze. From 10 trials in each of 23 subsequent days, five scores were used to evaluate alternating performance: T, the time (s) required to make a decision; S1, the number of alternations in sequence; S2, the number of differences between the observed sequence and an ideal alternating series; S3, the longest series of alternations in the series and S4, the longest series of perseverations in the series. All of the five scores showed a significant regression ($p \leqslant 0.002$) on time for the control group, and two of the five scores showed a significant ($p \leqslant 0.05$) regression on time for NADe-treated rats. The slopes were marginally less ($p = 0.05$) for NADe-treated rats, indicating a slower rate of learning.

These data provide convincing evidence of impairment of learning and memory in rats that had been treated with NADe, which is probably attributable to replacement of acetylcholine by the false transmitter AcNADe. They neither confirm nor deny the more general hypothesis on which these experiments were based, which would predict a progressive and probably irreversible loss of cholinergic function in animals on a NADe diet. To test this more explicitly, a group of rats (Group III in the Tables) was maintained on a NADe diet for 210 days before reinstitution of a normal diet for a further 60 days. Control groups were maintained on the NADe diet or a choline diet

(Groups I and II, respectively) for the entire duration of 270 days. As expected, NADe was rapidly replaced by choline in both free and phospholipid bound forms when its administration was discontinued; by 60 days free NADe and phospholipid-NADe in plasma had fallen to undetectable levels ($<1\%$). In the brain, NADe had fallen to 5.0% of phospholipids. Concomitantly, brain weight increased, although not to that of controls maintained on a choline diet throughout (Table 1).

Physiological and behavioral recovery were slower and generally incomplete. This is illustrated in Table 2, in which open field locomotor activity integrated over 24-h periods is shown. Rats on the NADe diet were significantly more active than controls on a normal diet at both 30 and 60 days. Group III, which had been returned from NADe to a normal diet, remained more active than controls but less active than the NADe group at both 30 and 60 days after the diet reversal. The difference from the NADe group was significant ($p = 0.038$) at 30 days but not at 60.

The inhibited avoidance test provided further evidence that memory was not fully restored when rats were returned to a normal diet after 210 days of treatment with NADe. As shown in Table 3, retention latency, when tested at regular intervals after diet reversal, was significantly less in animals on NADe, and did not increase significantly in the group that was restored to a normal diet.

Table 2. Open field locomotor activity expressed as light beam interruptions in a 24-h period

	Day of experiment	
Group	240	270
I	$11,431 \pm 861$ (9)	$12,618 \pm 1,333$ (9)
II	$6,595 \pm 530$ (12)	$6,843 \pm 485$ (12)
III	$8,103 \pm 585$ (9)	$10,248 \pm 750$ (9)

The groups received the dietary treatments described in Table 1.

Table 3. Memory measured by retention latency (seconds) after reversal of diets for group III

	Day of experiment		
Group	210	240	270
I	428 ± 104 (9)	306 ± 56 (9)	216 ± 49 (9)
II	767 ± 74 (12)	608 ± 81 (12)	829 ± 36 (12)
III	348 ± 81 (9)	460 ± 110 (9)	240 ± 95 (9)

The groups received the dietary treatments as stated in the legend in Table 1.

234

Summary and conclusions

Chronic administration to rats of a diet in which all choline is replaced by NADe, an unnatural choline analog, results in a classical hypocholinergic syndrome characterized by progressive loss of learning and memory, hyperkinesis, hyperreactivity and hyperalgesia. Discontinuation of the artificial diet results in rapid elimination of NADe from both free and phospholipid-bound pools in all tissues studied, but the behavioral effects recede more slowly and incompletely. These results are consistent with a model in which choline and NADe compete in both acetylcholine and phospholipid synthesis, resulting in selective vulnerability of cholinergic neurons. Histological studies are in progress to determine whether microanatomical changes are also consistent with this model.

Acknowledgements. The work described in this report was supported by USPHS grant MH 17691. The authors are grateful to Holly Batal for excellent editorial assistance and preparation of the manuscript.

Coyle, J. T., Price, D. L., and DeLong, M. R. (1983) Alzheimer disease: a disorder of cortical cholinergic innervation. Science 219: 1184–1190.

Crocker, A. D., and Russell, R. W. (1984) The up- and down method for the determination of nociceptive thresholds in rats. Pharmac. Biochem. Behav. 21: 133–136.

Davies, P., and Terry, R. D. (1981) Cortical somatostatin-like immunoreactivity in cases of Alzheimer's disease and senile dementia of the alzheimer type. Neurobiol. Aging 2: 9–14.

Jenden, D. J. (1986) The pharmacology of cholinergic mechanisms and senile brain disease. In: Scheibel, A. B., and Wechsler, A. P. (eds), The Biological Substrates of Alzheimer's Disease. M. A. B. Brazier, Academic Press, New York, UCLA Forum in Medical Sciences 27, pp. 205–215.

Karczmar, A. G. (1976) Central actions of acetylcholine, cholinomimetics, and related drugs. In: Hanin, I., and Goldberg, A. M. (eds), Biology of Cholinergic Function. Raven, New York, pp. 395–449.

Newton, M. W., and Jenden, D. J. (1985) Metabolism and subcellular distribution of N-amino-N, N-dimethylaminoethanol (N-aminodeanol) in rat striatal synaptosomes. J. Pharmac. exp. Ther. 235: 135–146.

Newton, M. W., Ringdahl, B., and Jenden, D. J. (1985a) Acetyl-N-aminodeanol: a cholinergic false transmitter in rat phrenic nerve-diaphragm and guinea-pig myenteric plexus preparations. J. Pharmac. exp. Ther. 235: 147–156.

Newton, M. W., Crosland, R. D., and Jenden, D. J. (1985b) In vivo metabolism of a cholinergic false precursor after dietary administration to rats. J. Pharmac. exp. Ther. 235: 157–161.

Olton, D. S., and Gage, F. H. (1976) Behavioral, anatomical and biochemical aspects of septal hyperreactivity. In: De France, J. P. (eds), The Septal Nuclei. Plenum, New York, pp 507–527.

Rossor, M. N. (1982) Neurotransmitters and CNS disease. Lancet 2: 1200–1204.

Russell, R. W., and Macri, J. (1978) Some behavioral effects of suppressing choline transport by cerebroventricular injection of hemicholinium-3. Pharmac. Biochem. Behav. 8: 399–403.

Silverman, R. W., Chang, A. S., and Russell, R. W. (1988) A microcomputer controlled system for measuring reactivity in small animals. Behav. Res. Methods, Instruments, and Computers 20: 495–499.

Sims, N. R., Bowen, D. M., Allen, S. J., Smith, C. C. T., Neary, D., Thomas, D. J., and Davison A. N. (1983) Presynaptic cholinergic dysfunction in patients with dementia. J. Neurochem. 40: 503–509.

Whitehouse, P. J., Price, D. L., Struble, R. G., Clark, A. W., Coyle, J. T., and DeLong, M. R. (1982) Alzheimer's disease and senile dementia: loss of neurons in the basal forebrain. Science 215: 1237–1239.

Wurtman, R. J., Blusztajn, J. K., and Maire, J.-C. (1985) 'Autocannibalism' of choline-containing membrane phospholipids in the pathogenesis of Alzheimer's disease—a hypothesis. Neurochem. Int. 7: 369–372.

Development of the septohippocampal projection *in vitro*

Beat H. Gähwiler, David A. Brown[a], Albert Enz[b] and
Thomas Knöpfel

*Brain Research Institute, August Forel-Str. 1, CH-8029 Zürich, Switzerland, [a]University
College, University of London, Gower Street, London WC1E 6BT, England, and [b]Sandoz
Ltd., CH-4001 Basel, Switzerland*

Summary. Slices were prepared from septal and hippocampal tissue and co-cultured for
periods up to one month. The presence of cholinergic neurons within the septal slices
was demonstrated by histochemical staining techniques for acetylcholinesterase or by Golgi-
like immunoperoxidase techniques with antibodies raised against the enzyme choline acetyltransferase. Cholinergic fibers originating in the septal explants started to grow radially in
all directions. By day 7, the first fibers were seen to reach their target, but maximal hippocampal ingrowth occurred between day 8 and 14 *in vitro*. Only those fibers
reaching the target were maintained, whereas cholinergic fibers growing in other directions
degenerate.

Electrophysiological studies showed that cholinergic fibers established functional cholinergic connections with hippocampal pyramidal cells. As a result of septal stimulation, two
different potassium currents were inhibited in pyramidal cells: a calcium-independent current,
I_M, and a calcium-dependent current, I_{AHP}, underlying spike afterhyperpolarization.

Application of nerve growth factor (NGF) strongly increased the number of cholinergic
fibers which invaded the hippocampal slices and raised the activities of the cholinergic
enzymes choline acetyltransferase and acetylcholinesterase, effects which were completely
blocked by anti-NGF antibodies. The response of septohippocampal co-cultures to NGF
depended on the time of application. During the first two weeks *in vitro*, NGF elicited
sustained increases in enzyme activities, whereas later administration of NGF produced
effects which were only maintained for several days.

Introduction

The hippocampus is innervated by fibers originating in the nuclei of
the medial septum and the adjacent vertical limb of the diagonal band
(Fibiger, 1982; Mellgren and Srebro, 1973; Mesulam et al., 1983; Milner
et al., 1983; Moske et al., 1973). The septohippocampal projection has
traditionally been one of the favorite objects for studying mechanisms
involved in neuronal development. First, the morphology of the projection is well characterized in terms of origin and target innervation.
Second, cholinergic neurons display a remarkable capacity for regeneration, as evidenced by studies using grafting techniques (Björklund et al.,
1979; Björklund and Stenevi, 1979; Kromer et al., 1981; Kromer et al.,
1983). Third, cholinergic projections in general are likely to play an
important role in many physiological processes such as learning and
cognitive functions (Seifert, 1983).

Several aspects of such a projection could be best studied in an *in vitro* preparation provided the cholinergic neurons of septal origin could be induced to innervate hippocampal tissue under culture conditions. In the present study, we will summarize the attempts of producing an *in vitro* analogue of the septohippocampal projection and describe the characteristic properties of this preparation (Finsen and Zimmer, 1985; Gähwiler and Brown, 1985; Gähwiler and Hefti, 1984; Zimmer et al., 1985). Slices of brain tissue cultured by our technique develop into a two-dimensional structure which retains most of the organotypic morphology, a feature which greatly facilitates tracing of fibers from one explant to the other (Gähwiler, 1988). In the second part, we will analyze whether one needs to postulate the existence of tropic factors which attract the growing cholinergic axons toward their target and investigate the possible influence of trophic factors such as nerve growth factor (NGF) on the development, establishment and maintenance of synaptic cholinergic connections.

Culturing techniques

For the cultivation of septum and hippocampus, we have used 5 to 6-day-old rats. The cultures were prepared by means of the roller-tube technique as previously described (Gähwiler, 1981, 1984a, b). In short, slices derived from septum and hippocampus were embedded side-by-side on glass coverslips. The situation at the time of explantation is illustrated in Figure 1A. The co-cultures were then placed in plastic tubes which were filled with 0.5 ml of a medium consisting of heat-inactivated horse serum (25%), basal medium (Eagle) (50%) and a balanced salt solution (25%). The cultures were fed twice per week. NGF-treated cultures received 100 ng/ml NGF (2.5S purified from mouse submaxillary glands) in normal medium either daily or twice weekly. Other cultures received twice weekly 1 μl of a sheep antiserum to NGF. This concentration of antiserum has been shown to inhibit the biological effect of 2 μg NGF on rat sympathetic neurons *in vitro* (Suda et al., 1978). After 1–30 days *in vitro*, the cultures were either used for electrophysiological recordings or taken for histochemical (Geneser-Jensen and Blackstad, 1972) and biochemical (Bradford, 1976; Fonnum, 1975; Potter, 1967) analysis. For the visualization of cholinergic neurons and fibers, the cultures were either stained immunohistochemically with antibodies against the cholinergic enzyme choline acetyltransferase (ChAT) (German et al., 1985) or histochemically stained for acetylcholinesterase (AChE), using acetylthiocholine (1.2 mg/ml) as a substrate and ethopropazine (0.06 mg/ml) as inhibitor of non-specific cholinesterases (Geneser-Jensen and Blackstad, 1972).

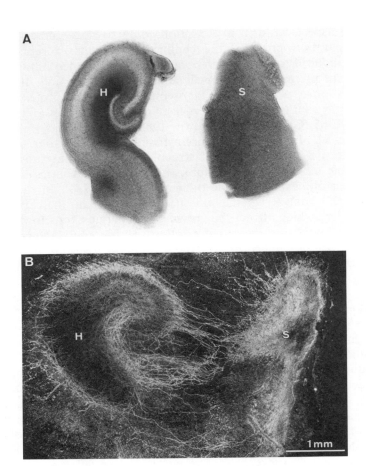

Figure 1. Septohippocampal co-culture. At the time of explantation, the two slices derived from septum (S) and hippocampus (H) are embedded side-by-side on a glass coverslip at a distance of approximately 1 mm (*A*). AChE-staining (dark-field optics) after 4 weeks *in vitro* reveals that the presence of AChE-positive fibers which have crossed the gap between the two explants (*B*).

Innervation of target tissue by cholinergic neurons

Development

In situ, the first AChE-positive fibers reach the hippocampus by postnatal day 3, but the majority of synapses are formed during the first two postnatal weeks (Milner et al., 1983). To study the temporal pattern of fiber growth *in vitro*, a number of co-cultures were stained for the presence of histochemically detectable AChE after variable time *in vitro*. During the first 5 days in culture, the slices appeared to slowly recover from dissection trauma and very few or no AChE-positive cell bodies or fibers could be detected within the septal slices. By day 6 *in vitro*, AChE-positive cell bodies were consistently observed in the septum, but stained processes were still missing. AChE-positive started to reach the the hippocampal target tissue by day 7, and their number strongly increased during the following days.

To investigate whether there are trophic factors (e.g. released by the target tissue) directing the growth of cholinergic axons, we examined the initial pattern of outgrowth from the septal slices. These studies revealed that cholinergic axons initially grew in all directions, including the direction opposite the hippocampal target. After about 2 weeks *in vitro*, the final pattern appeared to emerge, i.e. the fibers which had reached the hippocampal target became stabilized and all other fibers degenerated (Fig. 1B). We tend to conclude that there is initial random outgrowth of cholinergic fibers, but these studies nevertheless also suggest that there exist some additional cues which inform axons that they have grown in the right or wrong direction.

Although AChE has been shown to be a reliable marker for septal cholinergic neurons grown in cultures of dissociated cells (Hefti et al., 1985), some co-cultures were stained immunocytochemically for ChAT rather than AChE. In cultures grown in normal medium, faint staining of cell bodies could be detected within the septal slices. The staining intensity was, however, dramatically increased following treatment with NGF (for details see 'Trophic factors' section of this chapter) (Fig. 2). Under these conditions, ChAT-positive cell bodies as well as dendritic arborizations were clearly revealed within the septal slices (but not in the hippocampal tissue), and the very thin axonal processes could sometimes be followed all the way to their target within the hippocampal slice. The majority of cholinergic neurons had a soma diameter of approximately 30 μm and displayed two to four long dendrites.

Pharmacological modification

We tested in a total of 24 septohippocampal co-cultures whether abolishment of propagated electrical activity, chronic activation or blockade of cholinergic receptors or inhibition of cholinergic enzymes would influence the development of AChE-positive septohippocampal fibers. During their entire growth, the cultures were exposed to a medium containing either 10^{-7} M tetrodotoxin, 10^{-6} M carbachol, 10^{-5} M atropine or 10^{-6} M neostigmine. In the presence of tetrodotoxin, AChE-positive fibers still invaded hippocampal target tissue. Higher concentrations of the toxin could, however, not be tested because of its deleterious effect on the development of the hippocampal tissue, in particular of granule cells within the fascia dentata. Treatment

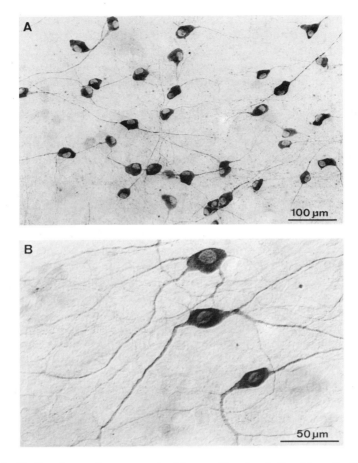

Figure 2. Cholinergic septal neurons stained immunohistochemically with antibodies raised against ChAT. *A* Group of ChAT-positive neurons viewed after 17 days in culture with bright field microscopy. *B* 20-day-old septal cholinergic neurons observed with interference contrast optics.

with muscarinic agonists or antagonists had no striking effect on hippocampal ingrowth by AChE-positive fibers. Fiber density tended, however, to be slightly increased by carbachol and decreased by atropine. Exposure of the cultures to neostigmine had no detectable influence on the pattern or density of hippocampal ingrowth by cholinergic fibers.

Selectivity of cholinergic innervation of target tissues

To test the specificity of growth of septal cholinergic fibers into their target areas, septal slices were co-cultured with slices of various areas which *in situ* lack a major cholinergic innervation. These areas included the cerebellum (Gähwiler and Hefti, 1984; Rimvall et al., 1985), the ventral mesencephalon and the hypothalamus (Gähwiler and Hefti, 1984). In the vast majority of such co-cultures, AChE-positive fibers remained restricted within the septal slices. In some cultures, a few AChE-positive fibers crossed the gap between the two slices but failed to invade the target tissue (Gähwiler and Hefti, 1984).

In a second set of experiments, we tested the ability of cholinergic neurons derived from different anatomical origin to grow AChE-positive fibers into adjacent hippocampal tissue. These areas included the nucleus basalis of Meynert, the spinal cord and striatal tissue. AChE-positive fibers originating in the cholinergic slices were regularly seen to invade this non-natural cholinoceptive target area (Gähwiler and Hefti, 1984, 1985). Since striatal cholinergic neurons are thought to be interneurons, we conclude that they have the capacity to develop into projection neurons when given appropriate target tissues.

These results indicate that under our culture conditions, AChE-positive fibers selectively invade cholinergic target areas. This effect is independent of the brain area from which the cholinergic neurons were derived suggesting that a selective signal from the target area is required for establishment of contacts between cholinergic neurons and cholinoceptive targets.

Functional aspects

For electrophysiological studies, the co-cultures were transferred to a temperature-controlled perfusion chamber where single CA3 hippocampal pyramidal cells were impaled with a microelectrode under visual control and then voltage-clamped.

Septal stimulation yielded both cholinergic and non-cholinergic components of functional septohippocampal innervation (Brown et al., 1986; Gähwiler and Brown, 1985). The cholinergic component, detectable in about half of the cultures, comprised a delayed, low-amplitude and very slow inward postsynaptic current (Fig. 4) which was enhanced and prolonged by 1 μM neostigmine and blocked by 0.1–1 μM atropine. In voltage-clamp experiments, current deflections produced by short voltage jumps were reduced during the slow excitatory postsynaptic currents (e.p.s.c.s) (Fig. 3A), indicating a fall in cell input conductance which was very similar to that seen during application of 1 μM muscarine (Fig. 3B). This fall in conductance seemed to stem at least in part from a reduction of the time-dependent current relaxations induced by the voltage jumps. Other tests suggested that these relaxations (see Fig. 3B) may have reflected de-activation and re-activation of a Ca^{2+}-independent K^+-current, I_M (Brown, 1988; Brown and

242

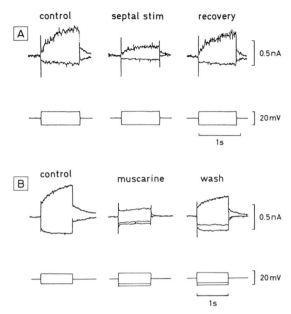

Figure 3. Effect of septal stimulation (*A*) and muscarine (*B*) on CA3 pyramidal cell: reduction of I_M. *A* A pyramidal cell was voltage-clamped at -32 mV and subjected to 1-s voltage jumps. Immediately after septal stimulation (in 1 μM neostigmine solution), the currents evoked by voltage commands were transiently reduced, indicating a fall in input conductance. The effect of septal stimulation was completely abolished by 0.1 μM atropine. *B* Effect of exogenously applied muscarine on clamp current in a CA3 pyramidal cell, superfused with a solution containing 1 μM tetrodotoxin to suppress indirect effects of muscarine. The record shows superimposed current responses to series of 0.5 s square voltage commands similar to those illustrated in *A*. Muscarine produced a sustained inward current (not shown) and reversibly reduced input conductance. Outward currents produced by depolarizing commands are probably composite currents containing components of the voltage-dependent current I_M and the Ca-dependent current(s) $I_{K(Ca)}$; inward relaxations induced by hyperpolarizing steps probably reflect deactivation of I_M since they persisted in Cd solution (see Gähwiler and Brown, 1985). These relaxations were abolished by septal stimulation or by muscarine, indicating that I_M was inhibited. (Record *A* from figure 4 of Gähwiler and Brown, 1985).

Adams, 1980), which had been previously described in hippocampal pyramidal cells (Halliwell and Adams, 1982).

In acute hippocampal slices, application of carbachol or electrical stimulation of cholinergic afferents have been reported to block a slowly decaying spike afterhyperpolarizations (Cole and Nicoll, 1984; Madison et al., 1987) that follow action potentials. The current underlying these hyperpolarizations is, in contrast to the calcium-independent current I_M, a calcium-activated potassium current termed I_{AHP} (Lancaster and Adams, 1986). Recent experiments carried out with septohippocampal co-cultures showed that septal stimulation reversibly reduced those spike afterhyperpolarizations in current-clamp recordings and the underlying outward-current I_{AHP} as observed in voltage-clamp experiments

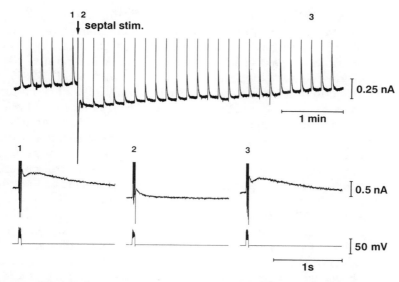

Figure 4. Effect of septal stimulation on CA3 pyramidal cell in septohippocampal co-culture: induction of slow inward current and reduction of I_{AHP}. *A* Continuous chart record shows the influence of septal stimulation (1-s train of stimuli at 40 Hz; 1 ms, 10 μA) on the slowly decaying outward currents which were induced by short command-voltage pulses (40 ms, 0.1 Hz, from −49 to −19 mV). *B* single tail currents (upper traces) and corresponding voltage records (lower traces) monitored before (*1), immediately after (*2) and 225 s after (*3) the septal stimulation.

(Fig. 4). During the slow e.p.s.c.s, I_{AHP} was often abolished. It appears, therefore, that in the hippocampus, cholinergic stimulation can inhibit two distinct K^+-currents which may both contribute to spike adaptation and repetitive firing as it is seen following synaptic release or exogenous application of acetylcholine.

Trophic factors

There is widespread evidence indicating that NGF is involved in the function of central cholinergic neurons. NGF, and the mRNA coding for NGF, are present in the rat brain and concentrated in cholinergic target areas (Korsching et al., 1985; Shelton and Reichardt, 1984; Whittemore et al., 1986; Whittemore et al., 1987). Receptors for NGF exist on cholinergic neurons of the forebrain (Richardson et al., 1986; Schwab et al., 1979; Seiler and Schwab, 1984) and the application of NGF leads to an increase in the activity of ChAT in central cholinergic neurons (Gähwiler et al., 1987; Hefti et al., 1984; Hefti et al., 1985; Honegger and Lenoir, 1982; Martinez et al., 1985; Mobley et al., 1986).

To test whether NGF affects the development of central cholinergic pathways, cultures were grown in normal medium or in medium containing either NGF or anti-NGF antibodies. In control cultures, about 70% of septal explants showed at least three or more AChE-positive fibers penetrating the adjacent hippocampal slice (Gähwiler et al., 1987). In contrast, this criterion for ingrowth was met in only 23% of co-cultures that were exposed to daily treatment with anti-NGF antiserum. Besides reducing hippocampal ingrowth of AChE-positive fibers, anti-NGF antibodies produced no other striking morphological alterations within the septal or hippocampal slices. In particular, septal neurons displayed their normal dendritic morphology as compared with control cultures, indicating that the effect of anti-NGF antiserum was specific and not due to a general toxicity of the antiserum (Gähwiler et al., 1987).

Application of NGF to the medium of cultured septal slices did not result in a diffuse, non-directed outgrowth of AChE-positive fibers as it is observed when peripheral sympathetic neurons are cultured in the presence of NGF (for review see Thoenen and Barde, 1980). However, when NGF was added to the culture medium, a striking increase was observed in the number of AChE-positive fibers which grew from the septum to the hippocampus (Gähwiler et al., 1987). Due to the large number of AChE-positive fibers in hippocampal slices grown in the presence of NGF, the histological effect could not be quantified. The biochemically determined activities of ChAT and AChE were, therefore, used. The dose-response-relation between NGF and ChAT activity is illustrated in Figure 5. ChAT (Fig. 6A) activities were elevated almost 10-fold in comparison with co-cultures. The effect was specific for

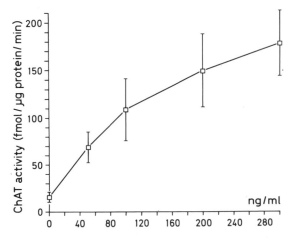

Figure 5. Dose-response relationship for NGF and ChAT activity. Values are mean +/− SEM, n = 10 each. Enzyme activity of septohippocampal culture was determined after 4 weeks *in vitro*. NGF was applied twice per week.

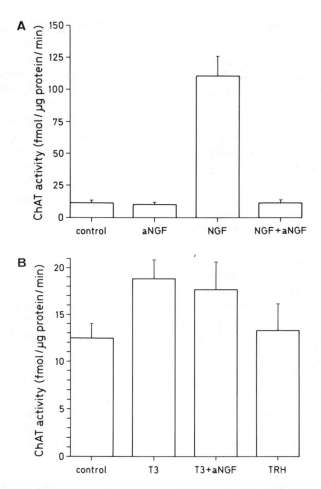

Figure 6. Effect of NGF (A) and T3 (B) on ChAT activity in septohippocampal cultures after 4 weeks *in vitro*. *A* Cultures were fed twice per week either control medium or medium containing NGF (100 ng/ml), anti-NGF antibodies or NGF together with anti-NGF antibodies. Values are means $+/-$ SEM, n = 26–40. NGF increased ChAT activity about 10-fold. While anti-NGF antibodies had no effect on basal ChAT levels, they completely abolished the effect of NGF. *B* T3 (500 ng/ml) also elevated ChAT activities, but compared with NGF its efficacy was considerably smaller and not abolished by anti-NGF antibodies. Values are means $+/-$ SEM, n = 31–41 each. TRH (1 μg/ml) was without any effect on the activity of ChAT.

cholinergic neurons, since it did not alter the protein content of the cultures. Moreover, the effect of NGF was completely blocked by simultaneous application of anti-NGF antibodies. Interestingly, anti-NGF antibodies failed to significantly alter basal activities of ChAT and AChE activities (Gähwiler et al., 1987; Martinez et al., 1985). The effects of NGF were qualitatively similar to those produced by thri-iodothyronine (T3) (Gähwiler et al., 1987; Hayashi and Patel, 1987).

Compared with NGF, the increase in ChAT (Fig. 6B) and AChE activity (not shown) were relatively small and declined with concentrations higher than 50 ng/ml. Since T3 had been reported to increase endogenous activities of NGF in the CNS, we checked whether anti-NGF would influence the action of T3 on the activities of cholinergic enzymes. As illustrated in Figure 6B, anti-NGF failed to influence T3-induced increases in ChAT activity, results which suggest that the effects of T3 are at least in this preparation not mediated by increased activities of endogenous NGF. In contrast to T3, thyrotropin releasing hormone (up to a concentration of 1 μg/ml) had no effect on the activity of ChAT.

Effect of NGF at different developmental stages

In the experiments described before, NGF was applied during a 4-week period and thereafter the effect assessed using morphological or biochemical criteria. These studies do not allow to determine whether the continuous presence of NGF is necessary. Moreover, it is not clear whether NGF receptors are functional only during certain periods of development or specifically expressed following axotomy as it necessarily occurs during preparation of the cultures. We therefore applied NGF either during a 4-week period or exposed the cultures to NGF during only one particular week. In all cases, ChAT and AChE activities were determined following four weeks in vitro.

In 96 co-cultures of septum and hippocampus, application of NGF during a 4-week period elevated the activity of ChAT from 36.2 to 259.4 fmole/μg protein/min and AChE activities from 8.5 to 24.9 pmole/μg protein/min (Fig. 7). These increases were considerably smaller when NGF was administered only during the first or second week, and no effect at all could be seen following application during the third week in vitro. Interestingly, both ChAT and AChE activities were again raised when NGF was applied during the fourth week in culture.

It therefore appears that, in the septohippocampal system, NGF can increase the activity of cholinergic enzymes by two different mechanisms. First, NGF might lead to an induction of cholinergic enzymes lasting less than two weeks. This hypothesis could explain the lack of an effect of NGF during the third week in vitro and would be consistent with previously reported time courses of the response to NGF. In developing rats, single intracerebroventricular injections of NGF produced a rise in ChAT activity which was sustained for at least 48 h and returned to control values within about ten days (Mobley et al., 1986). Second, the relatively strong increase in the activities of cholinergic enzymes produced by application of NGF during the first week in culture points to a long-lasting effect of NGF. It is conceivable that at

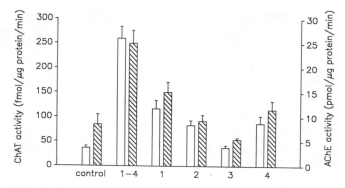

Figure 7. Effect of NGF on ChAT and AChE activities in septohippocampal cultures. The abscissa denotes the time of NGF application: during the entire 4-week period or only during the first, second, third or fourth week *in vitro*. Values are means $+/-$ SEM, n = 12.

this stage of development NGF enhances neuronal survival, as previously shown for septal neurons following fimbrial transsections (Hefti, 1986). Alternatively, the elevated activity of cholinergic enzymes might reflect the increase in the number of cholinergic fibers which have been observed following NGF treatment (Gähwiler et al., 1987). Experiments designed to distinguish between these possibilities are in progress in our laboratory.

Conclusions

The results obtained with slice cultures demonstrate the potential of this *in vitro* approach for studying the mechanisms involved in the establishment of specific neuronal connections between two anatomically remote brain areas, the septum and the hippocampus. Cholinergic nerve fibers originating in the septal slices could be shown to explore their environment, to grow into their target area and establish functional synaptic connections with pyramidal cells.

Cholinergic fiber growth was specific in the sense that contact was exclusively formed with cholinoceptive target areas whereas tissue which *in situ* lacks a major cholinergic innervation was not contacted by these fibers. Our studies with NGF strongly suggest that NGF is involved in the mechanisms establishing these pathways. The exact mechanism by which NGF promotes the establishment of the septohippocampal projection remains to be clarified. The presently available data do not support the view that NGF is a general stimulus for the growth of cholinergic fibers, but rather suggest that it is required for the stabilization and long-term maintenance of cholinergic connections. Analysis of co-cultures after several weeks *in vitro* in fact revealed that NGF strongly enhanced hippocampal ingrowth by cholinergic fibers, whereas no outgrowth of fibers in other directions was detected. Moreover, the ability of anti-NGF antibodies to partially prevent the establishment of septohippocampal projections *in vitro* suggests a possible role of NGF in the stabilization of connections after the axons have reached their targets.

Acknowledgements. The generous supply of anti-ChAT antibodies and NGF by Prof. L. Hersh and Dr R. Lindsay was greatly appreciated. We thank L. Rietschin for expert technical help. Supported by grant No. 3.534-086 from the Swiss National Science Foundation and the Dr Eric Slack-Gyr Foundation.

Björklund, A., Kromer, F., and Stenevi, U. (1979) Cholinergic reinnervation of the rat hippocampus by septal implants is stimulated by perforant path lesion. Brain Res. 173: 57–64.

Björklund, A., and Stenevi, U. (1979) Reformation of the severed septohippocampal cholinergic pathway in the adult rat by transplanted septal neurons. Cell Tissue Res. 185: 289–302.

Bradford, M. M. (1976) A rapid and sensitive method for the quantification of microgram quantities of protein utilizing the principle of protein-dye binding. Ann. Biochem. 72: 248–254.

Brown, D. (1988) M-currents: an update. TINS 7: 294–299.

Brown, D. A., and Adams, P. R. (1980) Muscarinic suppression of a novel voltage-sensitive K^+ current in a vertebrate neurone. Nature 283: 673–676.

Brown, D. A., Gähwiler, B. H., Marsh, S. J., and Selyanko, A. A. (1986) Mechanisms of muscarinic excitatory synaptic transmission in ganglia and brain. TIPS Suppl.: 66–71.

Cole, A. E., and Nicoll, R. A. (1984) Characterization of a slow cholinergic post-synaptic potential recorded *in vitro* from rat hippocampal pyramidal cells. J. Physiol. (Lond.) 352: 173–188.

Fibiger, H. C. (1982) The organization and some projections of cholinergic neurons of the mammalian forebrain. Brain Res. Rev. 4: 327–388.

Finsen, B., and Zimmer, J. (1985) Nerve connections between organotypic slice cultures of immature mouse and rat brain tissue. Acta physiol. scand. 124, Suppl: 542: 64.

Fonnum, F. J. (1975) A rapid radiochemical method for the determination of choline acetyltransferase. J. Neurochem. 24: 407–409.

Gähwiler, B. H. (1981) Organotypic monolayer cultures of nervous tissue. J. Neurosci. Meth. 4: 329–342.

Gähwiler, B. H. (1988) Organotypic cultures of neural tissue. TINS 11: 484–489.

Gähwiler, B. H. (1984a) Slice cultures of cerebellar, hippocampal and hypothalamic tissue. Experientia 40: 235–243.

Gähwiler, B. H. (1984b) Development of the hippocampus *in vitro*: cell types, synapses and receptors. Neuroscience 11: 751–760.

Gähwiler, B. H., and Brown, D. A. (1985) Functional innervation of cultured hippocampal neurones by cholinergic afferents from co-cultured septal explants. Nature 313: 577–579.

Gähwiler, B. H., Enz, A., and Hefti, F. (1987) Nerve growth factor promotes development of the rat septo-hippocampal cholinergic projection *in vitro*. Neurosci. Lett. 75: 6–10.

Gähwiler, B. H., Enz, A., and Hefti, F. (1987) NGF and T_3 increase activity of cholinergic enzymes in cultured septal slices. Soc. Neurosci. Abstr. 13: 1615.

Gähwiler, B. H., and Hefti, F. (1984) Guidance of acetylcholinesterase-containing fibres by target tissue in co-cultured brain slices. Neuroscience 13: 681–689.

Gähwiler, B. H., and Hefti, F. (1985) Striatal acetylcholinesterase-containing interneurons innervate hippocampal tissue in co-cultured slices. Dev. Brain Res. 18: 311–314.

Geneser-Jensen, F. A., and Blackstad, T. W. (1982) Distribution of acetylcholinesterase in the hippocampal region of the guinea pig. I. Entorhinal area, parasubiculum, and presubiculum. Z. Zellforsch. Mikrosk. Anat. 114: 460–481.

German, D. C., Bruce, G., and Hersh, L. B. (1985) Immunohistochemical staining of cholinergic neurons in the human brain using a polychlonal antibody to human choline acetyltransferase. Neurosci. Lett. 61: 1–5.

Halliwell, J. V., and Adams, P. R. (1982) Voltage-clamp analysis of muscarinic excitation in hippocampal neurons. Brain Res. 250: 71–92.

Hayashi, M., and Patel, A. J. (1987) An interaction between thyroid hormone and nerve growth factor in the regulation of choline acetyltransferase activity in neuronal cultures, derived from the septal-diagonal band region of the embryonic rat brain. Dev. Brain Res. 36: 109–120.

Hefti, F. (1986) Nerve growth factor promotes survival of septal cholinergic neurons after fimbrial transections. J. Neurosci. 6: 2155–2162.

Hefti, F., Dravid, A., and Hartikka, J. (1984) Chronic intraventricular injections of nerve growth factor elevate hippocampal choline acetyltransferase activity in adult rats with partial septo-hippocampal lesions. Brain Res. 293: 205–311.

Hefti, F., Hartikka, J., Gnahn, H., Heumann, R., and Schwab, M. (1985) Nerve growth factor increases choline acetyltransferase but not survival or fiber outgrowth of cultured fetal septal cholinergic neurons. Neuroscience 14: 55–68.

Honegger, P., and Lenoir, D. (1982) Nerve growth factor (NGF) stimulation of cholinergic telencephalic neurons in aggregating cell cultures. Dev. Brain Res. 3: 229–238.

Korsching, S., Auburger, G., Heumann, R., and Thoenen, H. (1985) Levels of nerve growth factor and its mRNA in the cental nervous system of the rat correlate with cholinergic innervation. EMBO J. 4: 1389–1393.

Kromer, L. F., Björklund, A., and Stenevi, U. (1981) Innervation of embryonic hippocampal implants by regenerating axons of cholinergic septal neurons in the adult rat. Brain Res. 210: 153–171.

Kromer, L. F., Björklund, A., and Stenevi, U. (1983) Intracephalic embryonic neural implants in the adult rat brain. I. Growth and mature organization of brainstem cerebellar and hippocampal implants. J. comp. Neurol. 218: 433–459.

Lancaster, B., and Adams, P. R. (1986) Calcium-dependent current generating the afterhyperpolarization of hippocampal neurons. J. Neurophysiol. 55: 1268–1282.

Madison, D. V., Lancaster, B., and Nicoll, R. A. (1987) Voltage clamp analysis of cholinergic action in the hippocampus. J. Neurosci. 7: 733–741.

Martinez, H. J., Dreyfus, Ch. F., Jonakait, G. H., and Black, I. B. (1985) Nerve growth factor promotes cholinergic development in brain striatal cultures. Proc. natl Acad. Sci. 82: 7777–7781.

Mellgren, S. I., and Srebro, B. (1973) Changes in acetylcholinesterase and distribution of degenerating fibers in the hippocampal region after septal lesions in the rat. Brain Res. 52: 19–36.

Mesulam, M.-M., Mufson, E. J., Wainer, B. H., and Levey A. I. (1983) Central cholinergic pathways in the rat: an overview based on an alternative nomenclature (Ch1–Ch6) Neuroscience 10: 1185–1201.

Milner, T. A., Ly, R., and Amaral, D. G. (1983) An anatomical study of the development of the septo-hippocampal projection in the rat. Dev. Brain Res. 8: 343–371.

Mobley, W. C., Rutkowski, J. L., Tennekoon, G. I., Gemski, J., Buchanan, K., and Johnston, M. V. (1986) Nerve growth factor increases choline acetyltransferase activity in developing basal forebrain neurons. Molec. Brain Res. 1: 53–62.

Mosko, S., Lynch, G., and Cotman, C. W. (1983) The distribution of septal projections to the hippocampus of the rat. J. comp. Neurol. 152: 163–174.

Potter, L. T. (1967) A radiometric microassay of acetylcholinesterase. J. Pharmac. exp. Ther. 156: 500–506.

Richardson, P. M., Verge Issa, V. M. K., and Riopella, R. J. (1986) Distribution of neuronal receptors for nerve growth factor in the rat. J. Neurosci. 6: 2312–2321.

Rimvall, K., Keller, F., and Waser, P. G. (1985) Development of cholinergic projections in organotypic cultures of rat septum, hippocampus and cerebellum. Dev. Brain Res. 19: 267–278.

Schwab, M., Otten, U., and Thoenen, H. (1979) Nerve growth factor (NGF) in the rat CNS: absence of specific retrograde axonal transport and tyrosine hydroxylase induction in locus coeruleus and substantia nigra. Brain Res. 168: 473–483.

Seifert, W., ed. (1983) Neurobiology of the Hippocampus. Academic Press, London.

Seiler, M., and Schwab, M. (1984) Specific retrograde transport of nerve growth factor (NGF) from neocortex to nucleus basalis in the rat. Brain Res. 300: 33–36.

250

Shelton, D. L., and Reichardt, L. F. (1984) Expression of the nerve growth factor gene correlates with the density of sympathetic innervation in effector organs. Proc. natl Acad. Sci. USA 81: 7951–7955.

Suda, K., Barde, Y. A., and Thoenen, H. (1978) Nerve growth factor in mouse and rat serum; correlation between bioassay and radioimmunoassay determination. Proc. natl Acad. Sci. USA 4042–4046.

Thoenen, H., and Barde, Y. A. (1980) Physiology of nerve growth factor. Physiol. Rev. 60: 1284–1335.

Whittemore, S. R., Ebendal, T., Lärkfors, L., Olson, L., Seiger, A., Strömberg, I., and Persson, H. (1986) Developmental and regional expression of β nerve growth factor messenger RNA and protein in the rat central nervous system. Proc. natl Acad. Sci. USA 83: 817–821.

Whittemore, S. R., Lärkfors, L., Ebendal, T., Holets, V. R., Ericsson, A., and Persson, H. (1987) Increased β-nerve growth factor messenger RNA and protein levels in neonatal rat hippocampus following specific cholinergic lesions. J. Neurosci. 7: 244–251.

Zimmer, J., Sunde, N., Sörensen, T., Jensen, S., Möller, A. G., and Gähwiler, B. H. (1985) The hippocampus and Fascia Dentata. An anatomical study of intracerebral transplants and intraocular and in vitro cultures. In: Björklund, A., and Stenevi, U. (eds), Neural Grafting in the Mammalian CNS. Elsevier Science Publishers, Amsterdam.

A role of basic fibroblast growth factor for rat septal neurons

C. Grothe, D. Otto, M. Frotscher* and K. Unsicker

Department of Anatomy and Cell Biology, University of Marburg, D-3550 Marburg, and
*Institute of Anatomy, University of Frankfurt, D-6000 Frankfurt 70, Federal Republic of Germany

Summary. The *in vitro* and *in vivo* relevance of basic fibroblast growth factor (bFGF) for rat septal neurons was studied and compared with the effects of nerve growth factor (NGF).

Implantation of gel foam soaked with saline, NGF or bFGF following fimbria fornix (FF) transection in adult rats showed that after 4 weeks the neuronal death in the medial septum of saline-treated rats (87% as compared to the unlesioned side) was reduced by NGF- or bFGF-treatment (NGF 0.3 µg: 71%; NGF 20 µg: 54%; bFGF 8 µg: 68%). These results indicate that both NGF and bFGF are able to sustain neurons in the medial septum after FF transection. Moreover, choline acetyltransferase (ChAT)-immunocytochemistry revealed that rescued neurons comprise a large proportion of the cholinergic population.

In cultured embryonic rat septal neurons seeded at high densities both NGF and bFGF significantly enhanced ChAT activity (7.5- and 3-fold, respectively) without affecting cell survival. In low density cultures both neurotrophic proteins increased the survival after 4 days. The portions of cholinergic and GABAergic neurons did not change after NGF- and bFGF-treatment (acetylcholinesterase cytochemistry, anti-GABA immunocytochemistry). These results show that i) NGF and bFGF promote survival of embryonic septal cholinergic and GABAergic neurons and may enhance ChAT activity, and ii) bFGF is a potent trophic factor for septal neurons *in vivo* and *in vitro*.

Introduction

The physiological role of nerve growth factor (NGF) for cholinergic magnocellular neurons of the basal forebrain has been extensively studied in the last year (for reviews see Korsching, 1986; Whittemore and Seiger, 1987). Central cholinergic neurons show a specific receptor-mediated retrograde transport of NGF (Johnson and Taniuchi, 1987; Seiler and Schwab, 1984). The highest levels of NGF are found in those regions of the central nervous system (CNS) where cell bodies or terminals of cholinergic neurons are located (Korsching et al., 1985). Furthermore, the highest levels of mRNA^{NGF} are present in the target areas of central cholinergic neurons (Korsching et al., 1985; Shelton and Reichardt, 1984). Following fimbria fornix (FF) transection (the septo-hippocampal pathway lesion paradigm) NGF protein levels transiently increase in the hippocampus, a target region of septal cholinergic neurons, and intracerebrally administered NGF rescues septal neurons (Gasser et al., 1986; Hefti, 1986; Korsching et al., 1985; Kromer, 1987; Williams et al., 1986). Moreover, exogenous treatment with NGF increases levels of choline acetyltransferase (ChAT) *in vivo* and *in vitro* (Gnahn et al., 1983; Hefti et al., 1985).

Another growth factor that has been proposed to act on CNS neurons is basic fibroblast growth factor (bFGF). Basic FGF occurs in the brain (Gospodarowicz et al., 1984) and is immunocytochemically demonstrable in CNS neurons including the hippocampus (Pettmann et al., 1986). *In vitro* bFGF promotes survival of embryonic peripheral and CNS neuron populations including chick ciliary, spinal cord (Unsicker et al., 1987), rat hippocampal (Walicke et al., 1986) and cortical neurons (Morrison et al., 1986). Furthermore, administration of bFGF to the transectioned optic and sciatic nerves protects a substantial number of retinal and sensory neurons from lesion-induced death (Sievers et al., 1987; Otto et al., 1987).

The present report summarizes the results from *in vivo* and *in vitro* studies that point out a potential role of bFGF for rat septal neurons.

Effects of bFGF and NGF on survival of septal neurons after fimbria fornix transection

Neurons of the medial septum (MS) and the vertical limb of the diagonal band innervate the hippocampus. Most of their projections are located in the fimbria fornix. In adult male Sprague-Dawley rats the lesion of the septohippocampal fibers was performed by a complete transection of the left fimbria fornix. The transection was located close to the rostral end of the septal pole of the hippocampus. A gel foam piece soaked with different additives (see below) was placed into the wound cavity. Four weeks after the transection animals were fixed and brain slices containing the septal area were processed for cresyl violet-staining or ChAT-immunocytochemistry (for details see Otto et al., 1989). Cell counts in cresyl violet-stained sections of the MS were performed in the following experimental groups: (1) unlesioned animals, (2) animals which had received an implant containing 8 µg bFGF, or (3) 20 µg NGF, or (4) 0.3 µg NGF, or (5) Dulbecco's modified Eagle's medium plus 0.25% BSA (DMEM/BSA). Unilateral fimbria fornix transection and implantation of gel foam containing DMEM/BSA revealed a neuronal cell loss of 87% as compared to the contralateral intact side. Implantation of gel foam soaked with bFGF significantly reduced neuron death to 68%. Similar data (71% neuron death) were obtained with 0.3 µg NGF. Implantation of gel foam containing 20 µg NGF reduced neuronal cell losses to 54% (Table 1). The number of cholinergic neurons decreased dramatically on the transectioned as compared to the intact side in DMEM/BSA-treated animals as revealed by choline acetyltransferase immunocytochemistry. Implantation of gel foam containing bFGF or NGF, respectively, largely prevented death of

Table 1. Percentages of surviving MS neurons 4 weeks after left FF lesion as compared to the unlesioned right side.

Experimental groups	(N)	Percent neurons ($\pm$ SEM)
Unlesioned	(4)	99.28 $\pm$ 2.43
8 μg bFGF	(7)	31.61 $\pm$ 0.33[+]
0.3 μg NGF$_1$	(3)	28.6 $\pm$ 2,26[+]
20 μg NGF$_2$	(4)	44.62 $\pm$ 4.44[+++]
DMEM/BSA	(6)	12.58 $\pm$ 0.87

[+]$p < 0.01$ compared to DMEM/BSA—data; [++]$p < 0.01$ compared to NGF$_1$—and bFGF—data. Cell counts were performed in cresyl violet-stained paraffin-sections of the MS. Data were analyzed using the U-Test according to Mann and Whitney.

cholinergic septal neurons, suggesting that rescued neurons comprise a large proportion of the cholinergic population. Measurements of neuronal cell size in lesioned and unlesioned MS showed that bFGF, in contrast to NGF, did not prevent a reduction in size of surviving neuronal cell bodies.

Effects of bFGF and NGF on septal neurons in culture

Cultures of dissociated septal neurons from embryonic day 18 were performed using two different culture media: i) serum-containing Leibowitz L15 medium which was optimal for quantitive determinations of ChAT activity (cf. Hefti et al., 1985), and ii) serum-free DMEM containing BSA and N1 supplements was used in experiments assessing neuronal survival because of the low non-neuronal cell proliferation and low neuronal survival in the absence of trophic additives (for details see Grothe et al., 1988). The cellular composition of the septal cultures is

Table 2. Cellular composition of septal cultures maintained at low density (17,000 cells/cm^2) in i) serum-containing L15 medium and ii) serum-free DMEM/N1/BSA for 14 h and 4 days. Since neurons did not survive in DMEM without trophic additives after 4 days, DMEM-data were taken from cultures where bFGF was supplemented. Values represent means from 3 experiments ($\pm$ SD). TT, tetanus toxin; NF, neurofilament; GalC, galactocerebroside; GFAP, glial fibrillary acid protein.

Culture conditions	Percent positive cells using neuronal or glial cell markers			
	TT/anti-NF	anti-GalC	anti-GFAP	Without immunoreaction
14 hours				
L15	50–70	—	—	30–50
DMEM	86 $\pm$ 4	—	—	14
4 days				
L15	50–70	—	—	30–50
DMEM	87 $\pm$ 8	< 5	< 10	—

254

Table 3. Effects of bFGF and NGF on ChAT activity of rat septal neurons in culture.

Culture conditions	ChAT activity (pmol/min/μg protein)
untreated	0.033 ± 0.017
bFGF 400 ng/ml	0.099 ± 0.013
NGF 50 ng/ml	0.25 ± 0.028

Septal neurons were cultured at 400,000 cells/cm^2 in serum-containing L15 medium for 4 days. Data were taken from 3 (n = 6; untreated and NGF) or 4 experiments (n = 8; bFGF), respectively ($\pm$ SEM).

compiled in Table 2. In high density cultures (400,000 cells/cm^2) addition of bFGF or NGF, respectively, to serum-containing L15 medium increased ChAT activity 3-fold (bFGF) and 7.5-fold (NGF) as compared to control cultures during a 4-day culture period (Table 3). Addition of trophic factors did not alter the protein concentrations suggesting that cell survival was not affected by these factors. Survival of septal neurons using serum-containing L15 medium was 4 times higher at high cell densities (33,000, 50,000 and 67,000 cells/cm^2) than at low cell density (17,000 cells/cm^2) after 4 days. Neuronal survival at

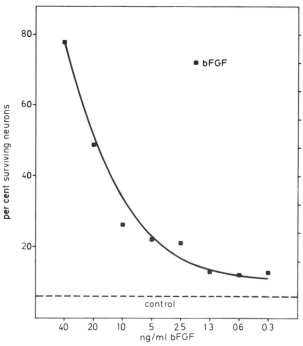

Figure 1. Dose dependence of the effect of bFGF on neuronal survival of low-density cultures in DMEM/N1/BSA after 4 days. Data represent the means from 4 wells of one representative experiments. The number of seeded cells was set as 100%.

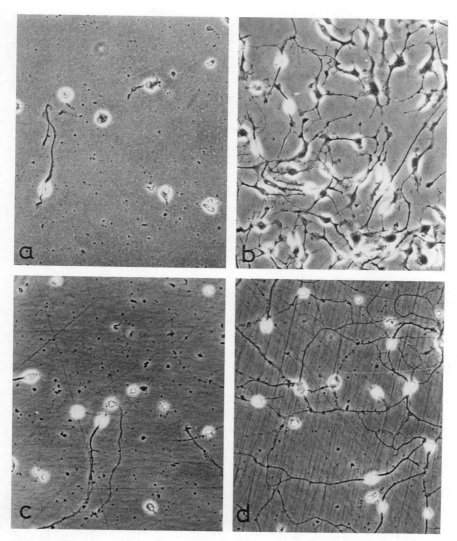

Figure 2. Phase contrast micrographs of representative low density cultures in (a) DMEM/ N1/BSA alone, (b) supplemented with bFGF (40 ng/ml), (c) serum-containing L15 medium alone, and (d) containing NGF (50 ng/ml). ×265.

high densities was not affected by antibodies to bFGF and/or NGF. Basic FGF or NGF supplements did not further increase neuronal survival in high density cultures.

In low density cultures addition of bFGF to serum-containing L15 medium or to serum-free DMEM/N1, respectively, enhanced neuronal survival in a dose-dependent manner after 4 days (Figs 1, 2). The effects

256

of bFGF were blocked by antibodies to bFGF. A survival promoting effect of NGF was consistently observed in cultures using serum-containing L15 medium, but did not occur in the serum-free DMEM/N1 (Fig. 2). The effects of bFGF and NGF were not additive. Immunocytochemical studies using neuronal (tetanus toxin, anti-neurofilament antibodies) and glial markers (antibodies to glial fibrillary acid protein and galactocerebroside) showed that the neuron–non-neuron ratio remained stable under all conditions employed. Further, different cell densities did not alter the cellular composition.

In low density cultures maintained for 4 days in serum-containing L15 medium the portions of cholinergic and GABAergic septal neurons did not change after bFGF- and NGF-treatment as shown by acetylcholinesterase cytochemistry and anti-GABA immunocytochemistry. This implies that both bFGF and NGF may promote survival of cholinergic and GABAergic neurons.

Discussion

Next to the transected optic and sciatic nerves, where bFGF reduces cell death of retinal and dorsal root ganglion cells, respectively (Sievers et al., 1987; Otto et al., 1987), the septohippocampal lesion paradigm is the third lesion model where a neurotrophic action of bFGF has been documented (see also Anderson et al., 1988). Moreover, bFGF and NGF support survival of embryonic septal cholinergic and GABAergic neurons in 4-day cultures. Like NGF, bFGF enhances ChAT activity in cultured septal neurons but to a lesser extent than NGF does. Based on the facts that bFGF (i) is immunocytochemically demonstrable in neurons of the hippocampus (Pettmann et al., 1986), (ii) prevents neuronal cell losses after FF transection (Anderson et al., 1988; Otto et al., 1989), and (iii) enhances survival and ChAT activity in cultured septal neurons (Grothe et al., 1988), we propose a possible *in vivo* role of bFGF for the maintenance and transmitter metabolism of septal neurons. Following the *in vitro* data which showed that NGF and bFGF do not act in an additive manner it seems that NGF and bFGF address largely overlapping or identical neuronal subpopulations. Such a redundancy in terms of trophic support is not unique and has also been made likely for dorsal root ganglion neurons (supported by NGF, ciliary neurotrophic (CNTF) and brain-derived neurotrophic factors), sympathetic neurons (NFG, CNTF) and ciliary ganglionic neurons (CNTF, bFGF, purpurin) (Barbin et al., 1984; Seidl et al., 1987; Schubert et al., 1986; Unsicker et al., 1987).

Acknowledgement. This work was supported by German Research Foundation Grants.

Anderson, K. J., Dam, D., Lee, S., and Cotman, C. W. (1988) Basic fibroblast growth factor prevents death of lesioned cholinergic neurons *in vivo*. Nature 332: 360–361.

Barbin, G., Manthrope, M., and Varon, S. (1984) Purification of the chick eye ciliary neurotrophic factor. J. Neurochem. 43: 1468–1478.

Gasser, U. E., Weskamp, G., Otten, U., and Dravid, A. R. (1986) Time course of elevation of nerve growth factor (NGF) content in the hippocampus and septum following lesions of the septohippocampal pathway in rats. Brain Res. 376: 351–356.

Gnahn, H., Hefti, F., Heumann, R., Schwab, M., and Thoenen, H. (1983) NGF-mediated increase of choline acetyltransferase (ChAT) in the neonatal forebrain; evidence for a physiological role of NGF in the brain? Dev. Brain Res. 9: 45–52.

Gospodarowicz, D., Cheng, J., Lui, G. M., Baird, A., and Böhlen, P. (1984) Isolation by heparin sepharose affinity chromatography of brain fibroblast growth factor: identity with pituitary fibroblast growth factor. Proc. natl Acad. Sci. USA 81: 6963–6967.

Grothe, C., Otto, D., and Unsicker, K. (1988) Basic fibroblast growth factor promotes *in vitro* survival and cholinergic development of rat septal neurons. Comparison with the effects of nerve growth factor. Neuroscience, in press.

Hefti, F. (1986) Nerve growth factor (NGF) promotes survival of septal cholinergic neurons after fimbrial transection. J. Neurosci. 6: 2155–2162.

Hefti, F., Hartikka, J., Eckenstein, F., Gnahn, H., Heumann, R., and Schwab, M. (1985) Nerve growth factor increases choline acetyltransferase but not survival or fiber outgrowth of cultured fetal septal cholinergic neurons. Neuroscience 14: 55–68.

Johnson, E. M., Jr, and Taniuchi, M. (1987) Nerve growth factor (NGF) receptors in the central nervous system. Biochem. Pharmac. 36: 4189–4195.

Korsching, S. (1986) The role of nerve growth factor in the CNS. Trends Neurosci. 9: 570–573.

Korsching, S., Auburger, G., Heumann, R., Scott, J., and Thoenen, H. (1985) Levels of nerve growth factor and its mRNA in the central nervous system of the rat correlate with cholinergic innervation. EMBO J. 4: 1389–1393.

Kromer, L. F. (1987) Nerve growth factor treatment after brain injury prevents neuronal death. Science 253: 214–216.

Morrison, R. S., Sharma, A., De Vellis, J., and Bradshaw, R. A. (1986) Basic fibroblast growth factor supports the survival of cerebral neurons in primary culture. Proc. natl Acad. Sci. USA 83: 7537–7541.

Otto, D., Unsicker, K., and Grothe, C. (1987) Pharmacological effects of nerve growth factor and fibroblast growth factor applied to the transectioned sciatic nerve on neuron death in adult rat dorsal root ganglia. Neurosci. Lett. 83: 156–160.

Otto, D., Frotscher, M., and Unsicker, K. (1989) Basic fibroblast growth factor and nerve growth factor administered in gel foam rescue medial septal neurons after fimbria fornix transection. J. Neurosci. Res. 22: 83–91.

Pettmann, B., Labourdette, G., Weibel, M., and Sensenbrenner, M. (1986) The brain fibroblast growth factor (FGF) is localized in neurons. Neurosci. Lett. 68: 175–180.

Schubert, D., LaCorbiere, M., and Esch, F. (1986) A chick neural retina adhesion and survival molecule is a retinol-binding protein. J. Cell Biol. 102: 2295–2301.

Seidl, K., Manthorpe, M., Varon, S., and Unsicker, K. (1987) Differential effects of nerve growth factor and ciliary neurotrophic factor on catecholamine storage and catecholamine synthesizing enzymes of cultured rat chromaffin cells. J. Neurochem. 49: 169–174.

Seiler, M., and Schwab, M. E. (1984) Specific retrograde transport of nerve growth factor (NGF) from neocortex to nucleus basalis in the rat. Brain Res. 300: 33–39.

Shelton, D. L., and Reichardt, L. F. (1984) Expression of β-nerve growth factor gene correlates with the density of sympathetic innervation in effector organs. Proc. natl Acad. Sci. USA 81: 7951–7955.

Sievers, J., Hausmann, B., Unsicker, K., and Berry, M. (1987) Fibroblast growth factors promote the survival of adult retinal ganglion cells after transection of the optic nerve. Neurosci. Lett. 76: 157–162.

Unsicker, K., Reichert-Preibsch, H., Schmidt, R., Labourdette, G., and Sensenbrenner, M. (1987) Acidic and basic fibroblast growth factors have neurotrophic functions for cultured periphereal and central nervous system neurons. Proc. natl Acad. Sci. USA 84: 5459–5436.

Walicke, P., Cowan, W. M., Ueno, N., Baird, A., and Guillemin, R. (1986) Fibroblast growth factor promotes survival of dissociated hippocampal neurons and enhances neurite extension. Proc. natl Acad. Sci. USA 83: 9231–9235.

Whittemore, S. R., and Seiger, A. (1987) The expression, localization and functional significance of β-nerve growth factor in the central nervous system. Brain Res. Rev. 12: 439–464.

Williams, L. R., Varon, S., Peterson, G. M., Wictorin, K., Fischer, W., Björklund, A., and Gage, F. H. (1986) Continuous infusion of nerve growth factor prevents basal forebrain neuronal death after fimbria-fornix transection. Proc. natl Acad. Sci. USA 83: 9231–9235.

Survival, growth and function of damaged cholinergic neurons

Fred H. Gage, Mark H. Tuszynski, Karen S. Chen,
David Armstrong and György Buzsáki

*Department of Neurosciences M-024, University of California at San Diego, La Jolla,
CA 92093, USA*

Summary. Recent progress has been made in defining the requirements for survival, growth and function of damaged cholinergic neurons of the central nervous system. In particular, the responsiveness of cholinergic neurons to nerve growth factor (NGF) in the regulation of development, cell survival, axon elongation, and response to injury has led to the formulation of the Neurotrophic Hypothesis, a unifying hypothesis of neuronal responsiveness to growth-promoting substances. NGF-mediated effects on cholinergic neurons in culture as well as in the septum, basal nucleus, striatum, and hippocampus, and the ability of NGF to prevent lesion-induced cell death and to ameliorate the effects of aging, provide the foundation for this work. A potential role for glia and microglia in mediating the effects of NGF is proposed.

Introduction

Throughout normal development the brain makes more neurons than are needed. Some neurons undergo cell death at times when their growing axons compete for target territories, reducing the final number of neurons to that found in the adult (Cowan et al., 1984). The extent of neuronal cell death during development can be affected by experimentally manipulating the target area of the developing neurons to add or subtract target tissue (Landmesser and Pilar, 1978; Hamburger and Oppenheimer, 1982). These observations have led several investigators to propose that developmental cell death and survival are regulated by proteins presumably supplied by the target territory, which have been called *neurotrophic factors* (NTF) (Thoenen and Bard, 1980). 'Trophic' refers to the ability of one tissue, cell, or protein to support and/or nourish another; thus a neurotrophic factor is a chemical or molecule that is made in any cell and supports the survival of, or nourishes, neurons. *Trophic* differs markedly from *tropic* which refers to the influence of one cell or tissue on the direction of movement or outgrowth of another; thus a neurotropic factor is a chemical or molecule that can influence the direction and/or growth of a neuronal axon (Table 1).

Table 1. Trophic and tropic effects of NGF on cholinergic neurons

Trophic effects upon:	Tropic effects upon:
Cholinergic: basal forebrain	Axon regeneration
Sympathetic neurons	Collateral sprouting
Sensory neurons of the dorsal roots	Reactive synaptogenesis

The Neurotrophic Hypothesis

The Neurotrophic Hypothesis (Fig. 1) postulates that 1) adult CNS neurons *in situ* are supported and regulated by their respective NTFs, 2) proper maintenance of these neurons depends on adequate supply and utilization of the NTFs, 3) an interference with the NTF support, or 'neurotrophic deficit', will result in defective performance or even degeneration and death of the target neurons, and 4) such trophic deficits may be the basis of degenerative central nervous system (CNS) diseases (e.g. Parkinson's, motor neuron, or Alzheimer's disease) or normal aging (Appel, 1981).

The Neurotrophic Hypothesis is supported in the CNS by 1) 'developmental neuronal death', in which the excessive number of neurons produced during development is decreased to accommodate the limited target cell number (Cowan et al., 1984); 2) 'retrograde neuronal degeneration', in which axotomized neurons cut off from their innervation target and surrounding glial cells undergo degeneration or even death (Pearson et al., 1983); and 3) 'pathological neuronal death', where specific populations of neurons degenerate and die (Appel, 1981; Bartus et al., 1982). One explanation commonly put forth for such neuronal death-inducing situations is that neurons normally depend for their continued health upon NTFs supplied by their target and associated glial cells, and that disruption in this trophic supply causes their death.

Nerve growth factor and the cholinergic neuron

Nerve growth factor (NGF) is currently the best characterized neurotrophic factor (NTF) (Barde et al., 1983; Berg, 1984; Thoenen et al., 1980; Ulrich et al., 1983). NGF supports the survival and axonal growth *in vitro* and *in vivo* of sensory and sympathetic neurons from the peripheral nervous system (PNS) (Gunderson, 1980) [neurotropic]. Furthermore NGF can attract and guide regenerating axons, whether provided to the neuron in a soluble or immobilized form (Gunderson, 1980), and may even guide axons whose neurons do not require NGF for survival (Collins, 1983) [neurotropic].

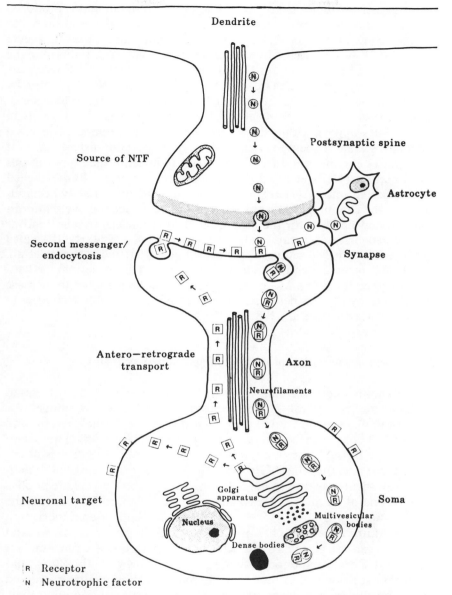

Figure 1. Schematic diagram of elements important to the neurotrophic hypotheses. These include: synthesis of neurotrophic factor (NTF) in target neurons or glia, and synthesis of receptors for the NTF in the dependent neuron. Secretion of NTF and anterograde transport of receptors to membrane surface. Binding and coupled interaction of NTF and receptor complex where this complex is retrogradely transported to the cell body where intracellular action is consummated. Arrows indicate directionality.

A number of studies have shown that NGF occurs in and is produced by the CNS. For example, mammalian CNS tissues have been shown to contain NGF messenger RNA by *in situ* hybridization (Ayer-Le Lievre et al., 1983), NGF antigen by immunohistochemical and radio-immune assays (Ayer-Le Lievre et al., 1983; Greene et al., 1980), NGF receptors by autoradiography (Richardson et al., 1986), and NGF protein by biological assays (Collins et al., 1983; Manthorpe et al., 1983; Nieto-Sampedro et al., 1983; Scott et al., 1984). The greatest NGF levels in CNS tissues appear within the target areas of the cholinoceptive basal forebrain systems (Sheldon et al., 1986), and NGF administered into rat brain raises choline acetyltransferase (ChAT) levels in the hippocampus and striatum (Hefti et al., 1984; Mobley et al., 1985). Radiolabelled NGF injected into target regions is taken up and retrogradely accumulated by the cholinergic neurons innervating them, such as the septal/diagonal band neurons for the hippocampus and nucleus basalis neurons for the neocortex (Schwab et al., 1979; Seiler et al., 1984). NGF is also produced by purified cultures of cerebral astroglial cells (Rudge et al., 1985) and accumulates in fluids surrounding rat brain lesions (Nieto-Sampedro et al., 1983). More recently NGF has been shown to increase ChAT activity in CNS neuronal cultures (Hefti et al., 1985; Honegger et al., 1982; Martinez et al., 1985).

Retrograde degeneration in the septohippocampal system

The cholinergic projection from the adult rat septum and diagonal band to the ipsilateral hippocampus has been a useful model for examining CNS plasticity (Fig. 2). Neurons of the medial septum and the vertical limb of the diagonal band project dorsally to the hippocampus mainly through the fimbria-fornix (Gage et al., 1983; Lewis et al., 1967; Gage and Björklund [HC vol. 3]). About 50% of the septal/diagonal band neurons sending fibers through the fimbria-fornix are cholinergic (Amaral et al., 1985; Wainer et al., 1985) and provide the hippocampus with about 90% of its total cholinergic innervation (Storm-Mathisen et al., 1974). The cholinergic neurons, axons and terminals can be visualized by acetylcholinesterase (AChE) (Butcher, 1983) and ChAT immunocytochemistry (Armstrong et al., 1983), and the terminal fields within the hippocampal formation can be quantified biochemically by measuring extracted ChAT activity (Fonnum, 1969, 1975).

Complete transection of the FF pathway in adult rats results in a rapid and consistent retrograde degeneration and death of many of the septal/diagonal band neurons (including the cholinergic ones) that originally contributed axons through this pathway (Cunningham, 1982; Grady et al., 1984; Hefti, 1986; Kromer and Cornbrooks, 1984; Wainer

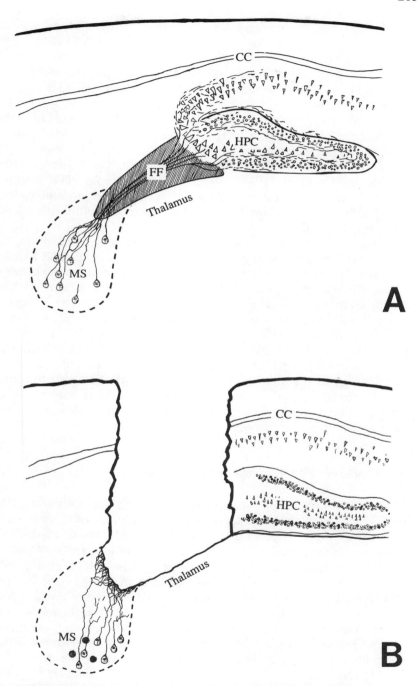

Figure 2. Schematic diagram of cholinergic septohippocampal pathway in the sagittal plane.
A Intact system. *B* 2 weeks following complete fimbria-fornix and overlying cortical aspiration.

et al., 1985) (Fig. 2). Markers of cell survival (retrogradely transported fluorescent dyes, Nissl stains) (Gage et al., 1986; Tuszynski et al., 1988), transmitter enzyme expression (AChE, ChAT) (Armstrong et al., 1987), and NGF-r expression on cells of the MS (Montero and Hefti, 1988; Tuszynski et al., 1988) demonstrate a loss of 70% to 90% of cells in this region.

One explanation for this axotomy-induced cell death is that the septal neurons become deprived of a critical supply of NTF provided possibly by the postsynaptic neurons or glial cells in the target areas of the hippocampus (Collins, 1985; Gage et al., 1986; Gnahn et al., 1983; Nieto-Sampedro et al., 1983). That this hippocampal NTF might be NGF or NGF-like is supported by the previously listed studies from several laboratories reporting an NGF presence within the septohippocampal system.

In addition to the lesion-induced degeneration described above, several laboratories have demonstrated that, in aged animals and humans, the cholinergic neurons of the basal forebrain are compromised (Bartus et al., 1982; Whitehouse et al., 1982). This compromise is reflected in cell shrinkage and in some cases cell loss of cholinergically marked neurons in the basal forebrain region, which seems to be correlated with a decrease in cognitive ability in animals (Fisher et al., 1987) as well as humans (Coyle et al., 1987). We initially reported and subsequently confirmed and extended the findings that transplantation of fetal cholinergic neurons to the hippocampus of aged rats pre-screened for cognitive impairments could result in substantial improvement in the previously impaired animal's behavior, and that this was in part mediated via the cholinergic system (Gage et al., 1986b).

Trophic effects of NGF on the cholinergic neuron

These observations raise the question whether exogenous administration of NGF to axotomized septal neurons might rescue them from the ensuing retrograde degeneration and death and thus allow them to regenerate their already cut axons or even new ones back to the hippocampal formation. Recently, three groups (Hefti et al., 1984; Kromer, 1987; Williams et al., 1986; Gage et al., 1988) have independently reported that the intraventricular administration of purified NGF into adult rats from the time of fimbria-fornix transection onward prevents the death of most of the axotomized cholinergic septum/diagonal band neurons (Fig. 3). Furthermore, it appears that even noncholinergic septal neurons (Panula et al., 1984) are destined to die (Gage et al., 1986) and may be saved by NGF administration (Williams et al., 1986). NGF administration also seems to prevent the degeneration of the cut septal cholinergic axons and/or to stimulate their regrowth, since a large

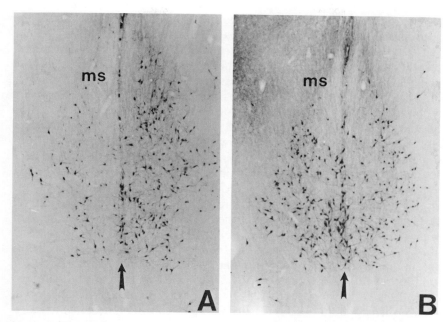

Figure 3. Photomicrograph of choline-acetyltransferase (ChAT) immunoreactivity in the medial septum. *A* 2 weeks following unilateral fimbria-fornix lesion with chronic vehicle infused in the lateral ventricle ipsilateral to the lesion. *B* 2 weeks following NGF infusion. ms, medial septal area. Arrow indicates midline.

number of AChE-positive fibers appear to form a neuroma-like structure proximal to the transection site (Gage et al., 1986).

More recently, based on our initial observations of the trophic effect of NGF on cholinergic neurons in the retrograde degeneration model described above, we have injected NGF into cognitively impaired aged rats and have found that all aged rats with intact NGF pumps showed an improvement in retention of a complex spatial learning task relative to matched, non-infused cognitively impaired rats (Fischer et al., 1987). In the same study, we observed that there was a significant increase in the size of the cholinergic neurons in the basal forebrain region on the side of the brain into which the NGF was intraventricularly infused.

NGF influences expression of its own receptors

The ability of neurons to respond to NGF seems to depend on the presence of cell surface receptors, which in the PNS appear to mediate the binding, internalization and transport from the terminals to the parent cell bodies (see Thoenen and Bard, 1985 for review). Such NGF receptors (NGFr) have been demonstrated also on the NGF-responsive cholinergic neurons in the CNS, both during development and in the adult animal (Taniuchi et al., 1986a; Richardson et al., 1986).

The cholinergic neurons of the striatum appear to represent a special case. In contrast to the neurons in the septal-diagonal band area and the nucleus basalis, these neurons possess very low or undemonstrable levels of NGFr in the adult, and their responsiveness to NGF has been reported to decline dramatically during postnatal development (Martinez et al., 1985; Mobley et al., 1985; Johnston et al., 1987). Nevertheless, our previous findings have shown that chronic infusion of NGF in the lateral ventricle of adult animals following FF lesions not only spared the medial septal neurons from degeneration, but also resulted in hypertrophy of the cholinergic neurons of the ipsilateral striatum (Gage et al., 1988). Similarly, chronic infusions of NGF into the lateral ventricle of aged rats ameliorated the age-related atrophy of the cholinergic neurons of the striatum, as well as the basal forebrain (Fischer et al., 1987). The *in vivo* effects of NGF on the striatal cholinergic neurons are in apparent contradiction with the lack of demonstrable NGFr immunoreactivity in these neurons. However, Taniuchi et al. (Taniuchi et al., 1986a, b) have recently reported that peripheral nerve damage will induce the expression of NGFr on Schwann cells within the denervated distal portion of the nerve, raising the possibility that the ability of striatal neurons to respond to NGF depends on the up-regulation of the NGFr and that this up-regulation is induced by the tissue damage. In a recent experiment we tested this hypothesis. Chronic NGF infusion into the adult neostriatum resulted in re-expression of the NGFr such that many cholinergic interneurons become immunoreactive for NGFr. This effect was seen also after striatal damage induced by infusion of vehicle alone, whereas infusion of anti-NGF serum partially inhibited the receptor's re-expression. Infusion of NGF, but not vehicle alone, dramatically increased the size and ChAT-immunoreactivity of these same cholinergic neurons (Gage et al., 1989).

These findings indicate that central cholinergic neurons which lose their NGFr during postnatal development will resume their NGF responsiveness when the tissue is damaged. Such a damage-induced mechanism may act to enhance the action of trophic factors, including NGF, released at the site of injury, and enhance the responsiveness of damaged CNS neurons to exogenously administered trophic factors.

Tropic effects of NGF on cholinergic neurons

In addition to its function of maintaining the normal integrity of the central cholinergic basal forebrain neurons and of peripheral sympathetic and neural crest-derived sensory neurons (neurotrophic), NGF has also been postulated to have a role in the growth of these neurons after damage to the nervous system (neurotropic) (Table 1). Two forms of axonal growth are commonly observed in the mature hippocampus: these are regeneration, or the regrowth of axons previously damaged, and collateral sprouting, or the new growth of remaining (intact) axons (Gage and Björklund, 1986b). The latter of these responses, collateral sprouting, occurs within the hippocampus by two distinct populations of neurons following FF transection. Superior cervical ganglion (SCG) derived sympathetic axons, which normally surround the hippocampal vasculature, undergo a robust sprouting response into the dentate gyrus and CA3 pyramidal cell region of the hippocampal parenchyma (Loy and Moore, 1977; Stenevi and Björklund, 1978). In addition, the magnocellular midline cholinergic neurons of the dorsal hippocampal formation sprout into the medial subiculum and CA1 pyramidal cell layers in response to FF lesion (Blaker et al., 1988). To test the postulate that NGF and its receptors have tropic effects on cholinergic axons in the hippocampus, and that the lesion-induced increase in NGF in the hippocampus may serve as a chemo-attractant of NGF-responsive axons towards the source of its production (Crutcher, 1987), we examined the FF lesioned hippocampal formation to determine whether the sprouting neurons stain positively for NGFr (Batchelor et al., in press). We found that the two populations of neurons that undergo collateral sprouting, namely the midline magnocellular cholinergic neurons of the dorsal hippocampus and the sympathetic neurons of the SCG, stain strongly for NGFr with a monoclonal antibody (Taniuchi et al., 1986a, b). In contrast, the small intrinsic cholinergic neurons of the hippocampus exhibited neither sprouting response nor staining for NGFr. In view of these results we suggest that the differing sprouting responses demonstrated by these three neuronal populations may be due to their responsiveness to NGF, as indicated by the presence or absence of NGFr.

NGF enhances regeneration of cholinergic axons

Attempts to restore this severed FF pathway have been made by grafting fetal tissue (Gage et al., 1988; Kromer et al., 1981) which may act as a bridge between the disconnected septohippocampal pathway. Increases in ChAT activity and AChE fiber innervation in the host hippocampus have been consistently reported in these studies, but the extent of reinnervation is small.

A possible explanation for the limited restoration of the cholinergic circuitry may be that the majority of the cholinergic neurons in the medial septum and diagonal band of Broca degenerate, become dysfunctional, and die within a month following the transection (Daitz et al., 1954; Gage et al., 1986; Armstrong et al., 1987). This observation, and the evidence of a link between NGF and cholinergic neurons (Heumann et al., 1985; Honegger et al., 1982; Sheldon et al., 1986), have prompted several groups to test and subsequently demonstrate the dependence of adult denervated cholinergic neurons on exogenous NGF for survival in the absence of endogenous NGF previously transported from the hippocampus (Gage et al., 1988; Hefti, 1986; Kromer, 1987; Williams et al., 1986).

These results lead to the prediction that the exogenous delivery of NGF could not only promote the survival of septal neurons but also would then promote the cholinergic axons to extend across a bridge of hippocampal fetal tissue placed in the FF cavity. Thus, in a recent study we combined the exogenous infusion of NGF to the lateral ventricle adjacent to the denervated septum with the simultaneous grafting of fetal hippocampal tissue to the fimbria-fornix cavity, as a set of procedures that may more fully and functionally restore the severed septohippocampal circuitry (Buzsàki et al., 1987). Imbedded in the design of this study were two additional related questions: 1) Does the transient two-week NGF infusion period which has been shown to result in significant cholinergic cell rescue have an enduring effect on the medial septal cells 6–8 months following termination of the NGF infusion? 2) Will fetal hippocampal grafts alone, in the absence of exogenous NGF infusion, support the survival of the axotomized cholinergic neurons of the medial septum?

A combination of intracerebral grafting and intraventricular infusion of NGF was used to attempt to reconstruct the cholinergic component of septohippocampal pathway following FF lesions. Four groups were used: lesion only (FF); lesion and fetal hippocampal graft (FF-HPC); lesion and NGF (FF-NGF); and lesion, graft and NGF (FF-HPC-NGF). Choline acetyltransferase immunoreactivity (ChAT-IR), acetylcholinesterase (AChE) fiber staining, and behavior-dependent theta electrical activity were used to assess the extent of pathway reconstruction. The NGF infusion only lasted the first two weeks following the FF lesion, while theta activity and histological analysis were conducted 6–8 months after the lesion. Only the FF-HPC-NGF group had long-term savings of ChAT-IR cells as compared to the FF and FF-HPC group. In addition the FF-HPC-NGF group had more extensive reinnervation of the hippocampus than any other group. Further, the FF-HPC-NGF group had the most complete evidence of behavior-dependent theta activity restoration. These results demonstrate clearly that a combination of short-term intraventricular NGF infusion and fetal hippocampal grafts can result in a more complete reconstruction of the damaged septohippocampal circuit.

A role for glia in mediating the tropic effects of NGF

Damage to the fimbria-fornix, and separately to the perforant path, leads to distinct and dramatic time-dependent increases in glial fibrillary acidic protein immunoreactivity (GFAP-IR) in specific areas of the hippocampal formation (Gage et al., 1988). Specifically, FF lesions resulted in an increase in the GFAP-IR in the pyramidal and oriens area of the CA3, as well as the inner molecular layer of the dentate gyrus. In addition, ipsilateral to the lesion, there was a rapid and robust increase in GFAP-IR in the dorsal lateral quadrant of the septum, but not in the medial region. Only after 30 days did the GFAP-IR reach the medial septum. Following perforant path lesions, there was a selective increase in GFAP-IR in the outer molecular layer of the dentate gyrus. Most of these changes were transient, and had disappeared by 30 days post lesion. We speculate that the increase in GFAP-IR in these target areas is a necessary requirement for the sprouting responses that are observed (Gage et al., 1988).

Considerable evidence supports the presumption that astrocytes can make and secrete neurotrophic factors that can subsequently support the survival and/or axonal outgrowth of a variety of central and peripheral neurons, *in vitro* and *in vivo* (Banker, 1980; Hatten and Liem, 1981; Liesi et al., 1984; Lindsay, 1977; Tarris et al., 1986). At present when a new putative factor is tested *in vitro* for its neurotrophic activity, rigorous controls must be used to establish that the neuronal population is not contaminated with glia, and that the presumed neurotrophic factor is not acting through the glial cell population *in vitro*. To date the evidence supports the notion that only reactive and/or proliferating astrocytes secrete trophic and tropic substances; thus it is essential to understand the signals for this activation *in vivo*. As stated above, microglial proliferation often precedes the astrocytic proliferation (Gall et al., 1979; Vijayan, 1983). Recently, Guilian and Baker (1985) showed that activated microglia secrete a substance which stimulates the proliferation of astroglia. Previously these same authors had shown that interleukin-1 (IL-1) could also stimulate astrocyte proliferation. Very recently it has been shown that IL-1 regulates the synthesis of NGF in non-neuronal cells of the damaged rat sciatic nerve (Lindholm et al., 1987), and that IL-1 is most likely secreted from activated macrophages in the vicinity of the damaged nerve (Korsching et al., 1985).

Working hypothesis of NGF effects

The results of data summarized in the previous sections suggest: 1) a role of IL-1 in the proliferation of astrocytes, 2) activated microglia and macrophages can secrete IL-1, and 3) IL-1 can activate NGF synthesis in non-neuronal cells. We have made the following suggestion for the outline of the events that lead to the NGF-sensitive sprouting responses in the hippocampus and septum following FF and PP lesions:

Perforant pathway damage induces terminal degeneration from the cells which are transected in the entorhinal cortex. This terminal degeneration activates the microglia to phagocytize in the restricted zone of terminal degeneration. These activated microglia subsequently release IL-1 into the surrounding environment, which in turn induces the proliferation of astrocytes. The activated astrocytes then secrete NGF into this region of the outer molecular layer of the dentate gyrus which results in the attraction of more cholinergic fibers that express NGFr on their membrane surfaces.

Following FF lesions, cholinergic terminals are disconnected from their cell bodies in the septum, and once again there is a terminal degeneration response, this time in the areas of heaviest cholinergic innervation, CA3 and dentate gyrus. This degeneration leads to a microglial proliferation, IL-1 secretion, and an astrocytic mitogenic reaction in CA3 and inner molecular layer of the dentate gyrus. The activated astrocytes secrete NGF and promote the ingrowth of NGFr-bearing sympathetic fibers of the superior cervical ganglia.

Concurrently in the septum, degeneration of terminals from the hippocampus to the dorsal lateral quadrant results in microglia reactivity, and subsequent IL-1 secretion, proliferation of glia, and NGF concentration. This in turn induces the growth of cholinergic fibers from the medial septum, which are undergoing retrograde degeneration, into the dorsolateral quadrant. This NGF source in the dorsal lateral quadrant is not sufficient to support all the medial cholinergic cells, but some of the medial septal cells are always spared even with a complete bilateral FF lesion. Meanwhile, the absence of reactive astrocytes in the medial septum soon after the FF lesion is the paramount reason for the death of these cells, because, as in development, these cells could survive in the presence of adequate glia-derived NGF.

This working hypothesis generates several specific testable predictions, which when examined should reveal more about the mechanism underlying survival, growth and function of damaged cholinergic neurons.

Acknowledgments. We thank Sheryl Christenson for typing the manuscript. The research presented in this manuscript was supported by NIA AG 06988, AG 05344, AG 08206, Office of Naval Research, California Department of Health Services, the Sandoz Foundation, and the J.D. French Foundation.

Amaral, D. G., and Kurz, J. (1985) An analysis of the origins of the cholinergic and noncholinergic septal projections to the hippocampal formation of the rat. J. comp. Neurol. 240: 37–59.

Appel, S. H. (1981) A unifying hypothesis for the cause of amyotrophic lateral sclerosis, parkinsonism, and Alzheimer disease. Ann. Neurol. 10: 400–405.

Armstrong, D. M., Saper, C. B., Levey, A. I., Wainer, B. H., and Terry, R. D. (1983) Distribution of cholinergic neurons in rat brain: Demonstrated by the immunocytochemical localization of choline acetyltransferase. J. comp. Neurol. 216: 53–68.

Armstrong, D. M., Terry, R. D., Deteresa, R. M., Bruce, G., Hersh, L. B., and Gage, F. H. (1987) Response of septal cholinergic neurons to axotomy. J. comp. Neurol. 264: 421–436.

Ayer LeLievre, C. S., Ebendal, T., Olsen, L., and Seiger, A. (1983) Localization of NGF-like immunoreactivity in rat neurons tissue. Med. Biol. 61: 296–304.

Banker, G. A. (1980) Tropic interactions between astroglial cells and hippocampal neurons in cultures. Science 209: 809-810.

Barde, Y. A., Edgar, D., and Thoenen, H. (1983) New neurotrophic factors. A. Rev. Physiol. 45: 601–612.

Bartus, R., Dean, R. L., Beer, C., and Lippa, A. S. (1982) The cholinergic hypothesis of geriatric memory dysfunction. Science 217: 408–417.

Batchelor, P. E., Armstrong, D. M., Blaker, S. M., and Gage, F. H. Nerve growth factor receptor and choline acetyltransferase colocalization in neurons within the rat forebrain: Response to fimbria-fornix transection. J. comp. Neurol., in press.

Berg, D. K. (1984) New neuronal growth factors. A. Rev. Neurosci. 7: 149–170.

Blaker, S. N., Armstrong, D. M., and Gage, F. H. (1988) Cholinergic neurons within the rat hippocampus: Response to fimbria-fornix transection. J. comp. Neurol. 272: 127–138.

Buzsaki, G., Bickford, R. G., Varon, S., Armstrong, D. M., and Gage, F. H. (1987) Reconstruction of the damaged septohippocampal circuitry by a combination of fetal grafts and transient NGF infusion. Soc. Neurosci. Abstr. 13: 568.

Butcher, L. L. (1983) Acetylcholinesterase histochemistry. Handbook of Chemical Neuroanatomy, Vol. 1. Elsevier, Amsterdam, pp. 1–49.

Collins, F., and Crutcher, K. A. (1985) Neurotrophic activity in the adult rat hippocampal formation: Regional distribution and increase after septal lesion. J. Neurosci. 5: 2809–2814.

Collins, F., and Dawson, A. (1983) An effect of nerve growth factor on parasympathetic neurite outgrowth. PNAS 80: 2091–2094.

Cowan, W. M., Fawcett, J. W., O'Leary, D. D., and Stanfield, B. B. (1984) Regressive events in neurogenesis. Science 225: 1258–1265.

Coyle, J. T., Price, P. H., and Delong, M. R. (1983) Alzheimer's disease: A disorder of cortical cholinergic innervation. Science 219: 1184–1189.

Crutcher, K. A. (1987) Sympathetic sprouting in the central nervous system: A model for studies of axonal growth in the mature mammalian brain. Brain Res. Rev. 12: 203–233.

Cunningham, T. J. (1982) Naturally occurring neuron death and its regulation by developing neural pathways. Int. Rev. Cytol. 74: 163–186.

Daitz, H. M., and Powell, T. P. S. (1985) Studies on the connexions of the fornix system. J. Neurol. Neurosurg. Psychiat. 7: 75–82.

Fawcett, J. W., O'Leary, D. D. M., and Cowan, W. M. (1984) Activity and the control of ganglian cell death in the rat retina. PNAS 81: 5589–5593.

Fischer, W., Gage, F. H., and Björklund, A. (1988) Degenerative changes in forebrain cholinergic nuclei correlate with cognitive impairments in aged rats. Eur. J. Neurosci., in press.

Fischer, W., Wictorin, K., Björklund, A., Williams, L. R., Varon, S., and Gage F. H. (1987) Amelioration of cholinergic neuron atrophy and spatial memory impairment in aged rats by nerve growth factor. Nature 329: 65–68.

Fonnum, F. (1984) Topographical and subcellular localization of choline acetyltransferase in the rat hippocampal region. J. Neurochem. 24: 407–409.

Fonnum, F. (1969) Radiochemical micro assays for the determination of choline acetyltransferase and acetylcholinesterase activities. J. Biochem. 115: 465–472.

Gage, F. H., Batchelor, P., Chen, K. S., Chin, D., Higgins, G. A., Koh, S., Deputy, S., Rosenberg, M. B., Fischer, W., and Björklund, A. (1989) NGF receptor reexpression and NGF-mediated cholinergic neuronal hypertrophy in the damaged adult neostriatum. Neuron 2: 1177–1184.

Gage, F. H., and Björklund, A. (1986) Enhanced graft survival in the hippocampus following selective denervation. Neuroscience 17: 89–98.

Gage, F. H., and Björklund, A. (1986a) Neural grafting in the aged rat brain. Ann., Res. Physiol. 48: 447–459.

Gage, F. H., and Björklund, A. (1986b) Cholinergic septal grafts into the hippocampal formation improve spatial learning and memory in aged rats by an atropine sensitive mechanism. J. Neurosci. 2837–2847.

Gage, F. H., Björklund, A., Stenevi, U., and Dunnett, S. B. (1983) Functional correlates of compensatory collateral sprouting by aminergic and cholinergic afferents in the hippocampal formation. Brain Res. 268: 39–47.

Gage, F. H., Wictorin, K., Ficher, W., Williams, L. R., Varon, S., and Björklund, A. (1986) Life and death of cholinergic neurons: In the septal and diagonal band region following complete fimbria fornix transection. Neuroscience 19: 241–255.

Gage, F. H., Armstrong, D. M., Williams, L. R., and Varon, S. (1988) Morphologic response of axotomized septal neurons to nerve growth factor. J. comp. Neurol. 269: 147–155.

Gage, F. H., Blaker, S. N., Davis, G. E., Engvall, E., Varon, S., and Manthorpe, M. (1988) Human amnion membrane matrix as a substratum for axonal regeneration in the central nervous system. Exp. Brain Res., in press.

Gage, F. H., Olinechek, P., and Armstrong, D. M. (1988) Astrocytes are important for NGF-mediated hippocampal sprouting. Exp. Neurol., in press.

Gall, C., Rose, G., and Lynch, G. (1979) Proliferative and migratory activities of glial cells in the partially deafferented hippocampus. J. comp. Neurol. 183: 539–550.

Giulian, D., and Baker, T. J. (1985) Peptides released by ameloid microglia regulate atroglial proliferation. J. Cell Biol. 101: 2411–2415.

Gnahn, H., Hefti, F., Heumann, R., Schwab, M. E., and Thoenen, H. (1983) NGF-mediated increase in choline acetyltransferase (ChAT) in the neonatal rat forbrain; evidence for physiological role of NGF in the brain? Dev. Brain Res. 9: 45–52.

Grady, S., Reeves, T., and Steward, O. (1984) Time course of retrograde degeneration of the cells or origin of the septohippocampal pathway after fimbria-fornix transections. Soc. Neurosci. Abstr. 10: 463.

Greene, L., and Shooter, E. M. (1980) The nerve growth factor: Biochemistry, synthesis, and mechanism of action. A. Rev. Neurosci. 3: 353–402.

Gundersen, R. W., and Barrett, J. N. (1980) Characterization of the turning response of dorsal root neurites toward nerve growth factor. J. Cell Biol. 87: 546–554.

Hamburger, V., and Oppenheim, R. W. (1982) Naturally occurring neuronal death in vertebrates. Neurosci. Com. 1: 55–68.

Hattan, M. E., and Liem, R. H. K. (1981) Astroglial cells provide a template for the positioning of developing cerebellar neurons in vitro. J. Cell Biol. 90: 622–630.

Hefti, F. (1986) Nerve growth factor (NGF) promotes survival of septal cholinergic neurons after fimbrial transection. J. Neurosci. 6: 2155–2162.

Hefti, F., Dravid, A., and Hartikka, J. J. (1984) Chronic intraventricular injections of nerve growth factor elevate hippocampal choline acetyltransferase activity in adult rats with partial septohippocampal lesions. J. Brain Res. 293: 305–311.

Hefti, F., Hartikka, J. J., Echenstein, F., Gnahn, H., Heumann, R., and Schwab, M. (1985) Nerve growth factor increases choline acetyltransferase but not survival or fiber outgrowth of cultured fetal septal cholinergic neurons. Neuroscience 14: 55–68.

Heumann, R., Korsching, S., and Thoenen, H. (1987) Changes of nerve growth factor synthesis in nonneuronal cells in responses to sciatic nerve transection. J. Cell Biol. 104: 1623–1631.

Honegger, P., and Lenoir, D. (1982) Nerve growth factor (NGF) stimulation of cholinergic telencephalic neurons in aggregating cell cultures. Dev. Brain Res. 3: 229–238.

Kromer, L. F., Björklund, A., and Stenevi, U. (1981) Regeneration of the septohippocampal pathways in adult rats is promoted by utilizing embryonic hippocampal implants was bridges. Brain Res. 210: 173–200.

Kromer, L. F. (1987) Nerve growth factor treatment after brain injury prevents neuronal death. Science 235: 214–216.

Kromer, L. R., and Cornbrooks, C. (1984) Laminin and a Schwann cell surface antigen present within transplants of cultured CNS cells co-localize with CNS axons regenerating in vivo. Soc. Neurosci. Abstr. 10: 1084.

Landmeser, L., and Pilar, G. (1978) Interactions between neurons and their targets during in vivo synaptogenesis. Fedn Proc. 37: 2016–2021.

Lewis, P. R., Shute, C. C. D., and Silver, A. (1967) Confirmation from choline-acetylase of a massive cholinergic innervation to the rat hippocampus. J. Physiol. 191: 215–224.

Liesi, P., Kaakkola, S., Dahl, D., and Vaheri, A. (1984) Laminin is induced in astrocytes of adult brain injury. EMBO J. 683–686.

Lindholm, D., Heumann, R., Leyer, M., and Thoenen, H. (1987) Interleukin-1 regulates synthesis of nerve growth factor in non-neuronal cells of rat sciatic nerve. Nature 330: 658–660.

LIndsay, R. M. (1979) Adult rat brain astrocytes support survival of both NGF-dependent and NGF-insensitive neurons. Nature 282: 80–82.

Loy, R., and Moore, R. Y. (1977) Anomalous innervation of the hippocampal formation by peripheral synpathetic axons following mechanical injury. Exp. Neurol. 57 (2): 645–650.

Manthorpe, M., Nieto-Sampedro, M., Skaper, S. D., Barbin, G., Longo, F. M., Lewis, E. R., Cotman, C. W., and Varon, S. (1983) Neurotrophic activity in brain wounds of the development rat. Correlation with implant survival in the wound cavity. Brain Res. 267: 47–56.

Martinez, H. J., Dreyfus, C. F., Jonakait, G. M., and Black, I. B. (1985) Nerve growth factor promotes cholinergic development in brain striatal cultures. PNAS 82: 7777–7781.

Mobley, W. C., Rutkowski, J. L., Tennekoon, G. I., Buchanan, K., and Johnston, M. W. (1985) Choline acetyltransferase activity in striatum of neonatal rats increased by nerve growth factor. Science 229: 284–287.

Montero, C. N., and Hefti, F. (1988) Rescue of lesioned septal cholinergic neurons by nerve growth factor: Specificity and requirement for chonic treatment. J. Neurosci. 8: 2986–2999.

Nieto-Sampedro, M., Manthorpe, M., Barbin, G., Varon, S., and Cotman, C. W. (1983) Injury-induced neuronotrophic activity in adult rat brain: Correlation with survival delayed implants in the wound cavity. J. Neurosci. 3: 2219–2229.

Panula, P., Revuelta, A. V., Cheney, D. L., Wu, J.-Y., and Costa, E. (1984) An immunohistochemical study in the location of GABAergic neurons in rat septum. J. comp. Neurol. 222: 69–80.

Pearson, R. C. A., Gatter, K. C., and Powell, T. P. S. (1983) Retrograde cell degeneration in the basal nucleus of monkey and man. Brain Res. 261: 321–326.

Richardson, P. M., Verge Isse, V. M. K., and Riopelle, R. J. (1986) Distribution of neuronal receptors for nerve growth factor in the rat. J. Neurosci. 6: 2312–2321.

Rudge, J. S., Manthorpe, M., and Varon, S. (1985) The output of neuronotrophic and neurite promoting agents from rat brain astroglial cells: A microculture method for screening potential regulatory molecules. Brain Res. 19: 161–172.

Schwab, M. E., Otten, U., Agid, Y., and Thoenen, H. (1979) Nerve growth factor (NGF) in the rat CNS: Absence of specific retrograde axonal transport and tyrosine hydroxylase induction in locus coeruleus and substantia nigra. Brain Res. 168: 473–483.

Scott, S. M., Tarris, R., Eveleth, D., Mansfield, H., Wiechsel, M. E., and Fischer, D. A. (1918) Bioassay detection of mouse nerve growth factor (mNGF) in the brain of adult mice. J. Neurosci. Res. 6: 653–658.

Seiler, M., and Schwab, M. E. (1984) Specific retrograde transport of nerve growth factor (NGF) from cortex to nucleus basalis in the rat. Brain Res. 300: 33–39.

Sheldon, D. L., and Reichardt, L. F. (1986) Studies on the expression of the beta-nerve growth factor (NGF) gene in the central nervous system; level and regional distribution of NGF and mRNA suggest that NGF functions as a trophic factor for several distinct populations of neurons. PNAS 83: 2714–2718.

Stenevi, U., and Björklund, A. (1978) Growth of vascular sympathetic axons into the hippocampus after lesions of the septohippocampal pathway: A pitfall in brain lesion studies. Neurosci. Lett. 7: 219–224.

Storm-Mathisen, J. (1974) Choline acetyltransferase and acetylcholinesterase in fascia dentata following lesions of the entorhinal afferent. J. Brain Res. 80: 119–181.

Taniuchi, M., and Johnson, E. M. (1985) Characterization of the binding properties and retrograde axonal transport of monoclonal antibody directed against the rat nerve growth factor receptor. J. Cell Biol. 101: 1100–1106.

Taniuchi, M., Schweizer, J. B., and Johnson, E. M. (1986) Nerve growth factor receptor molecules in rat brain. Proc. natl Acad. Sci. USA 83: 1950–1954.

Theonen, H., and Barde, Y. A. (1980) Physiology of nerve growth factor. Physiol. Rev. 60: 1284–1335.

Ullrich, A., Gray, A., Berman, C., and Dull, T. J. (1983) Human beta-nerve growth factor gene sequence highly homologous to that of mouse. Nature 303: 821–825.

Tarris, R. H., Wieschsel, M. E. Jr, and Fisher, D. A. (1986) Synthesis and secretion of a nerve growth-stimulating factor by neonatal mouse astrocyte cells in vitro. Ped. Res. 20: 367–372.

274

Tuszynski, M. H., Buzsàki, G., Stearns, G., and Gage, F. H., (1988) Septal cell death following fimbria/fornix transection, and hippocampal cholinergic regeneration following nerve growth factor infusion plus grafting of synthetic and neuronal bridges. Soc. Neurosci. Abstr.

Vijayan, V. K. (1983) Lysosomal enzyme changes in young and aged control and entorhinal-lesioned rats. Neurobiol. Aging 4: 13–23.

Wainer, B. H., Levey, A. I., Rye, D. B., Mesulam, M., and Mufson, E. J. (1985) Cholinergic and non-cholinergic septohippocampal pathways. Neurosci. Lett. 54: 45–52.

Whitehouse, P. J., Price, D. L., Struble, R. G., Clark, A. W., Coyle, J. T., and Delong, M. R. (1982) Alzheimer's disease and senile dementia: Loss of neurons in the basal forebrain. Science 215: 1237–1239.

Williams, L. R., Varon, S., Peterson, G. M., Wictorin, K., Fisher, W., Björklund, A., and Gage, F. H. (1986) Continuous infusion of nerve growth factor prevents basal forebrain neuronal death after fimbria-fornix transection. Proc. natl Acad. Sci. USA 83: 9231–9235.

Restoration of cholinergic circuitry in the hippocampus by foetal grafts

D. J. Clarke and A. Björklund*

Department of Pharmacology, South Parks Road, Oxford OX1 3QT, England, and
*Department of Medical Cell Research, Lund, Sweden

Summary. The pathway from medial septum to hippocampus is one of the major and most well-documented cholinergic connections in the rodent brain. Interruption of this pathway by either direct destruction of the cells of origin in the medial septum or by transection of the fimbria-fornix, the fibre tract along which the septohippocampal axons traverse, results in a virtually complete depletion of cholinergic markers within the hippocampal formation. Previous experiments have shown that grafts of foetal rat septal-diagonal band region placed into the denervated hippocampus can restore acetylcholinesterase (AChE) fibre density to 85–90% of control values (Björklund et al., Acta physiol. scand. Suppl. 522 (1983) 49–58).

More recently, it has been demonstrated using the more specific technique of choline acetyltransferase (ChAT) immunocytochemistry in combination with electron microscopy that septal grafts are also able to restore the cholinergic connectivity at the synaptic level in the dorsal hippocampal formation. However, we have demonstrated that this restoration of both AChE and ChAT fibre density represents a specific mechanism and that the source of the foetal cholinergic neurons is crucial to the extent of reinnervation and pattern of connectivity achieved.

In aged rats, judged as being behaviourally impaired with respect to their spatial memory, there appears to be an intrinsic denervation of the septohippocampal pathway such that the hippocampus is depleted of cholinergic markers. In these cases, transplantation can again restore cholinergic innervation but without the requirement of prior denervation by a fimbria-fornix transection—grafts are placed into the intact hippocampus. Results show that the grafts survive well in the aged, intact hippocampus and are able to ameliorate the behavioural impairments, perhaps by the formation of substantial numbers of cholinergic synapses between the graft and host brain.

In conclusion, therefore, neural grafting of cholinergic neurons of appropriate type and origin is able to reinnervate the hippocampal formation previously denervated either by mechanical transection of the fimbria-fornix or as a result of an age-dependent deterioration.

Introduction

The neuronal pathway from the medial septum to the hippocampal formation has attracted much interest over recent years due to its involvement in long-term potentiation (Robinson and Racine, 1982) and behavioural processes such as learning and memory (Brito et al., 1983; Meck et al., 1984). Mapping of this connection using anterograde tracing techniques (Chandler and Crutcher, 1983; Meibach and Segal, 1977) has revealed a very precise pattern of termination within both the dentate gyrus and the hippocampus itself, such that the fibres appear to form laminae adjacent to the granule and pyramidal cell layers and in the outer third of the molecular layer.

The use of acetylcholine as a transmitter in at least a proportion of this pathway (Wainer et al., 1985) is now well documented (Fibiger, 1982; Mellgren and Srebro, 1973). When the hippocampal formation is stained with markers for the cholinergic system, e.g. acetyl cholinesterase (AChE) (Bakst and Amaral, 1984; Storm-Mathisen, 1970) or muscarinic binding site agonists (Wheal and Miller, 1980), the same laminar distribution of afferents is observed as in the anterograde tracing studies. With the advent of more specific and sensitive immuno-cytochemical techniques, antibodies to the synthetic enzyme of acetyl-choline, choline acetyltransferase (ChAT) have been used to further study the cholinergic termination of the septohippocampal pathway at both the light and electron microscopical levels (Clarke, 1985; Frotscher and Léranth, 1985).

The cholinergic system has also aroused much interest in the field of neural transplantation. Interruption of the septohippocampal choliner-gic system in rats as a result of either experimental manipulation or by an age-related degenerative process results in behavioural impair-ment with respect to learning and memory (Nilsson et al., 1987; Lippa et al., 1980). Several studies using grafts of cholinergic-rich tissue taken from the developing septal-diagonal band region of rat foetuses have demonstrated a behavioural recovery of memory impairments in either young (Dunnett et al., 1982, 1985; Low et al., 1982; Pallage et al., 1986: Segal et al., 1987; Nilsson et al., 1987; Arendt et al., 1988) or aged rats (Gage et al., 1984; Gage and Björklund, 1986). These studies also demonstrated that, at the light microscopical level, the grafts were extending AChE-positive fibres into the host hippocampus, thus indicating that the grafts could be making connections with the host brain which may, in part at least, be responsible for the behavioural ameliorations.

Thus it has been ascertained that grafts of foetal septum can amelio-rate the behavioural impairments resulting from a reduced cholinergic input to the hippocampus, but it is not yet known if any cholinergic neurons, regardless of their origin, would serve the same purpose. The literature, with respect to other transmitter-specific systems, seems somewhat contradictory. Foster et al. (1985, 1987) working with the serotonin system and Kromer (1983) with respect to the cholinergic system have suggested that the origin of the foetal tissue is important to the specificity of growth and connectivity of transplant to host when placed in correct or incorrect target zones. However, Lewis and Cotman (1983) and Gibbs et al. (1986) provided evidence to support the oppo-site view. In a recent paper (Nilsson et al., 1988) we began a systematic study to examine this question in the hippocampus using AChE histo-chemistry.

The aims of the studies presented in this chapter were firstly, to ascertain whether cholinergic grafts exert their effects on the animal's behaviour by forming new functional synaptic connections with the host hippocampus; secondly, to ascertain to what extent these novel contacts resemble those found in control animals and finally, to see if the origin of the foetal cholinergic neurones played any role in the extent of reinnervation and synapse formation seen in the host brain.

Cholinergic grafts in the hippocampus

In all of the grafting experiments presented in this chapter, the suspension graft procedure of Björklund et al. (1983) was used. For septal grafts, the tissue of the developing septum and diagonal band region was dissected from E14–E16 rat foetuses (crown-to-rump length 12–16 mm) and injected at two sites in the dorsal hippocampal formation. In the experiments using different types of cholinergic graft, developing cholinergic neurones were taken from the developing nucleus basalis magnocellularis (NBM) of E15 embryos, the pontomesencephalic tegmentum of the brain stem and the striatal primordium of E14 embryos and the spinal cord from E13–E14 embryos and compared with the innervation properties of septal grafts.

It has been demonstrated that there is enhanced graft survival if the grafts are placed into a lesioned host brain (Gage et al., 1986b). In these experiments, the animals with septal grafts or cholinergic grafts from a variety of sites received aspirative lesions of the fimbria-fornix in the same surgical session as the grafting. The aged animals did not receive any lesion prior to grafting but were lesioned only 3 days prior to perfusion. These aged rats were assessed as being behaviourally impaired as a result of their performance in the Morris Water Maze. Following transplantation surgery, all groups of animals were allowed to survive 14–17 weeks prior to perfusion and histochemical analysis with either AChE histochemistry or ChAT immunocytochemistry.

The results from each group of animals will be discussed separately.

Normal cholinergic circuitry in hippocampus

With the advent of the sensitive immunocytochemical techniques using the ChAT antibodies, it has been possible to determine the precise pattern of termination of the cholinergic septohippocampal pathway (Clarke, 1985) and it can be examined at both the light and electron microscopical levels. These studies supported the earlier experiments using AChE histochemistry, such that the bulk of the ChAT-positive fibres formed laminar patterns at the interface of the granule cell layer and molecular layer in the dentate gyrus (Fig. 1A) and adjacent to the pyramidal cell layer in the hippocampus. The cholinergic fibres are fine

278

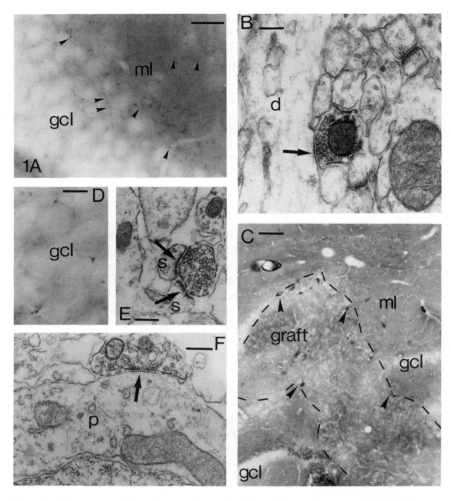

Figure 1. *A* Low power light micrograph showing the position and appearance of ChAT-positive fibres (arrowheads) in the dentate gyrus in normal rats. These fibres form a dense supra-granular band adjacent to the granule cell layer (gcl) at the base of the molecular layer (ml). *B* Electron micrograph showing the appearance of a ChAT-immunoreactive bouton in normal rat dentate gyrus. The bouton is forming a symmetrical synaptic contact (arrow) with a dendritic shaft (d). *C* Light micrograph showing the position and extent of a septal suspension graft within the dentate gyrus. The granule cell layer (gcl) is broken by the graft. Numerous ChAT-positive neurons (arrowheads) are seen in the graft, often aligned at the graft-host interface (broken lines). ml = Molecular layer. *D* Light micrograph of the position and appearance of ChAT-positive fibres derived from the septal graft. These fibres (arrowheads) are intermingled within the granule cell layer (gcl) itself and do not form a supra-granular band as in normal animals (see 1*A*). *E* Electron micrograph of a ChAT-positive bouton of graft origin forming synaptic specializations (arrows) with two spine heads (s). *F* Electron micrograph of a ChAT-positive bouton of graft origin forming a symmetrical synapse (arrow) onto the perikaryon (p) of a granule cell within the granule cell layer. Scales: *A*, 25 μm; *B*, *E* and *F*, 0.25 μm; *C*, 75 μm; *D*, 15 μm.

Table 1. Percentage distribution of postsynaptic targets of ChAT-positive boutons in control, young grafted, and aged grafted hippocampus

Postsynaptic target	Control		Young grafted		Aged grafted	
	N	%	N	%	N	%
Dendritic shafts	51	64.6	31	23.7	85	64.9
Dendritic spines	17	21.5	12	9.1	28	21.4
Small dendrites or spines	5	6.3	—	—	11	8.4
Perikarya	4	5.1	88	67.2	7	5.3
Non-identified	2	2.5	—	—	—	—
Total	79	100.0	131	100.0	131	100.0

calibre and possess numerous varicosities along their length. Very few varicosities were found within the cell layers themselves but appeared associated with the dendrites of the granule cells and both the apical and basal dendrites of the pyramidal cells. A few cholinergic interneurons were also found in stratum oriens of hippocampus (Frotscher and Leranth, 1985) and in the hilar region just below the granule cell layer.

At the electron microscopical level, the varicosities seen in the light microscope could be identified as terminal boutons by their electron dense appearance, due to the deposition of reaction product around the synaptic vesicles (Fig. 1B). These boutons were generally small (0.4–0.7 μm diameter) and formed symmetrical synaptic specializations. The predominant post-synaptic target of these ChAT-immunoreactive boutons was dendritic shafts (65%; see Table 1) and only a very small percentage (5%) were in contact with neuronal perikarya.

Young, denervated grafted animals

Grafts were readily visible in these animals, usually found within the hippocampal formation and occasionally extending into the choroidal fissure (Fig. 1C). Numerous intensely ChAT-immunoreactive perikarya were seen, aligned predominantly around the perimeters of the graft at the interface of the graft and host tissue. Within the grafts, also, a dense plexus of fine, ChAT-positive fibres could be visualized. It was often possible to trace these fibres from the graft into the surrounding host neuropil where they appeared to terminate predominantly within the granule cell layer of the dentate gyrus (Fig. 1D). No supragranular band of fibre staining was apparent as in the control animals.

At the EM level, the anomalous distribution was also seen. The majority of synaptic specializations were made onto neuronal perikarya (Fig. 1F) within the granule cell layer and not onto their dendrites (see

Table 1). Although the majority of these contacts were of the symmetrical type, a considerable number of ChAT-immunoreactive boutons formed asymmetric synapses with, especially, dendritic spines (Fig. 1E), but also, to some extent with dendritic shafts and neuronal perikarya. Thus, new contacts were made between the graft and host tissue but these new connections were somewhat anomalous in both their post-synaptic target and pattern of termination, as compared to control animals (Clarke et al., 1986a; Anderson et al., 1986).

The possible reason for the change in post-synaptic target of these graft-derived cholinergic axons are twofold. Firstly, the findings may reflect an ontogenetic effect since in other cortical areas, a greater percentage of the synaptic contacts in immature animals are axo-somatic (Mates and Lund, 1983) and it is believed that these additional contacts may be eliminated or relocated during development (Purves and Lichtman, 1980). The alternative hypothesis is that the ingrowing cholinergic axons from the graft are occupying vacant synaptic sites which are due to the removal of both cholinergic and non-cholinergic septohippocampal afferents (Baisden et al., 1984; Vincent and McGeer, 1981) by the large fimbria-fornix transection.

Aged, impaired grafted animals

These animals were previously assessed as being behaviourally impaired with respect to their spatial memory in the Morris Water Maze. The animals did not receive a fimbria-fornix lesion prior to grafting, but only 3 days prior to perfusion to remove any remaining intrinsic cholinergic inputs to the hippocampus. This ensured that any cholinergic innervation in hippocampus was probably of graft origin (Clarke et al., 1986b).

At the light microscopical level, grafts were again readily visible and contained numerous ChAT-positive neurones and fibres. The fibres were traced from the graft and seen to ramify in the supragranular band region of the dentate gyrus, adjacent to the granule cell layer. This pattern of termination was similar to that observed in the young, control animals. At the EM level, the results again matched those of the young, control animals and were markedly dissimilar from those of the young, grafted animals, (see Table 1), in that the predominant post-synaptic target was dendritic shafts. Very few contacts onto the granule cell bodies were observed.

These results suggest, therefore, that in these aged animals which did not receive a prior denervating lesion the cholinergic innervation of the hippocampus from the septal graft is 'normal'. This may be caused by an intrinsic age-related degeneration of the septohippocampal cholinergic pathway and a filling of the vacated synaptic sites by new cholinergic axons. Thus a more precise and selective synaptogenesis may be occurring in these aged, behaviourally impaired animals and the connections formed be more functionally appropriate than those anomalous ones formed in the young, grafted animals with prior fimbria-fornix denervation.

These findings are supported to some extent by the literature. It has been reported that in the dentate gyrus of senescent animals there is a 35% decrease in synaptic density (Geinisman et al., 1978; Bondareff, 1980), which may thus reflect an age-related denervation of the hippocampus. However, the old brain still remains plastic enough to accept and incorporate ingrowing axons from the graft which may thus fulfil a necessary role in reinnervating a partially denervated target and thus restore the animal's behavioural ability to learn and remember.

Specificity of reinnervation from cholinergic grafts of diverse origin

In the previous experiments, it has been demonstrated that cholinergic grafts are able to ameliorate behavioural impairments and form new synaptic connections within the host hippocampus. The aims of the present experiments (Nilsson et al., 1988) were to ascertain if the source of the cholinergic neurons was important to the extent of reinnervation achieved in the hippocampus (as assessed by AChE histochemistry) and whether the hippocampus would incorporate all types of ingrowing cholinergic axon into its circuitry, or would there be a more selective mechanism involved.

At the light microscopical level, AChE histochemistry allowed good visualization of the grafts and the extent of fibre innervation. Graft size varied substantially; the largest being spinal cord grafts and the smallest those of striatal origin. NBM grafts were also relatively large whilst septal and brain stem grafts were of approximately equal size and lay in the middle of the size range. However, the extent of reinnervation was even more dramatic (see Fig. 2). Only septal grafts were shown to achieve 'normal' values of reinnervation, with NBM and brain stem grafts achieving approximately half the innervation scores of the contralateral intact side and striatum and spinal cord grafts not reinnervating the host hippocampus to any real extent. Each graft type seemed also to adopt its own pattern of reinnervation, irrespective of its extent of fibre ingrowth into the hippocampus. The septal grafts (as was discussed in the young, grafted section) hyperinnervated the cell body layers but otherwise mimicked the fibre distribution seen in control animals. NBM grafts appeared to preferentially innervate the hilar region of the dentate gyrus; brain stem grafts, the molecular layer of CA3, as did striatal grafts to a limited extent, and spinal cord grafts, the CA2 region of hippocampus (unpublished observation).

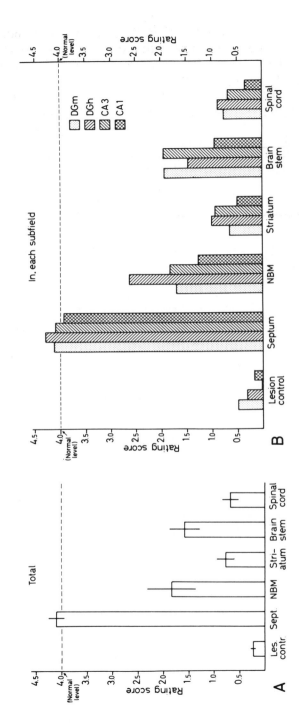

Figure 2. Extent of AChE-positive innervation in the hippocampus of the lesion-control animals together with those receiving the 5 types of cholinergic grafts. Rating scores along the y axis are obtained as a mean of 5 standardized sections extending throughout the dorsal hippocampus. 0 = no AChE-positive fibres; 1 = sparse; 2 = moderate; 3 = near normal; 4 = normal; 5 = hyperinnervation. *A* Total amount of AChE innervation (all subfields pooled). Vertical bars are +/- SEM. *B* Extent of reinnervation in different subfields of the hippocampus. DGm = dentate gyrus, molecular layer; DGh = dentate gyrus, hilus; CA3 = layer CA3 off hippocampus; CA1 = layer CA1 of hippocampus.

Thus, the light microscopical results suggest that there is, indeed, a distinct specificity of cholinergic innervation of the hippocampus from cholinergic grafts. Only the grafts taken from the septal region, the indigenous origin of the hippocampal cholinergic input, were able to reinnervate the denervated hippocampus to any extent resembling the normal cholinergic input. NBM grafts were probably the next most successful graft type, perhaps due to the intermingling of the cortically projecting basal forebrain cholinergic systems to which the septo-hippocampal and NBM-cortical pathways belong. Dunnett et al. (1986) reported similar findings in that septal graft provided the most extensive AChE reinnervation of hippocampus while, conversely, NBM grafts acted preferentially in neocortex; in both cases the appropriate graft type provides the most complete reinnervation. Similar results have been demonstrated by Foster et al. (1985, 1987) using medullary and mesen-cephalic raphé tissue grafted into both their appropriate and inappropri-ate target regions.

These studies have now been extended to the electron microscopical level using ChAT-immunocytochemistry (paper in preparation). The preliminary results suggest that some synaptic specializations are seen in the hippocampus from all graft types but that only septal grafts and to a limited extent NBM grafts, form any significant number of contacts with host hippocampal neuronal elements. The distribution of post-synaptic targets of the ChAT-immunoreactive boutons from septal grafts appears virtually identical as previously described, although fibres from NBM grafts appear to preferentially contact dendritic spines within the hilar region. From all other graft types, the extent of synaptic connectivity is minimal.

Discussion

Neural transplantation of foetal tissue of appropriate transmitter type and origin is thus providing a means of repairing a damaged central nervous system. Similar results have been described in other transmitter systems e.g. dopaminergic (Freund et al., 1985; Mahalik et al., 1985; Nishino et al., 1986; Clarke et al., 1988a), serotoninergic (Beebe et al., 1979; Anderson et al., 1986) and noradrenergic systems (Björklund et al., 1979). In all cases, the plasticity of the adult brain has clearly been demonstrated. The hippocampus is well known as being a plastic structure which undergoes synaptogenesis and collateral sprouting in response to partial deafferentation (Gasser and Dravid, 1987; Cotman and Nadler, 1978; Gage et al., 1983; Hoff et al., 1982). It was previously believed that synaptogenesis occurred only during development and that no further synapses could be made or accepted into the fully-developed neuronal circuitry of the adult brain. However, in the light of the results showing compensatory collateral sprouting and those following trans-plantation of foetal tissue in the hippocampus, new ideas concerning central nervous system plasticity are emerging. Despite the ability of the cholinergic system to sprout following lesions, it has been conclusively demonstrated that the cholinergic neurons of the hippocampus cannot compensate for the loss of the septohippocampal system (Frotscher, 1987). It seems that the adult brain is indeed capable of accepting new synapses into its already-formed pattern of connectivity and, in some cases (Clarke et al., 1988a, b; Sotelo and Alvarado-Mallart, 1987; Peschanski and Isacson, 1987; Albert and Das, 1984; Matsumoto et al., 1985) can itself form synaptic contacts with newly-introduced neuronal elements. This ability of the brain to incorporate new circuitry also extends to the senescent brain as demonstrated by the results presented in this chapter. These novel synaptic specializations also appear to be functional connections as assessed by the electrophysiological data from the grafting of septal tissue to the adult hippocampus (Segal et al., 1985).

The results also demonstrate that to enable complete reinnervation of the cholinergic system to the denervated hippocampus, it is important to have grafted cholinergic neurones from the appropriate origin—the medial septum. The ability of the host brain to differentiate between cholinergic neurones of different origins may be dependent on trophic factors being released from the denervated target area which attract the ingrowing cholinergic axons. The different types of cholinergic fibre may react to a greater or lesser extent to these trophic factors. Recognition of their appropriate targets within the hippocampal laminae are proba-bly also reliant on some sort of trophic interaction. It is still unclear, however, as to what precise mechanisms are involved in guiding the ingrowing axons to their targets, although nerve growth factor (NGF) has recently been implicated in playing an important role in the cholin-ergic septohippocampal pathway (Hefti et al., 1985; Fischer et al., 1988). It is apparent from these experiments that foetal grafts are acting as more than 'mini-pumps' supplying transmitter to the target since they are able to produce a precise point-to-point pattern of reconnectivity at the synaptic level; connections which are functional electrophysiologi-cally. There is a high degree of specificity as to the origin of the cholinergic tissue grafted, suggesting that properties of neurones beyond the transmitter type are essential for optimal performance in grafting of cholinergic grafting experiments.

Acknowledgements. I would like to thank my collaborators Ola Nilsson and Fred H. Gage without whose help this work would not have been possible. Special thanks go to Laura Packwood and Jill Lloyd for expert technical and typographical assistance.

Albert, E. N., and Das, D. C. (1984) Neocortical transplant in the rat brain: An ultrastruc-tural study. Experientia 40: 294–298.

Anderson, J. K., Gibbs, R. B., Salvaterra, P. M., and Cotman, C. W. (1986) Ultrastructural characterisation of identified cholinergic neurons transplanted into the hippocampal formation of the rat. J. comp. Neurol. 249: 279–292.

Arendt, T., Allen, Y., Sinden, J., Schugens, M. M., Marchbanks, R. M., Lantos, P. L., and Gray, J. A. (1988) Cholinergic-rich brain transplants reverse alcohol-induced memory deficits. Nature 323: 448–450.

Baisden, R. H., Woodruff, M. L., and Hoover, D. B. (1984) Cholinergic and non-cholinergic septo-hippocampal projections: a double-label horseradish peroxidase-acetylcholinesterase study in the rabbit. Brain Res. 290: 146–151.

Bakst, I., and Amaral, D. G. (1984) The distribution of acetylcholinesterase in the hippocampal formation of the monkey. J. comp. Neurol. 225: 344–371.

Beebe, B. K., Møllgård, A., Björklund, A., and Stenevi, U. (1979) Ultrastructural evidence of synaptogenesis in the adult rat dentate gyrus from brainstem implants. Brain Res. 167: 391–395.

Björklund, A., Segal, M., and Stenevi, U. (1979) Functional reinnervation of rat hippocampus by locus coeruleus implants. Brain Res. 170: 409–426.

Björklund, A., Stenevi, U., Schmidt, R. H., Dunnett, S. B., and Gage, F. H. (1983a) Intracerebral grafting of neuronal cell suspensions. I. Introduction and general methods of preparation. Acta physiol. scand. Suppl. 522: 1–8.

Bondareff, W. (1980) Changes in synaptic structure affecting neural transmissions in the senescent brain. In: Oota, K., and Makinodan, M. (eds), Proceedings of the Naito International Symposium of Aging. Raven Press, New York.

Brito, G. N. O., Davis, B. J., Stopp, L. C., and Stanton, M. E. (1983) Memory and the septohippocampal cholinergic system in the rat. Psychopharmacology 81: 315–320.

Chandler, J. P., and Crutcher, K. A. (1983) The septohippocampal projection in the rat: an electron microscopic horseradish peroxidase study. Neuroscience 10: 685–696.

Clarke, D. J. (1985) Cholinergic innervation of the rat dentate gyrus: A light and electron microscopial study using a monoclonal antibody to choline acetyltransferase. Brain Res. 360: 349–354.

Clarke, D. J., Gage, F. H., and Björklund, A. (1986a) Formation of cholinergic synapses by intrahippocampal septal grafts as revealed by choline acetyltransferase immunocytochemistry. Brain Res. 360: 151–162.

Clarke, D. J., Gage, F. H., Nilsson, O. G., and Björklund, A. (1986b) Grafted septal neurones form cholinergic synaptic connections in the dentate gyrus of behaviourally impaired aged rats. J. comp. Neurol. 252: 483–492.

Clarke, D. J., Brundin, P., Strecker, R. E., Nilsson, O. G., Björklund, A., and Lindvall, O. (1988a) Human fetal dopamine neurons grafted in a rat model of Parkinson's Disease: Ultrastructural evidence for synapse formation using tyrosine hydroxylase immunocytochemistry. Exp. Brain Res. 73: 115–126.

Clarke, D. J., Dunnett, S. B., Isacson, O., Sirinathsinghji, D. J. S., and Björklund, A. (1988) Striatal grafts in rats with unilateral neostriatal lesions—I. Ultrastructural evidence of afferent synaptic inputs from the host nigrostriatal pathway. Neuroscience 24: 791–801.

Cotman, C. W., and Nadler, J. V. (1978) Reactive synaptogenesis in the hippocampus. In: Cotman, C. W. (ed.), Neuronal Plasticity. Raven Press, New York, pp. 227–271.

Dunnett, S. B., Low, W. C., Iversen, S. D., Steveni, U., and Björklund, (1982) Septal transplants restore maze learning in rats with fornix-fimbria lesions. Brain Res. 251: 335–348.

Dunnett, S. B., Toniolo, G., Fine, A., Ryan, C. N., Björklund, A., and Iverson, S. D. (1985) Transplantation of embryonic ventral forebrain neurones to the neocortex of rats with lesions of nucleus basalis magnocellularis. II. Sensorimotor and memory effects. Neuroscience 16: 787–797.

Dunnett, S. B., Whishaw, I. Q., Bunch, S. T., and Fine A. (1986) Acetylcholine-rich neuronal grafts in the forebrain of rats: Effects of environmental enrichment, neonatal noradrenaline depletion, host transplantation site, and regional source of embryonic donor cells on graft size and acetylcholinesterase-positive fibre outgrowth. Brain Res. 378: 357–373.

Fibiger, H. C. (1982) The organization and some projections of cholinergic neurons of the mammalian forebrain. Brain Res. Rev. 4: 327–388.

Fischer, W., Wictorin, K., Björklund, A., Williams, L. R., Varon, S., and Gage, F. H. (1987) Amelioration of cholinergic neuron atrophy and spatial memory impairment in aged rats by nerve growth factor. Nature 329: 65–68.

Foster, G. A., Schultzberg, M., Björklund, A., Gage, F. H., and Hökfelt, T. (1985) Fate of embryonic mesencephalic and medullary raphé neurons transplanted to the striatum, hippocampus or spinal cord of the adult rat: Analysis of 5-hydroxytryptamine, substance-P and thyrotropine-releasing hormone-immunoreactive cells. In: Björklund, A., and Stenevi, U. (eds), Neural Grafting in the Mammalian CNS. Elsevier, Amsterdam, pp. 179–190.

Foster, G. A., Schultzberg, M., Gage, F. H., Björklund, A., Hökfelt, T., Cuello, A. C., Verhofstad, A. A. J., and Visser, T. J. (1987) Transmitter expression and morphological development of embryonic medullary and mesencephalic raphe neurons after transplantation to the adult rat central nervous system. II. Grafts to the striatum. Exp. Brain Res. 60: 427–444.

Freund, T. F., Bolam, J. P., Björklund, A., Stenevi, U., Dunnett, S. B., Powell, J. F., and Smith, A. D. (1985) Efferent synaptic connections of grafted dopaminergic neurons reinnervating the host neostriatum: A tyrosine hydroxylase immunocytochemical study. J. Neurosci. 5: 603–616.

Frotscher, M. (1987) Cholinergic neurons in the rat hippocampus do not compensate for the loss of septohippocampal cholinergic fibers. Neurosci. Lett. 87: 18–22.

Frotscher, M., Léranth, Cs. (1985) Cholinergic innervation of the rat hippocampus as revealed by choline acetyltransferase immunocytochemistry. A combined light and electron microscopic study. J. comp. Neurol. 239: 237–246.

Gage, F. H., and Björklund, A. (1986a) Cholinergic septal grafts into the hippocampal formation improve spatial learning and memory in aged rats by an atropine sensitive mechanism. J. Neurosci. 17: 89–98.

Gage, F. H., and Björklund, A. (1986b) Enhanced graft survival in the hippocampus following selective denervation. Neuroscience 17: 89–98.

Gage, F. H., Björklund, A., Stenevi, U., Dunnett, S. B., and Kelly, P. A. T. (1984) Intrahippocampal septal grafts ameliorate learning impairments in aged rats. Science 225: 533–536.

Gage, F. H., Dunnett, S. B., Stenevi, U., and Björklund, A. (1983) Intracerebral grafting of neuronal cell suspensions. VIII. Survival and growth of implants of nigral and septal cell suspensions in intact brain of aged rats. Acta physiol. scand. (Suppl). 522: 67–75.

Gasser, U. E., and Dravid, A. R. (1987) Noradrenergic, serotonergic and cholinergic sprouting in the hippocampus that follow partial or complete transection of the septohippocampal pathway: Contributions of spared inputs. Exp. Neurol. 96: 352–364.

Geinisman, Y., Bondareff, W., and Dodge, J. T. (1978) Dendritic atrophy in the dentate gyrus of the senescent rat. Am. J. Anat. 152: 321–330.

Gibbs, R. B., Anderson, K., and Cotman, C. W. (1986) Factors affecting innervation in the CNS: Comparison of three cholinergic cell types transplanted to the hippocampus of adult rats. Brain Res. 383: 362–366.

Hefti, F., Hartikka, J., Eckenstein, F., Gnahn, H., Henmann R., and Schwab, M. (1985) Nerve growth factor increases choline acetyltransferase but not survival or fiber outgrowth of cultured fetal septal cholinergic neurons. Neuroscience 14: 55–68.

Hoff, S. F., Scheff, S. W., Bernardo, L. S., and Cotman, C. W. (1982) Lesion-induced synaptogenesis in the dentate gyrus of aged rats. I. Loss and reacquisition of normal synaptic denisty. J. comp. Neurol. 205: 246–252.

Kromer, L. F. (1983) Cholinergic transplants exhibit specificity in reinnervating the adult rodent hippocampus. Soc. Neurosci. Abstr. 9: 372

Lewis, E. R., and Cotman, C. W. (1983) Neurotransmitter characteristics of brain grafts: Striatal and septal tissues form the same laminated input to the hippocampus. Neuroscience 8: 57–66.

Lippa, A. S., Pelham, R. W., Beer, B., Critchett, D. J., Dean, R. L., and Bartus, R. T. (1980) Brain cholinergic dysfunction and memory in aged rats. Neurobiol. Aging 1: 13–19.

Low, W. C., Lewis, P. R., Bunch, S. T., Dunnett, S. B., Thomas, S. R., Iverson, S. D., Björklund, A., and Stenevi, U. (1982) Functional recovery following transplantation of embryonic septal nuclei into adult rats with septohippocampal lesions: The recovery of functions. Nature 300: 260–262.

Mahalik, T. J., Finger, T. E., Strömberg, I., and Olsen, L. (1985) Substantia nigra transplants into denervated striatum of the rat: Ultrastructure of graft and host interconnections. J. comp. Neurol. 240: 60–70.

Mates, S. L., and Lund, J. S. (1983) Developmental changes in the relationship between type 2 synapses and spiny neurons in the monkey visual cortex. J. comp. Neurol. 221: 98–105.

Matsumoto, A., Murakami, S., Arai, Y., and Osanai, M. (1985) Synaptogenesis in the neonatal preoptic area grafted into the aged brain. Brain Res. 347: 363–367.

Meck, W. H., Church, R. M., and Oton, D. S. (1984) Hippocampus, time, and memory. Behav. Neurosci. 98: 3–22.

Meibach, R. C., and Siegal, A. (1977) Efferent connections of the septal area in the rat: an analysis utilizing retrograde and anterograde transport methods. Brain Res. 119: 1–20.

Mellgren, S. I., and Srebro B. (1973) Changes in acetylcholinesterase and distribution of degenerating fibres in the hippocampal region after septal lesion in the rat. Brain Res. 52: 19–35.

Nilsson, O. G., Clarke, D. J., Brundin, P., and Björklund, A. (1988) Comparison of growth and reinnervation properties of cholinergic neurons from different brain regions grafted to the hippocampus. J. comp. Neurol. 268: 204–222.

Nilsson, O. G., Shapiro, M. L., Gage, F. H., Olton, D. A., and Björklund, A. (1987) Spatial learning and memory following fimbria-fornix transection and grafting of fetal septal neurons to the hippocampus. Exp. Brain Res. 67: 195–215.

Nishino, H., Ono, T., Takahashi, J., Kimura, M., Shiosaka, S., Yamasaki, H., Hatanaka, H., and Tohyama, M. (1986) The formation of new neuronal circuit between transplanted dopamine neurons and non-immunoreactive axon terminals in the host rat caudate nucleus. Neurosci. Lett. 64: 13–16.

Pallage, V. G., Toniolo, G., Will, B., and Hefti, F. (1986) Long-term effects of nerve growth factor and neural transplants on behavior of rats with medial septal lesions. Brain Res. 386: 197–208.

Peschanski, M., and Isacson, O. (1988) Fetal homotypic transplants in the excitotoxically neuron depleted thalamus. 2. Electron microscopy. J. comp. Neurol., in press.

Purves, D., and Lichtman, J. W. (1980) Elimination of synapses in the developing nervous system. Science 210: 153–157.

Robinson, G. B., and Racine, R. J. (1982) Heterosynaptic interactions between septal and entorhinal inputs to the dentate gyrus: long-term potentiation effects. Brain Res. 249: 162–166.

Segal, M., Björklund, A., and Gage, F. H. (1895) Transplanted septal neurons make viable cholinergic synapses with a host hippocampus. Brain Res. 336: 302–307.

Segal, M., Greenberger, V., and Milgram, H. W. (1987) A functional analysis of connections between grafted septal neurons and a host hippocampus. Prog. Brain Res. 71: 349–357.

Sotelo, C., and Alvarado-Mallart, R. M. (1987) Reconstruction of the defective cerebellar circuitry in adult pcd mutant mice by Purkinje cell replacement through transplantation of solid embryonic implants. Neuroscience 20: 1–22.

Storm-Mathisen, J. (1970) Quantitative histochemistry of acetylcholinesterase in rat hippocampal region correlates with histochemical staining. J. Neurochem. 17: 739–750.

Vincent, S. R., and McGeer, E. G. (1981) A substance P projection to the hippocampus. Brain Res. 215: 349–351.

Wainer, B. H., Levey, A. I., Rye, D. B., Mesulam, M. M., and Mufson, E. J. (1985) Cholinergic and non-cholinergic septohippocampal pathways. Neurosci. Lett. 54: 45–52.

Wheal, H. V., and Miller, J. J. (1980) Pharmacological identification of acetylcholine and glutamate excitatory systems in the dentate gyrus of the rat. Brain Res. 182: 145–155.

Effects of colchicine treatment on the cholinergic septohippocampal system

A. Represa and Y. Ben-Ari

INSERM U 29, 123 Bd. Port Royal, 75014 Paris, France

Summary. The influence of hippocampal target cells on the septohippocampal cholinergic system was studied using immunocytochemical and autoradiographic procedures. The destruction of dentate granular cells by colchicine injection promotes a significant increase in the density of acethylcholinesterase staining and cholinergic-muscarinic receptors in the zone denervated by the terminal axons of granular cells, which supports the hypothesis of a proliferation of cholinergic fibers in CA3. In the septal region the number of choline acetyl transferase positive cells was significantly lower (by 23%) ipsilaterally to the colchicine injection as compared to the contralateral side; when the hippocampus is almost completely destroyed by colchicine treatment this cell loss is more important (by over 50%). The present results agree with those of earlier studies and suggest that target-derived trophic factors are important for the maintenance of the basal forebrain cholinergic system and that the fascia dentata provides a significant source of such factors.

Introduction

Recent studies have suggested that target-derived trophic factors are important for the maintenance of the basal forebrain cholinergic projection system (Hsiang et al., 1987; Sofroniew et al., 1983, 1986) and that nerve growth factor (NGF) may be important in this trophic regulation (Whittemore et al., 1987; Heacock et al., 1986). Indeed it has been observed that transection of the fimbria-fornix, which contains the septohippocampal cholinergic fibers, results in retrograde degeneration of the parent cells bodies in the medial septum/vertical limb of diagonal band complex (MS/VLDB; Gage et al., 1986). This cell loss can be substantially attenuated by infusion of NGF into the lateral ventricle (Hefti, 1986; Kromer, 1987; Williams et al., 1986).

The major hippocampal target of the septal cholinergic fibers are the dentate granular cells, which are particularly enriched in mRNA for NGF (Ayer-LeLievre et al., 1988; Korsching et al., 1985). Therefore a selective lesion of this region should produce prominent changes in the septohippocampal cholinergic system. A local injection of the axonal transport and mitotic blocker agent colchicine produces a rather selective destruction of the granular neurons of fascia dentata (Goldschmidt and Stewart, 1980). Although the mechanism of this effect is not presently elucidated, it constitutes a useful tool to selectively destroy fascia dentata and to investigate the physiological and behavioral consequences of lesions. Using this procedure, Drust and Crawford (1985) have found an increase in acetylcholinesterase-staining in the zone denervated by the loss of mossy fiber (the lucidum layer of CA3), which suggests a compensatory mechanism.

The aim of the present study was 1) to confirm whether colchicine treatment induces a collateral sprouting of cholinergic fibers by a combined acetylcholinesterase (AChE) staining and autoradiographic study of muscarinic-cholinergic receptors; 2) to determine whether colchicine destruction of granular cells promotes a transynaptic loss of cholinergic neurons in medial septum/vertical limb of diagonal band complex by immunocytochemistry, using a monoclonal antibody against cholineacetyltransferase (ChAT).

Material and methods

Male wistar rats (180–200 g) were used throughout these experiments. They had access to food and water *ad libitum*, and were housed in individual cages under diurnal lighting conditions (from 08.00 to 20.00 h). For lesion experiments, animals were anesthetized with Equithesin (Jensen Salsbury, 3 ml/kg). Two injections/animal of colchicine (Sigma) were performed unilaterally with a microsyringe under stereotaxic guidance at the following coordinates according to the atlas of Albe Fessard et al. (1971); 5.4 mm anterior to lambda, 1.3 mm lateral to the sagittal suture and 7 mm below the interaural line ($1.8 \mu g/0.45 \mu l$ dosage/volume over 18 min) and 2.9 mm anterior, 3.3 mm lateral and 6.4 mm below ($2 \mu g/0.5 \mu l$, in 20 min). In a parallel study from our laboratory (Robain et al., 1989) this procedure was found to destroy over 85% of granular cells and the mossy fibers 8 days after the injection; furthermore, there was a good preservation of the GABAergic neurons of fascia dentata and CA3 pyramidal region.

For immunocytochemistry the animals were perfused under anesthesia with sodium pentobarbitone (20 mg/kg), with ice-cold normal saline followed by a fixative containing 3% paraformaldehyde and 0.01% glutaraldehyde in 0.05 M phosphate buffer (pH 7.4), then perfused with the same solution without glutaraldehyde and finally with a solution containing 10% sucrose in 0.1 M phosphate buffer. The brains were removed and immersed in 0.1 M phosphate buffer containing 30% sucrose for 1 or 2 days. Coronal sections (30 μm) were cut on a cryostat. The septal sections were stained by immunocytochemistry with a monoclonal antibody against ChAT from rat-mouse hybridoma (Boehringer Mannheim GmBH, FRG). The peroxidase-antiperoxidase technique was applied. Blanks were obtained incubating sections with nonimmune normal rat serum instead of the primary antibody. The hippocampal sections were stained with the AChE-staining method described by Shute and Lewis (1961); ethopropazine (30 μM) was used to inhibit the unspecific esterases. Additional sections were stained with cresyl violet.

The autoradiographic study of muscarinic-cholinergic receptors was performed with a previously described technique (Ben-Barak and Dudai, 1979) with minor modifications. Briefly, unfixed brain sections (20 μm) were preincubated for 30 min at 25°C in 50 mM Tris-HCl to remove endogenous competitive ligands. Then they were incubated at room temperature in the same buffer containing 1 nM [^{3}H]-QNB (NEN, 46 Ci/mmol) for 2 h. Additional sections were incubated in presence of 100 μM atropine sulphate to determine the non-specific binding. After rinsing and drying, the slices were put in an X-ray cassette and apposed to an [^{3}H]-sensitive film concomitantly with plastic standards (Amersham). Alternate sections were Nissl-stained or AChE-stained to compare the distribution of muscarinic binding sites and AChE-stained fibers in the similar sections.

Results and discussion

Cholinergic sprouting after colchicine treatment

One month after colchicine treatment there was an almost complete disappearance of dentate granular cells (by over 90%), as well as of Timm-stained mossy fibers. In contrast the number of pyramidal cells in the region CA3 remained unchanged. As shown in Figure 1B, the colchicine treatment induces a decrease of cholinesterase-stained fibers in the molecular layer of fascia dentata; this is in keeping with the atrophy and shrinkage of the fascia dentata. In contrast, in the CA3 area of Ammon's horn there was an enhancement of AChE elements. In agreement with the earlier study of Drust and Crawford (1985), we found that this increase is more conspicuous in the stratum lucidum of CA3, the region normally occupied by mossy fiber terminals which in control cases are poorly stained by this procedure (Fig. 1A).

The quantitative study of the autoradiographies obtained with ^{3}H-QNB reveal that the density of muscarinic binding sites significantly decreases by over 50% in the fascia dentata of colchicine-treated hippocampus as compared to the control hippocampus (Fig. 2). The present results therefore suggest that muscarinic binding sites are preferentially located postsynaptically on dentate granular cells; this confirms a previous report of Yamamura and Snyder (1974), showing that muscarinic binding sites are not lost in hippocampus after a lesion of septohippocampal fibers. In the lucidum layer of CA3 (Fig. 2), the mean density of muscarinic binding sites significantly increases by 30% as compared to the control cases. Both the AChE-staining enhancement and the rise of muscarinic binding sites in stratum lucidum of CA3

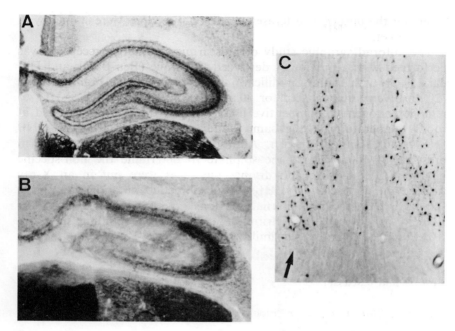

Figure 1. Effects of colchicine treatment on cholinergic septohippocampal system. In *A* and *B* acetylcholinesterase staining of hippocampus; note in colchicine-treated animal (*B*) the increased density of AChE staining in the stratum oriens and the formation of a clear-cut new band of AChE staining in the stratum lucidum of CA3. In *C* the medial septum shows a great number of ChAT positive cells; in the injected side (arrow) there is a loss in the number of immunoreactive cells as compared to the contralateral side.

support the hypothesis of a cholinergic sprouting of septohippocampal fibers to innervate the apical dendrite of CA3 pyramidal cells and compensate for the innervation deficits produced by mossy fibers degeneration.

Loss of ChAT positive cells after intrahippocampal treatment of colchicine

As shown in Figures 1C and 3, unilateral colchicine injection resulted, 4 weeks after treatment, in a loss of 23% of the number of total ChAT positive cell bodies in the MS/VLDB complex as compared to the contralateral side. Cholinergic neurons located on the horizontal limb of the diagonal band were not affected by the degranulation (i.e. 43 ± 7 and 49 ± 6 in ipsi- and contralateral side, respectively). The degeneration of ChAT positive cells induced by destruction of granular cells is clearly much smaller than that found by Gage et al. (1986) following a transection of fimbria-fornix (by over 60%). In some additional cases

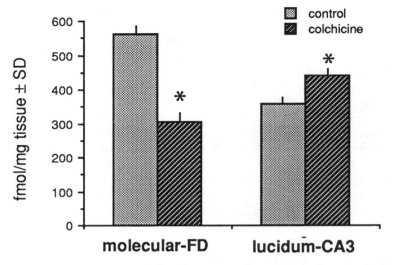

Figure 2. Density of muscarinic binding sites (mean values ±SD) in the molecular layer of fascia dentata (FD) and lucidum layer of CA3 in controls (n = 6) and colchicine-treated (n = 5) rats. Quantifications were performed in at least 12 sections/animal by microdensito-metry; * indicates a significative difference (p < 0.001, Anova).

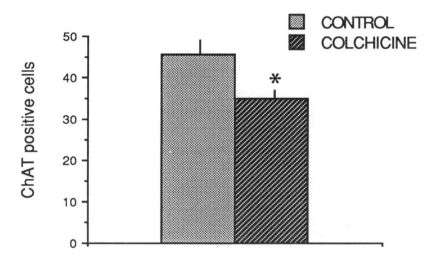

Figure 3. Number of ChAT positive cells ±SD in the medial septum/vertical limb of diagonal band complex. In colchicine-treated hippocampus as compared to the contralateral side (n = 7). Quantification of total ChAT positive neurons in the ipsilateral and contralateral side was performed in at least 8 separate sections located between 8.7 and 9.5 mm anterior to lambda, an area where the neurons of medial septum (MS)/vertical limb of diagonal band (VLDB) project ot the hippocampus (Amaral and Kurz, 1985). * indicates p < 0.001, Anova.

(n = 4) in which colchicine treatment induced a almost complete destruction of the hippocampus, the number of ChAT positive cells in MS/VLDB decreased by over 50% (from 48 ± 6 in controls to 22 ± 4, number of total cells/section $\pm$ SD). In conclusion, the present result clearly shows that the cell death in the septal region is induced by a selective destruction of their target structures. In a recent report, Nakagawa and Ishihara (1988) found that hippocampal extracts from colchicine-treated rats promote neurotrophic activity in cultured cholinergic cells; this neurotrophic activity could explain both, the sprouting of cholinergic fibers and the relative sparing of septal cholinergic cells as compared to fimbria-fornix lesioned rats. In contrast, when the hippocampus is almost completely destroyed by colchicine we found that the retrograde degeneration of cholinergic cells is quite similar to that found after axotomy of cholinergic fibers, supporting the idea that the extent of retrograde damage is directly proportional to the population of target neurons in the hippocampus which remain available for rearrangement of cholinergic connections. It is well known that septal cholinergic cells degenerate in Alzheimer's disease (Terry, 1985), and it has been speculated that this may be caused by a loss of specific neurotrophic factors (Hefti, 1986). Moreover, a dendritic degeneration of dentate-granular cells has been found in senile dementia patients (Flood et al., 1987; Ruiter and Uylings, 1987). We suggest now that cholinergic deficits in senile dementia of the Alzheimer type could result from a default of the hippocampal target cells to supply the necessary cholinergic neurotrophic factors. Moreover, we have found in Alzheimer hippocampi no sign of sprouting of mossy fibers in spite of the extensive denervation of their fascia dentata (Represa et al., 1988); this observation indicates a lack of hippocampal plasticity in senile dementia of the Alzheimer type.

Albe-Fessard, D., Stutinsky, F., and Liboudan, S. (1971) Atlas stéréotaxique du diencéphale du rat blanc. Editions du Centre National de la Recherche Scientifique, Paris 1971.

Amaral, D. G., and Kurz, J. (1985) An analysis of the origins of the cholinergic and noncholinergic septal projections to the hippocampal formation of the rat. J. comp. Neurol. 240: 37–59.

Ayer-LeLievre, C., Olson, L., Ebendal, T., Seiger, Å., and Persson, H. (1988) Expression of the β-Nerve Growth Factor gene in hippocampal neurons. Science 240: 1339–1341.

Ben-Barak, J., and Dudai, Y. (1979) Cholinergic binding sites in rat hippocampal formation: properties and ontogenesis. Brain Res. 166: 245–257.

Drust, E. G., and Crawford, I. L. (1985) Enhanced acetylcholinesterase staining in hippocampal area CA3 after lesion of granule cells by infusion of colchicine. Brain Res. Bull. 14: 9–14.

Flood, D. G., Buell, S. J., Horwitz, G. J., and Coleman, P. D. (1987) Dendritic extent in human dentate gyrus granule cells in normal aging and senile dementia. Brain Res. 402: 205–216.

Gage, F. H., Wictorin, K., Fischer, W., Williams, L. R., Varon, S., and Björklund, A. (1986) Retrograde cell changes in medial septum and diagonal band following fimbria-fornix transection: quantitative temporal analysis. Neuroscience 19: 241–255.

Goldschmidt, R. B., and Steward, O. (1980) Preferential neurotoxicity of colchicine for granule cells of the dentate gyrus of the adult rat. Proc. natl Acad. Sci. USA 77: 3047–3051.

294

Heacock, A. M., Schonfeld, A. R., and Katzman, R. (1986) Hippocampal neurotrophic factor: characterization and response to denervation. Brain Res. 363: 299–306.

Hefti, F. (1986) Nerve growth factor promotes survival of septal cholinergic neurons after fimbrial transections. J. Neurosci. 6: 2155–2162.

Hsiang, J., Wainer, B. H., Shalaby, I. A., Hoffmann, P. C., Heller, A., and Heller, B. R. (1987) Neurotrophic effects of hippocampal target cells on developing septal cholinergic neurons in culture. Neuroscience 21: 333–343.

Korsching, S., Auburger, G., Heumann, R., Scott, J., and Thoenen, H. (1985) Levels of nerve growth factor and its mRNA in the central nervous system of the rat correlates with cholinergic inervation. EMBO J. 4: 1389–1393.

Kromer, L. F. (1987) Nerve growth factor treatment after brain injury prevents neuronal death. Science 235: 214–216.

Nakagawa, Y., and Ishihara, T. (1988) Enhancement of neurotrophic activity in cholinergic cells by hippocampal extract prepared from colchicine-lesioned rats. Brain Res. 439: 11–18.

Represa, A., Duyckaerts, C., Tremblay, E., Hauw, J. J., and Ben-Ari, Y., (1988) Is senile dementia of the Alzheimer type associated with hippocampal plasticity? Brain Res. 457: 355–359.

Robain, O., Represa, A., Jardin, L., and Ben-Ari, Y. Selective destruction of the mossy fibers and granule cells with preservation of the GABAergic network in the inferior region of the rat hippocampus after colchicine treatment. J. comp. Neurol., in press.

Ruiter, J. P., and Uylings, H. B. M. (1987) Morphometric and dendritic analysis of fascia dentata granule cells in human aging and senile dementia. Brain Res. 402: 217–229.

Shute, C. C. D., and Lewis, P. R. (1961) The use of cholinesterase techniques combined with operative procedures to follow nervous pathways in the brain. Bibl. Anat. Basel 2: 34–49.

Sofroniew, M. V., Isacson, O., and Björklund, A. (1986) Cortical grafts prevent atrophy of cholinergic basal nucleus neurons induced by excitotoxic cortical damage. Brain Res. 378: 409–415.

Sofroniew, M. V., Pearson, R. C. A., Eckenstein, F., Cuello, A. C., and Powell, T. P. S. (1983) Retrograde changes in cholinergic neurons in the basal forebrain of the rat following cortical damage. Brain Res. 289: 370–374.

Terry, R. D. (1985) Alzheimer's disease. In: Davis, R. L., and Robertson, D. M. (eds), Textbook of Neuropathology. Williams and Wilkins, Baltimore, pp. 824–841.

Whittemore, S. R., Lärkfors, L., Ebendal, T., Holets, V. R., Ericsson, A., and Persson, H. (1987) Increased β-nerve growth factor messenger RNA and protein levels in neonatal rat hippocampus following specific choline lesions. J. Neurosci. 7: 247–251.

Williams, L. R., Varon, S., Peterson, G. M., Wictorin, K., Fischer, W., Björklund, A., and Gage, F. H. (1986) Continuous infusion of nerve growth factor prevents basal forebrain neuronal death after fimbria fornix transection. Proc. natl Acad. Sci. USA 83: 9231–9235.

Yamamura, H. I., and Snyder, S. H. (1974) Postsynaptic localization of muscarinic cholinergic receptor binding in rat hippocampus. Brain Res. 78: 320–326.

Effect of early visual pattern deprivation on development and laminar distribution of cholinergic markers in rat visual cortex

Christiane Walch, Reinhard Schliebs and Volker Bigl

Paul Flechsig Institute for Brain Research, Department of Neurochemistry, Karl Marx University, Karl-Marx-Städter-Str. 50, DDR- 7039 Leipzig, German Democratic Republic

Summary. In order to evaluate the role of cholinergic cortical mechanisms in the shaping of visual cortical plasticity in more detail the present paper summarizes recent studies on the laminar distribution of muscarinic acetylcholine receptors, choline acetyltransferase, and sodium-dependent high-affinity choline uptake sites during postnatal ontogenesis of the visual cortex of monocularly deprived rats using autoradiographic techniques as well as quantitative biochemical methods after separating the different cortical layers by a cryocut technique. The data are correlated to the laminar distribution of cholinergic fibers within the visual cortex as studied by the immunohistochemical visualization of choline acetyltransferase.

The laminar distribution of cholinergic receptor binding in the visual cortex changes during ontogenesis. In adult rats, the highest muscarinic acetylcholine receptor density is found in layer I.

The activity of the choline acetyltransferase is rather uniformly distributed in all cortical layers. Adult activity values are reached at the age of 25 days. In adult rats the enzyme activity is highest in layer V.

In all visual cortical layers the highest ^{3}H-hemicholinium-3 binding to choline uptake sites during the postnatal period studied is already detectable at the age of 10 days, then binding decreases sharply until day 25 at which age it nearly equals the value found in the adult brain. Binding sites exhibit highest density in layers I and IV of the adult rat visual cortex.

Monocular deprivation resulted in significant changes in all three parameters studied with different cortical laminae preferentially affected. The data suggest that the normal laminar development of the modulatory function of cholinergic transmission in the rat visual cortex depends on the presence of physiological light stimulation.

The functional maturation of the visual system depends on adequate visual stimulation during a certain critical period of susceptibility in the ontogenetic development of the brain. The experience-dependent modification of neuronal response properties in the visual cortex during this critical period of early postnatal development by abnormal visual experience (e.g. total exclusion of light or monocular visual deprivation) has been extensively documented with both electrophysiological as well as neuroanatomical methods. Much less is known, however, about the underlying biochemical mechanisms or loci at which these modifications occur. To unravel possible changes in synaptic efficacy during postnatal development following different conditions of visual deprivation various neurotransmitter systems has been studied and transient or permanent changes have been reported in the development of α- and β-adrenergic (Aurich et al., 1989; Kasamatsu and Shirokawa, 1985; Schliebs et al., 1982b), serotoninergic (Aurich et al., 1985), cholinergic (Schliebs et al.,

1982a) as well as benzodiazepine receptors (Schliebs et al., 1986b) and glutamate binding sites (Schliebs et al., 1986a). In rats most of these changes are confined to subcortical visual structures. Only the alteration of cholinergic receptor binding is restricted to cortical areas.

Electrophysiological studies have demonstrated that along with noradrenergic modulatory influences acetylcholine is involved in the facilitation of experience-dependent changes in the neuronal response properties of the kitten visual cortex (for references see Bear and Singer, 1985). Besides its role in shaping visual cortical plasticity acetylcholine modulates visually evoked neuronal responses and is involved in the state-dependent modification of cortical responsiveness (for references, see Sillito et al., 1985). In addition, a recent study suggests that the cholinergic cortical innervation from the nucleus basalis Meynert (NbM) strongly affects cortical morphogenesis (Höhmann et al., 1986).

In contrast to other neocortical areas the rat visual cortex receives a dual cholinergic innervation (Bigl et al., 1982; Carey and Rieck, 1987): the medial cholinergic cortical pathway from the nucleus of the diagonal band of Broca, and the nucleus preopticus magnocellularis innervating the medial portion of the visual cortex, the lateral aspects receiving cholinergic fibers via a lateral pathway from the rostral portions of nucleus basalis and the substantia innominata (see Eckenstein et al., 1988). Immunochemical studies on the intracortical distribution of choline acetyltransferase (ChAT)-containing fibers have shown that the cholinergic afferents from the nucleus basalis are distributed throughout the cortex with a more or less marked laminar pattern which varies between different cortical fields and between animal species (see de Lima and Singer, 1986). Within the visual cortex of rats ChAT-positive fibers are concentrated in layer I and the upper part of layer V. Additionally, a very dense band of ChAT-positive terminals is observed in lower layer IV which is not observed in any other cortical area (Eckenstein et al., 1988; Lysakowski et al., 1986). Like other neurotransmitter receptors the muscarinergic and nicotinergic acetylcholine receptors show a laminar distribution within the cerebral cortex (McDonald et al., 1987). In the cat visual cortex a reversal in the laminar pattern of muscarinic acetylcholine receptors during the critical period was described (Shaw et al., 1984). Similar age-dependent changes in the laminar distribution were detected in developmental studies of rat cerebral cortex (Krisst and Kasper, 1983; Rotter et al., 1979). Little is known about developmental changes in the laminar pattern of other cholinergic neurochemical markers in rat visual cortex and about the role of visual experience for the development of the laminar distribution of cholinergic parameters. In view of the fact that monocular visual deprivation has been shown to alter mAChR density in the rat visual cortex (Schliebs et al., 1982a) it was of interest to study the influence of monocular deprivation on the laminar development of cholinergic parameters in the visual cortex to reveal whether the development of the normal laminar pattern of cholinergic parameters in the visual cortex is driven or altered by visual experience. The present paper summarizes our recent results on the postnatal development of the laminar distribution of muscarinic acetylcholine receptor (mAChR), choline acetyltransferase (ChAT) activity, and high-affinity choline uptake sites in rat visual cortex (Schliebs et al., 1989; Walch and Schliebs, 1989).

The experiments were carried out on rats of the strain BD III at the ages of 10, 15, 25, and 90 days. They were raised in the laboratory's own animal house under a 12-h light/dark cycle and received food and water ad libitum. In visual deprivation experiments, rats were subjected to unilateral eyelid suture at the age of 11 days (monocular deprivation) under anesthesia, and killed at the age of 25 days.

The visual cortex comprising cortical areas 17, 18, and 18a, was isolated, frozen on dry ice, and trimmed precisely into 2×3 mm blocks by means of a tissue chopper, in order to obtain sections of equal size. The different cortical layers were then separated by cutting serial crystat sections of 10 μm thickness parallel to the cortical surface (Brückner et al., 1983). The thickness of each layer was estimated at every age studied from cresyl violet stained saggital sections.

For binding studies, slices were thaw-mounted onto glass slides and stored at $-10°C$ until used. Incubation of 5 slide-mounted tissue sections collected from one individual cortical layer was carried out in humid chambers by flooding the sections with an incubation buffer containing the appropriate radioligand: ^{3}H-quinuclidinylbenzilate (QNB) or ^{3}H-hemicholinium-3 (HCh3) for assaying muscarinic acetylcholine receptors (mAChR) or high-affinity choline uptake sites, respectively. The amount of radioligand bound to the tissue sections was assessed by scraping the tissue slices from the slides and measuring the radioactivity by liquid scintillation counting. For ChAT assay, all slices from one individual cortical layer were collected, homogenized in buffer, and immediately used for the biochemical analysis. ChAT activity was determined radiochemically using ^{3}H-acetyl-CoA as cosubstrate.

For autoradiographic experiments, the labelled and dried tissue sections were apposed to Tritium-sensitive film (Ultrofilm, LKB). Autoradiograms were analyzed with a computer-assisted imaging device, MAGISCAN 2A (Joyce and Loebel, U.K.) as described previously (Schliebs et al., 1989).

The developmental changes in the laminar distribution of mAChR in the rat visual cortex are displayed in Figure 1. The density of mAChR in layer I rises sharply from day 10 through day 25 and afterwards continues to increase more moderately until adulthood. In the layers II to VI receptor density increases slightly from day 10 to day 15, then it rises sharply until day 25. From this age onwards receptor binding decreases (layers II–V) or remains stable (layer VI). The K_d-values

298

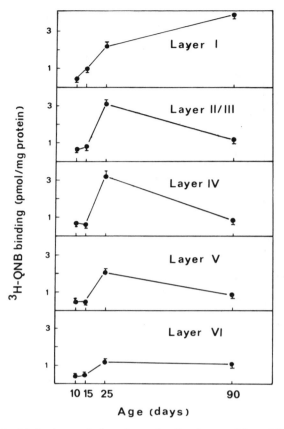

Figure 1. Postnatal development of maximum density of muscarinic acetylcholine receptors in the layers of rat visual cortex. The different cortical layers were separated by cutting serial cryostat sections of 10 μm thickness. Slices from one individual layer were thaw-mounted onto glass slides and incubated at room temperature with varying concentrations of ^{3}H-QNB ranging between 0.1 and 15 nM. The amount of radioligand bound to the tissue section was assessed by scraping the tissue slices from the slides and measuring the radioactivity via liquid scintillation counting. Maximum receptor densities were obtained from saturation experiments by fitting experimental data to the saturation curve using nonlinear least squares regression analysis. The data given represent the mean ± SEM from four separate saturation experiments each performed in duplicate.

obtained from Scatchard plots change only slightly during development (data not shown).

Quantitative receptor autoradiography performed in coronal slices of the adult rat brain revealed that ^{3}H-QNB binding to mAChR predominates in visual cortical layer I, but high levels of mAChR binding occurred also in the remaining layers with mAChR levels higher in layer II/III than in deeper layers (Schliebs et al., 1989). This distribution is in good agreement with the laminar pattern obtained by ^{3}H-QNB binding

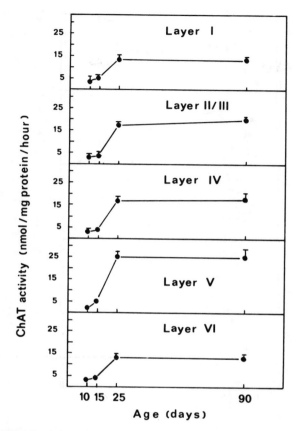

Figure 2. Developmental changes of ChAT activity in individual layers of rat visual cortex. Each point represents the mean ± SEM from 3 to 14 separate determinations each performed in duplicate.

in tissue slices from individual visual cortical layers as shown in Figure 1.

The developmental changes in the laminar pattern of ChAT activity are shown in Figure 2. In all layers ChAT activity develops slowly between 10 and 15 days and rises sharply from postnatal day 15 up to day 25 at which age the activity levels found in adult rats are almost reached. This developmental pattern is consistent with that found in homogenates of the visual cortex in rats.

The developmental profiles of ³H-HCh3 binding to choline uptake sites in the individual cortical layers of visual cortex are displayed in Figure 3. In all cortical layers, the highest ³H-HCh3 binding during the postnatal period studied was already detectable at the age of 10 days. The developmental pattern is qualitatively similar in all cortical layers:

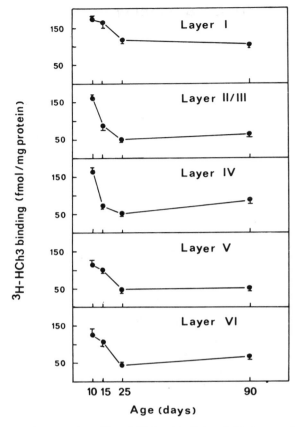

Figure 3. Developmental profiles of ³H-hemicholinium-3 binding to choline uptake sites in individual layers of rat visual cortex. Each point represents the mean ± SEM from 7 to 12 separate determinations (see also legend to Fig. 1).

it decreases more or less sharply between 10 and 25 days postnatally at which age the adult binding levels are reached. In the adult rat, ³H-HCh3 binding is highest in layer I followed by layer IV. Layer II/III, V and VI show only binding values amounting to about 50% of those measured in layer I.

Unilateral eyelid closure from postnatal day 11 until day 25 resulted in significant alterations in the laminar distribution of mAChR in the visual cortex both ipsilateral and contralateral to the closed eye. As a consequence of monocular deprivation the density of mAChR in the contralateral visual cortex was decreased in layer IV and layer V but increased in layer VI in comparison to age-matched controls; in the ipsilateral visual cortex it was reduced in layer II/III and layer IV but strongly enhanced in layer V (Fig. 4).

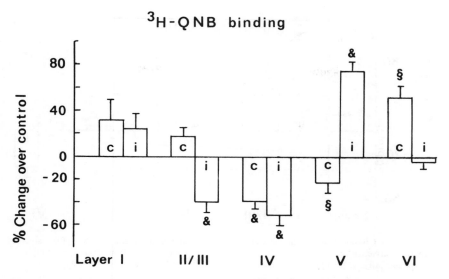

Figure 4. % Change over control in muscarinic acetylcholine receptor density in individual layers of rat visual cortex following monocular deprivation from 11 through 25 days postnatally. §, $p < 0.05$; &, $p < 0.01$ vs. control, two-tailed Student's t-test.

ChAT activity was increased by monocular deprivation only in the visual cortex ipsilateral to the closed eye (layers IVa and VI, Fig. 5).

As a consequence of monocular deprivation until day 25, ^{3}H-HCh3 binding was significantly higher in layer II/III of both ipsi- and contralateral visual cortex by about 30% and decreased in the ipsilateral layer VI in comparison to control values (Fig. 6).

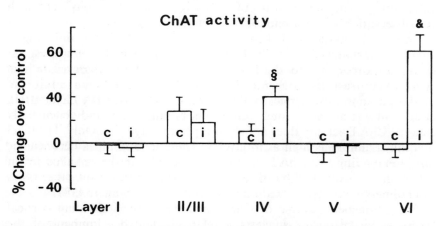

Figure 5. % Change over control in ChAT activity in individual layers of rat visual cortex following monocular deprivation from 11 through 25 days postnatally. §, $p < 0.05$; &, $p < 0.01$ vs. control, two-tailed Student's t-test.

302

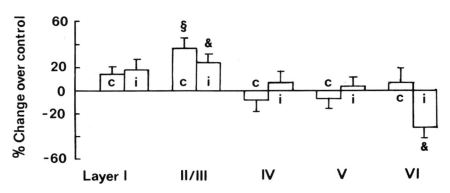

Figure 6. % Change over control in ^{3}H-hemicholinium-3 binding to choline uptake sites in individual layers of rat visual cortex following monocular deprivation from 11 through 25 days postnatally. §, p < 0.05; &, p < 0.01 vs. control, two-tailed Student's t-test.

These data obtained with a more quantitative and controlled approach confirm previous studies in cat and rat (Krisst and Kasper, 1983; Rotter et al., 1979; Shaw et al., 1984) and demonstrate that also in the rat visual cortex the laminar distribution of mAChR is changed during the ontogenetic development: whereas at 25 days of age mAChR binding is highest in layer IV and layer II/III of the visual cortex, in the adult animal mAChR density in layer I is two times higher compared to the remaining cortical layers. No similar changes are seen in the intracortical distribution of ChAT activity (as a measure of cholinergic fibers and neurons) or ^{3}H-HCh3 binding (as a marker of cholinergic terminals or varicosities). The high density of mAChR in layer I is consistent with the observation of an immunohistochemically high density of cholinergic fibers in this layer (Eckenstein et al., 1988; de Lima and Singer, 1986) and is matched by the high amount of ^{3}H-HCh3 binding sites, but does not correspond to the lower activity of .ChAT observed in this layer. In contrast, the concentration of ChAT-positive fibers in layer IV of the immunohistochemically demonstrated visual cortex (Eckenstein et al., 1988) is not correlated to mAChR binding. Also radiochemically assayed ChAT activity does not show a peak in layer IV. Only ^{3}H-HCh3 binding is almost as high as in layer I. These incongruities between the mAChR binding and ChAT activity observed also in the visual cortex of cats might be explained by the assumption that in this layer other types of cholinergic receptors predominate (Stichel and Singer, 1987).

The ontogenetic course of ChAT activity in the individual cortical layers of visual cortex suggests a relatively late development of the cholinergic system which seems to be terminated after the third week of postnatal life. At that time, ^{3}H-HCh3 binding sites also reach nearly adult values. The decrease in choline uptake sites occurring between 10 and 25 days might then be regarded as degradation of those uptake sites not effectively involved in the ongoing formation of cholinergic synaptic contacts in the individual visual cortical layers.

Disturbed visual input during the early period of life affects differentially the development of the laminar pattern of mAChR, ChAT activity, and choline uptake sites, suggesting that the development of the modulatory function of cholinergic transmission in the distinct visual cortical layers is mainly driven by the presence of physiological light stimulation. These changes in the laminar pattern of presynaptic cholinergic structures and their corresponding postsynaptic target sites following monocular deprivation support the view that inadequate visual experience differentially affects the maturation of the cholinergic neurons and afferent fiber system in the visual cortical layers. The exact mechanisms by which cholinergic afferents from the NbM and cholinergic interneurons facilitate experience-dependent response properties of visual cortical neurons during a critical period of ontogenetic development and which role the age-related laminar distribution of cholinergic parameters plays for these processes remains to be unravelled.

Acknowledgement. This study was supported by a grant of the Ministry of Sciences and Technology of G.D.R.

Aurich, M., Schliebs, R., and Bigl, V. (1985) Serotoninergic receptors in the visual system of light-deprived rats. J. Dev. Neurosci. 3: 285–290.
Aurich, M., Schliebs, R., Stewart, M. G., Rudolph, E., Fischer, H.-D., and Bigl, V. (1989) Adaptive changes in the central noradrenergic system in monocularly deprived rats. Brain Res. Bull. 22: in press.
Bear, M. F., and Singer, W. (1985) Modulation of visual cortical plasticity by acetylcholine and noradrenaline. Nature (Lond.) 320: 172–176.
Bigl, V., Woolf, N. J., and Butcher, L. L. (1982) Cholinergic projections from the basal forebrain to frontal, parietal, temporal, occipital, and cingulate cortices: a combined fluorescent tracer and acetylcholinesterase analysis. Brain Res. Bull. 8: 727–741.
Brückner, G., Braulke, T., Müller, L., and Biesold, D. (1983) A method for the dissection of the embryonic cerebral cortex into individual layers. An application to biochemical studies of glycan metabolism. J. Neurosci. Meth. 7: 215–226.
Carey, R. G., and Rieck, R. W. (1987) Topographic projections to the visual cortex from the basal forebrain in the rat. Brain Res. 424: 205–215.
De Lima, A. D., and Singer, W. (1986) Cholinergic innervation of the cat striate cortex: A choline acetyltransferase immunocytochemical analysis. J. comp. Neurol. 250: 324–338.
Eckenstein, F. P., Baughman, R. W., and Quinn, J. (1988) An anatomical study of cholinergic innervation in rat cerebral cortex. Neuroscience 25: 457–474.
Höhmann, C. F., Brooks, A. R., Oster-Granite, M. L., and Coyle, J. T. (1986) Neonatal basal forebrain lesions in mice induce abnormal cortical morphogenesis. Soc. Neurosci. Abstr. 12: 1582.
Kasamatsu, T., and Shirokawa, T. (1985) Involvement of β-adreno-receptors in the shift of ocular dominance after monocular deprivation. Exp. Brain Res. 59: 507–514.
Krisst, D. A., and Kasper, E. K. (1983) High density of cholinergic muscarinic receptors accompanies high intensity acetylcholin-esterase-staining in layer IV of infant rat somatosensory cortex. Dev. Brain Res. 8: 373–376.

304

Lysakowski, A., Wainer, B. H., Rye, D. B., Bruce, G., and Hersh, L. B. (1986) Cholinergic innervation displays strikingly different laminar preferences in several cortical areas. Neurosci. Lett. 64: 102–108.

McDonald, J. K., Speciale, S. G., and Parnavelas, J. G. (1987) The laminar distribution of glutamate decarboxylase and choline acetyltransferase in the adult and developing visual cortex of the rat. Neuroscience 21: 825–832.

Rotter, A., Field, P. M., and Raisman, G. (1979) Muscarinic receptors in the central nervous system of the rat. III. Postnatal development of binding of [³H]propylbenzilycholine mustard. Brain Res. Rev. 1: 185–205.

Schliebs, R., Bigl, V., and Biesold, D. (1982) Development of muscarinic cholinergic receptor binding in the visual system of monocularly deprived and dark reared rats. Neurochem. Res. 7: 1181–1198.

Schliebs, R., Burgoyne, R. D., and Bigl, V. (1982b) The effect of visual deprivation on β-adrenergic receptors in the visual centres of rat brain. J. Neurochem. 38: 1038–1043.

Schliebs, R., Kullmann, E., and Bigl, V. (1986) Development of glutamate binding sites in the visual structures of the rat brain. Effect of visual pattern deprivation. Biomed. biochim. Acta 45: 495–506.

Schliebs, R., Rothe, T., and Bigl, V. (1986b) Dark-rearing affects the development of benzodiazepine receptors in the cental bisual structures of rat brain. Dev. Brain Res. 24: 179–185.

Schliebs, R., Walch, C., and Stewart, M. G. (1989) Laminar pattern of cholinergic and adrenergic receptors in rat visual cortex using quantitative receptor autoradiography. J. Hirnforsch. 30: in press.

Shaw, C., Needler, M. C., and Cynader, M. (1984) Ontogenesis of muscarinic acetylcholine binding sites in cat visual cortex: reversal of specific laminar distribution during the critical period. Dev. Brain Res. 14: 295–299.

Sillito, A. M., Salt, T. E., and Kemp, J. A. (1985) Modulatory and inhibitory processes in the visual cortex. Vision Res. 25: 375–381.

Stichel, C. C., and Singer, W. (1987) Quantitative analysis of the choline acetyltransferase-immunoreactive axonal network in the cat primary visual cortex: II- Pre- and postnatal development. J. comp. Neurol. 258: 99–111.

Walch, C., and Schliebs, R. (1989) Age-dependent changes in the laminar distribution of cholinergic markers in rat visual cortex. Neurochem. Int. 14: in press.

The role of muscarinic acetylcholine receptors in ocular dominance plasticity

Qiang Gu and Wolf Singer

Max-Planck-Institut für Hirnforschung, D-6000 Frankfurt 71, Federal Republic of Germany

Summary. During a critical period of postnatal development neuronal connections in the visual cortex are susceptible to experience-dependent modifications. In normally reared kittens the majority of neurons respond to visual stimulation of either eye. A few days of monocular deprivation, however, are sufficient to render most cortical neurons unresponsive to visual stimuli presented to the deprived eye. Among other factors the cholinergic projection to striate cortex has been identified as having a permissive role in this use-dependent modification of synaptic transmission. In order to analyze further the influence of acetylcholine in cortical plasticity, we tested whether the blockade of muscarinic or nicotinic receptors interfered with ocular dominance plasticity. At four weeks of age kittens had one eyelid sutured closed and osmotic minipumps implanted, which delivered scopolamine (1 nmol/h) or hexamethonium (1 or 10 nmol/h) into the striate cortex of one hemisphere and vehicle solution (saline) into the other. After one week, ocular dominance distributions were determined in area 17 with single unit recording. In the control hemispheres, most neurons became unresponsive to the deprived eye, while in the scopolamine-treated hemispheres most neurons remained binocular. In contrast to the effects of scopolamine, the intracortical infusion of hexamethonium had no effect on ocular dominance plasticity. These results demonstrate that blockade of muscarinic, but not nicotinic receptors renders kitten striate cortex resistent to the effects of monocular deprivation.

Introduction

In the 1960s, Hubel and Wiesel demonstrated in their pioneering studies that the visual cortex of kittens is highly susceptible to experience-dependent modifications of its structural and functional architecture (Hubel and Wiesel, 1963, 1970; Wiesel and Hubel, 1963, 1965). Before visual experience becomes effective in influencing the connectivity of cortical neurons, most cells in the visual cortex of kittens respond to visual stimulation through either eye. With monocular deprivation, however, the large majority of cortical cells lose the ability to respond to the deprived eye. These changes of ocularity occur only during a critical period of early development, which in kittens lasts about 3 months.

Kasamatsu and Pettigrew proposed that noradrenaline (NA) has a permissive role in this ocular dominance plasticity. In a series of experiments they demonstrated that destruction of the noradrenergic projection by injection of 6-hydroxydopamine (6-OHDA) into the lateral ventricle or visual cortex reduced ocular dominance plasticity in the kitten visual cortex (Kasamatsu and Pettigrew, 1976, 1979). These effects of 6-OHDA could be reversed by infusing NA directly into the visual cortex (Pettigrew and Kasamatsu, 1978; Kasamatsu et al., 1979, 1981). However, several independent investigations have indicated that ocular dominance changes can be induced despite NA depletion. In these investigations 6-OHDA was either injected prior to monocular deprivation or the noradrenergic input to cortex was blocked by means other than local 6-OHDA application (Bear and Daniel, 1983; Bear et al., 1983; Daw et al., 1984, 1985; Adrien et al., 1985; Trombley et al., 1986).

This controversy was resolved by the demonstration that lesions affecting both the noradrenergic and the cholinergic input to striate cortex abolish ocular dominance plasticity, while lesions of either projection alone are ineffective and that 6-OHDA did not only destroy noradrenergic fibers but also had an anticholinergic effect (Bear and Singer, 1986). This suggested that in addition to NA, acetylcholine (ACh) also had a permissive function in ocular dominance plasticity.

In order to determine further the role of ACh in experience-dependent neuronal plasticity, we tested whether the local blockade of cholinergic transmission interfered with ocular dominance plasticity, and if so, whether the effects were mediated by muscarinic or nicotinic receptors. To this end we studied the effects of intracortical infusion of the muscarinic and nicotinic receptor blockers scopolamine and hexamethonium on ocular dominance plasticity.

Methods

After induction of general anesthesia with ketamine (20 mg/kg) and xylazine (10 mg/kg) i.m. four-week-old kittens were monocularly deprived by suturing the right eyelid closed. At the same time two osmotic minipumps (Alzet, model 2001) were implanted subcutaneously. These were connected with polyethylene tubing to 27-gauge cannulae, which were inserted 2–2.5 mm deep into the primary visual cortex of each hemisphere near the area centralis representation (Horsley-Clarke Coordinates $P - 5$, $L + 2$). The pumps delivered either scopolamine (1 nmol/h) or hexamethonium (1 or 10 nmol/h) into the left hemisphere and vehicle solution (saline) into the other hemisphere. After 6–8 days, the ocular dominance distributions of neurons were determined at various distances from the infusion cannulae.

For the electrophysiological experiments, the kittens were again anesthetized with a mixture of 20 mg/kg ketamine (5%) and 10 mg/kg xylazine (5%) i.m. Once all surgical interventions had been terminated anesthesia was maintained by artificial ventilation with nitrous oxide (70% N_2O, 30% O_2), supplemented by an i.v. infusion of pentobarbital (0.7 mg/kg/h). Relaxation was obtained with i.v. infusion of

0.5 mg/kg/h hexacarbacholine-bromide. To compensate for the loss of fluid, the kittens were also infused with a levulose-Ringer solution (4%) through an orally inserted gastric catheter (3 ml/h). Electroencephalogram, heart rate, body temperature and the CO_2 concentration in the expired air were monitored continuously. Body temperature was maintained at 37.5°C and end-tidal CO_2 concentration was adjusted between 3.5 and 4.0%. The nictitating membranes were retracted with Neosynephrine and the pupils dilated with atropine. Contact lenses with artificial pupils covered the corneae, the refractive state was determined by a refractometer and appropriate spectacle lenses were selected to focus the retina on a tangent screen positioned 114 cm in front of the kitten's eye plane. Retinal landmarks were plotted on the screen with the help of a fundus camera. Single unit activity was recorded with micropipettes containing 1.5 M potassium-citrate which had DC-resistances between 5 MΩ and 30 MΩ. The signals were fed to a conventional amplifier chain and a window discriminator. The output of the latter was monitored together with the AC-signal both visually and acoustically.

Single units were sampled from area 17 on the dorsal crest of the lateral gyrus in an area extending 2 to 4 mm anterior to the infusion cannula. To minimize sampling artifacts, the electrodes were inclined in the sagittal plane so that the recording trajectories had an angle of 30° with respect to the cortical surface.

Receptive fields were analyzed using hand-held light stimuli with slits whose length, width and orientation could be varied. The responses of the cells to stimulation of each eye were classified according to their vigor in a 3-group rating scale. Weak reactions just detectable with conventional hand-mapping were rated in class 1. Clear and reproducible responses to hand-held stimuli that could vary in time were rated in class 2. Reproducible and vigorous responses were assigned to class 3. The appropriate ocular dominance class was selected on a scale of 5 classes. Groups 1 and 5 contain monocular cells activated exclusively by one eye; groups 2 and 4 refer to cells activated by both eyes with one eye being dominant; group 3 comprises binocular cells that are equally activated by both eyes. Cells that could not be driven by a visual stimulus were classified separately as unresponsive.

After electrophysiological recording kittens were perfused with physiological saline, followed by 4% formaldehyde in 0.1 M phosphate buffer (pH 7.4). After postfixation and cryoprotection with 30% sucrose 50-μm thick sections of striate cortex were cut on a cryotome and Nissl-stained for anatomical analysis of the infusion site.

In order to estimate the distribution of scopolamine concentration in the experimental cortex and at the recording site, respectively, osmotic minipumps containing 1 mM scopolamine and 166.7 μCi/ml 1-(N-methyl-^{3}H)-scopolamine were implanted in one kitten as usual for 7 days. Subsequently after administration of a lethal dose of Nembutal the osmotic minipumps were removed and blocks of the visual cortex were taken out, rapidly frozen in cooled 2-methylbutane, which was kept at a temperature of $-35°C$. Parasagittal 12-μm thick sections were then prepared with a cryostat at $-17°C$. These were dried and covered with an ^{3}H-hyperfilm. Similarly, 12-μm thick sections were prepared from a tritium-labelled polymer standard block with known values of tissue equivalent tritium concentration (nCi ^{3}H per mg grey matter) and exposed for the same period of time on the same film. Densitometric analysis of the autoradiographs was performed with a digital image processing system. Tissue concentrations of scopolamine in the perfused cortex were assessed by comparing grey levels at various distances from the infusion site with those of the standards. On the basis of the assumptions that 166.7 mCi tritium corresponds to 1 mmol scopolamine and that the grey matter of visual cortex has a specific weight of about 1.1 g/ml, the respective scopolamine concentrations could be determined from the optical density of the autoradiographs.

To make sure that the distribution of tritium activity in the visual cortex actually reflects tritiated methylscopolamine and not pharmacologically inactive decomposition products, we performed additional control experiments using thin layer chromatography to identify the nature of the radioactve compounds. An osmotic minipump containing 1 mM scopolamine and 100 μCi/ml tritiated methylscopolamine was implanted in a kitten as described above. After 7 days the kitten was sacrificed, the osmotic minipump removed and a small block of visual cortex (approximately 20 mg) near the infusion site was taken out, and homogenized after addition of 10 μl 10 mM scopolamine. After centrifugation 2 μl of the supernatant were analyzed on a precoated silicagel-plate. In parallel, 2 μl of the original solution containing 100 μCi/ml tritiated methylscopolamine and 10 mM scopolamine were applied on the same plate. The samples were eluted in a solution consisting of n-butanol, acetic acid, H_2O and ethanol (80 : 10 : 30 : 20). The strips of each sample were then cut into 1-cm segments for scintillation counting.

Results

The ocular dominance distributions in Figure 1A/B summarize the results of 3 kittens that were monocularly deprived for 6, 7 and 8 days, respectively, and had the minipumps removed prior to recording. Units were sampled at a distance of 2–4 mm from the infusion cannulae. In the hemispheres infused with vehicle solution more than one third of the neurons became unresponsive to the deprived eye and the ocular dominance distributions were strongly biased towards the open eye (Fig. 1A). In the scopolamine-treated hemispheres by contrast, the majority of the

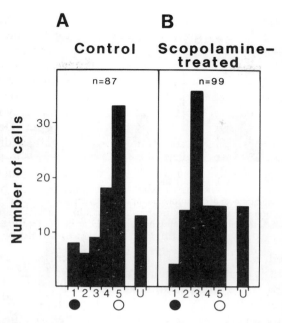

Figure 1. Ocular dominance distributions of the control (*A*) and scopolamine-treated (*B*) hemispheres in 3 kittens, that were monocularly deprived for 6, 7 and 8 days, respectively.

neurons remained binocular with cells driven equally well by either eye being most frequent (Fig. 1B). No significant differences existed with respect to the percentage of cells unresponsive to light.

In order to check the possibility of scopolamine blocking the light responses of cortical neurons and preventing use-dependent modifications simply by abolishing activity, we analyzed single cell responses in one kitten with the scopolamine infusing pump in place. Ninety cells were sampled in 3 tracks at distances of 1.9–3.7 mm from the infusion cannula 8 days after implantation and while the pump was still in action. These control recordings showed that neurons remained responsive to light stimuli at distances from the infusion site where ocular dominance shifts were readily suppressed (Fig. 2A, B, C).

In order to test whether the nicotinic receptor blocker, hexamethonium, has a similar effect on ocular dominance plasticity, we intracortically infused hexamethonium at doses of 1 and 10 nmol/h in kitten striate cortex. After one week of monocular deprivation the ocular dominance distributions of neurons recorded at a distance of 2–4 mm from the infusion site were strongly biased toward the open eye (Fig. 3B) and not significantly different from the distributions found in the control hemispheres after one week of monocular deprivation (Fig. 3A).

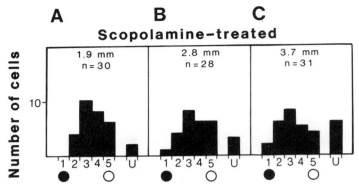

Figure 2. Ocular dominance distributions in another kitten visual cortex after 8 days of monocular deprivation and scopolamine treatment (1 nmol/h) at different distances from the infusion cannula. Scopolamine was continuously infused during the recordings.

Nissl-stained serial sections of the occipital cortex revealed that a spherical necrosis of about 1–2 mm in diameter had formed in all kittens around the tips of the infusion cannulae both in the drug treated and in the control hemispheres. Outside this area tissue had a normal appearance suggesting that lesions were confined essentially to the well-demarcated necrosis around the tips of the cannulae. Since these necroses were

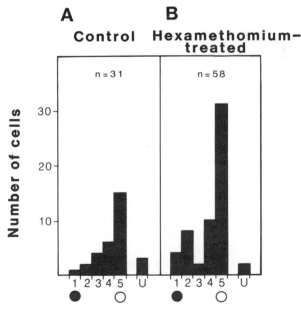

Figure 3. Ocular dominance distributions of the control (A) and hexamethonium-treated (B) hemispheres in 2 kittens, that were monocularly deprived for 7 days and infused with 1 and 10 nmol/h hexamethonium.

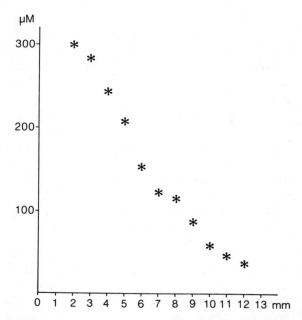

Figure 4. The calculated distribution of scopolamine concentration from the cannula after 7 days of infusion (1 nmol/h).

similar in both hemispheres they are probably due to mechanical and osmotic disturbances rather than to specific pharmacological effects.

For the interpretation of the present results it is crucial to estimate the steady state concentrations of scopolamine at the recording site. The autoradiographic analysis revealed an exponential decay of radioactivity with increasing distance from the infusion site. To determine to what extent the radioactivity measured after 7 days of infusion was due to scopolamine or to breakdown products, we measured the fraction of radioactivity actually associated with scopolamine in one additional control kitten that had been infused for 7 days with tritiated scopolamine. This analysis revealed that 92% of total radioactivity was contained in the scopolamine fraction. Thus, at our usual recording location (2–4 mm from the infusion site) the scopolamine concentration was calculated to be in the range of 200–300 μM (Fig. 4).

Discussion

The present results indicate that chronic blockade of muscarinic receptors prevents use-dependent modifications of ocular dominance such as occur normally after monocular deprivation. This corroborates the conjecture that ACh has a permissive effect on use-dependent plasticity of kitten visual cortex (Bear and Singer, 1986). Moreover, the present findings indicate that this effect is mediated by muscarinic receptors and that blockade of muscarinic transmission is alone capable of preventing ocular dominance changes. Our control experiments indicate that the drug concentrations at the recording site were sufficient to block muscarinic receptors and that this blockade did not abolish the activity of cortical neurons. This suggests that the effects of scopolamine were due to its selective action on muscarinic receptors.

Nicotinic receptors have been shown to be localized presynaptically in the kitten striate cortex (Prusky et al., 1988). Our present data indicate that these nicotinic receptors are not involved in the control of ocular dominance plasticity.

Our results bear on the much-debated question of the relative contribution of noradrenergic and cholinergic mechanisms in the gating of visual cortex plasticity. Lesions of the cholinergic or the noradrenergic projection to striate cortex when restricted to either system alone did not block ocular dominance plasticity even when they caused a substantial depletion of noradrenergic (Bear and Daniels, 1983; Bear et al., 1983; Daw et al., 1984, 1985; Adrien et al., 1985; Trombley et al., 1986) or cholinergic (Bear and Singer, 1986) markers. However, combined lesions were effective (Bear and Singer, 1986) suggesting a synergistic action of ACh and NA in cortical plasticity. By contrast, pharmacological blockade of β-receptors alone was reported to be solely sufficient to prevent ocular dominance plasticity (Kasamatsu and Shirokawa, 1985; Shirokawa and Kasamatsu, 1986). Taken together with the present results this suggests that neither the noradrenergic nor the cholinergic system by itself is sufficient to maintain plasticity when muscarinic or β-adrenergic transmission is blocked reliably. We propose as the most likely explanation for the ineffectiveness of isolated lesions that the ensuing depletion of ACh and NA may not have been sufficient. Incomplete destruction of the respective subcortical projections or persisting intrinsic sources of NA and ACh may be considered as causes. In addition, receptor supersensitivity may have developed and amplified the effectiveness of the remaining transmitters. There is evidence that chronic disruption of noradrenergic (Sporn et al., 1976; U'Prichard et al., 1980; Harik et al., 1981) and cholinergic transmission (Westlind et al., 1981; Sato et al., 1987) induce hypersensitivity in the corresponding postsynaptic receptors.

We have proposed two possibilities to account for the synergy of ACh and NA in supporting cortical plasticity: One common action of both neuromodulators is that they reduce potassium permeability (Halliwell and Adams, 1982; Madison and Nicoll, 1982) and hence both could enhance depolarization in response to visual input (Sillito, 1983). Thereby ACh and NA increase the probability that retinal input reaches the activation threshold of the NMDA-receptor mechanism, the latter acting as a link in use-dependent modifications of synaptic transmission in striate cortex (Kleinschmidt et al., 1987). Another possibility is that ACh and NA, by virtue of their influence on the second messengers inositol triphosphate and cAMP, directly gate cellular processes that are involved in synaptic modifications.

Adrien, J., Blanc, G., Buisseret, P., Frégnac, Y., Gary-Bobo, E., Imbert, M., Tassin, J. P., and Trotter, Y. (1985) Noradrenaline and functional plasticity in kitten visual cortex: A re-examination. J. Physiol. 367: 73–98.

Bear, M. F., and Daniels, J. D. (1983) The plastic response to monocular deprivation persists in kitten visual cortex after chronic depletion of norepinephrine. J. Neurosci. 3: 407–416.

Bear, M. F., Paradiso, M. A., Schwartz, M., Nelson, S. B., Carnes, K. M., and Daniels, J. D. (1983) Two methods of catecholamine depletion in kitten visual cortex yield different effects on plasticity. Nature 302: 245–247.

Bear, M. F., and Singer, W. (1986) Modulation of visual cortical plasticity by acetylcholine and noradrenaline. Nature 320: 172–176.

Daw, N. W., Robertson, T. W., Rader, R. K., Videen, T. O., and Coscia, C. J. (1984) Substantial reduction of cortical noradrenaline by lesions of adrenergic pathway does not prevent effects of monocular deprivation. J. Neurosci. 4: 1354–1360.

Daw, N. W., Videen, T. O., Rader, R. K., Robertson, T. W., and Coscia, C. J. (1985) Substantial reduction of noradrenaline in kitten visual cortex by intraventricular injections of 6-hydroxydopamine does not always prevent ocular dominance shifts after monocular deprivation. Exp. Brain Res. 59: 30–35.

Halliwell, J. V., and Adams, P. R. (1982) Voltage-clamp analysis of muscarinic excitation in hippocampal neurons. Brain Res. 250: 71–92.

Harik, S. I., Duckrow, R. B., LaManna, J. C., Rosenthal, M., Sharma, V. K., and Banerjee, S. P. (1981) Cerebral compensation for chronic noradrenergic denervation induced by locus ceruleus lesion: Recovery of receptor binding, isoproterenol-induced adenylate cyclase activity, and oxidative metabolism. J. Neurosci. 1: 641–649.

Hubel, D. H., and Wiesel, T. N. (1963) Receptive fields of cells in striate cortex of very young, visually inexperienced kittens. J. Neurophysiol. 26: 994–1002.

Hubel, D. H., and Wiesel, T. N. (1970) The period of susceptibility to the physiological effects of unilateral eye closure in kittens. J. Physiol. 206: 419–436.

Kasamatsu, T., and Pettigrew, J. D. (1976) Depletion of brain catecholamines: Failure of ocular dominance shift after monocular occlusion in kittens. Science 194: 206–209.

Kasamatsu, T., and Pettigrew, J. D. (1979) Preservation of binocularity after monocular deprivation in the striate cortex of kittens treated with 6-hydroxydopamine. J. comp. Neurol. 185: 139–162.

Kasamatsu, T., Pettigrew, J. D., and Ary, M. (1979) Restoration of visual cortical plasticity by local microperfusion of norepinephrine. J. comp. Neurol. 185: 163–182.

Kasamatsu, T., Pettigrew, J. D., and Ary, M. (1981) Cortical recovery from effects of monocular deprivation: Acceleration with norepinephrine and suppression with 6-hydroxydopamine. J. Neurophysiol. 45: 254–266.

Kasamatsu, T., and Shirokawa, T. (1985) Involvement of β-adrenoreceptors in the shift of ocular dominance after monocular deprivation. Exp. Brain Res. 59: 507–514.

Kleinschmidt, A., Bear, M. F., and Singer, W. (1987) Blockade of 'NMDA' receptors disrupts experience-dependent plasticity of kitten striate cortex. Science 238: 355–358.

Madison, D. V., and Nicoll, R. A. (1982) Noradrenaline blocks accommodation of pyramidal cell discharge in the hippocampus. Nature 299: 636–638.

Pettigrew, J. D., and Kasamatsu, T. (1978) Local perfusion of noradrenaline maintains visual cortical plasticity. Nature 271: 761–763.

Prusky, G. T., Shaw, C., and Cynader, M. S. (1988) The distribution and ontogenesis of [^{3}H] nicotine binding sites in cat visual cortex. Dev. Brain Res. 39: 161–176.

Sato, H., Hata, Y., Hagihara, K., and Tsumoto, T. (1987) Effects of cholinergic depletion on neuron activities in the cat visual cortex. J. Neurophysiol. 58: 781–794.

Shirokawa, T., and Kasamatsu, T. (1986) Concentration-dependent suppression by β-adrenergic antagonists of the shift in ocular dominance following monocular deprivation in kitten visual cortex. Neuroscience 18: 1035–1046.

314

Sillito, A. M. (1983) Plasticity in the visual cortex. Nature 303: 477–478.

Sporn, J. R., Harden, T. K., Wolfe, B. B., and Molinoff, P. B. (1976) β-adrenergic receptor involvement in 6-hydroxydopamine-induced supersensitivity in rat cerebral cortex. Science 194: 624–626.

Trombley, P., Allen, E. E., Soyke, J., Blaha, C. D., Lane, R. F., and Gordon, B. (1986) Doses of 6-hydroxydopamine sufficient to deplete norepinephrine are not sufficient to decrease plasticity in the visual cortex. J. Neurosci. 6: 266–273.

U'Prichard, D. C., Reisine, T. D., Mason, S. T., Fibiger, H. C., and Yamamura, H. I. (1980) Modulation of rat brain α- and β-adrenergic receptor populations by lesion fo the dorsal noradrenergic bundle. Brain Res. 187: 143–154.

Westlind, A., Grynfarb, M., Hedlund, B., Bartfai, T., and Fuxe, K. (1981) Muscarinic supersensitivity induced by septal lesion or chronic atropine treatment. Brain Res. 225: 131–141.

Wiesel, T. N., and Hubel, D. H. (1963) Single-cell responses in striate cortex of kittens deprived of vision in one eye. J. Neurophysiol. 26: 1003–1017.

Wiesel, T. N., and Hubel, D. H. (1965) Comparison of the effects of unilateral and bilateral eye closure on cortical unit responses in kittens. J. Neurophysiol. 28: 1029–1040.

Acetylcholine-dopamine balance in striatum: Is it still a target for antiparkinsonian therapy?

P. Calabresi, A. Stefani, N. B. Mercuri and G. Bernardi

Clinica Neurologica, Dipartimento di Sanitá Pubblica II Universitá di Roma, Tor Vergata, Via O. Raimondo, I-00173 Roma, Italy

Introduction

In the basal ganglia, the balance between acetylcholine (ACh) and dopamine (DA) levels has been considered of main importance for the control of motor activity. Considering that an imbalance in favor of ACh is associated with Parkinson's disease, the parkinsonian symptoms can be reduced either by elevating DA levels with DA-mimicking drugs or by reducing the ACh effects with ACh receptor antagonists. In the striatum, the interaction between the local intrinsic cholinergic system and the dopaminergic afferents from the substantia nigra (pars compacta) occurs at different levels. At presynaptic level, DA tonically inhibits ACh release in the striatum (Lehmann and Langer, 1983). On the other hand, ACh is able to increase DA release in this structure (Raiteri et al., 1984). At a postsynaptic level, ACh and DA may interact either in the proximal region (perikarya and proximal dendrites) or at more distal level (spines of distal dendrites) of the medium size spiny neurons. The location of both cholinergic and dopaminergic synapses on the neck of the spines receiving glutamatergic inputs may result in the modulation of the excitatory cortical signals (Freund et al., 1984; Izzo and Bolam, 1988). In addition, it has been shown that either DA or ACh may affect their own release by acting on autoreceptors located respectively on nigrostriatal neurons and intrinsic cholinergic aspiny neurons (Chesselet, 1984). As a consequence of these interactions, DA and ACh may differentially influence the neuronal firing and modulate the output signals from the striatum to other structures of the basal ganglia. However, the situation is much more complicated, since the neostriatum can be divided into at least two fundamental compartments, the striosomes (referred to by various workers as patches, islands, or cell clusters) and the matrix (Gerfen, 1984; Gerfen, 1985; Gerfen et al., 1987; Graybiel and Ragsdale, 1978; Nastuk and Graybiel, 1985; Penny et al., 1988).

Striosomes are distinct islands that are relatively low in ACh-esterase staining, while the surrounding matrix is relatively rich in that enzyme.

Recently, it has been shown that most transmitters and related receptors are preferentially accumulated in either the matrix or the patches (Bolam et al., 1988; Gerfen et al., 1987; Izzo et al., 1987)

Experimental and clinical observations on dopaminergic system

The existence of at least two subtypes of brain DA receptors is now widely accepted (Stoof and Kebabian, 1981; Stoof and Kebabian, 1984). The D1 subtype is defined as positively linked to adenylate cyclase while the D2 subtype is negatively coupled or uncoupled to this enzyme (Stoof and Kebabian, 1981; Stoof and Kebabian, 1984). Until recently, it was generally recognized that dopaminergic D2 receptors mediate most of the behavioral and therapeutic effects of neuroleptics and DA agonists. In fact, the extrapyramidal motor alterations by neuroleptics are thought to be connected to increases in striatal cholinergic neuronal activity subsequent to the blockade of inhibitory DA D2 receptors located on striatal cholinergic interneurons (Grigoriadis and Seeman, 1984).

However, recently, it has been shown that D1 and D2 receptors cooperate in the modulation of basal ganglia activity (Braun et al., 1986; Saller and Salama, 1986). In fact, in the motor and behavioral model of Parkinson's disease (chronic reserpine pretreatment, 6-OHDA nigral lesions) it has been observed that the concomitant administration of D1 and D2 agonists can better exert a therapeutic effect in comparison with the results obtained with the administration of D1 or D2 agonists alone (Arnt, 1985; Arnt and Hyttel, 1986). These behavioral observations fit well with the electrophysiological data obtained in our laboratory. We observed that bath application of selective D1 agonists decreases postsynaptic membrane excitability of neostriatal cells (mainly spiny neurons) intracellularly recorded *in vitro* (Calabresi et al., 1987). DA, as well as the D1 selective agonist SKF 38393 and the cAMP analogue 8-Br-cAMP, decreases neuronal excitability in a voltage-dependent manner. This inhibitory effect is due to a decrease of the membrane rectification in the depolarizing direction and it is caused by the activation of postsynaptic D1 receptors. D2 agonists fail to produce detectable electrophysiological effects in control slices. On the other hand, D2 agonists (quinpirole, bromocriptine, lysuride) decrease excitatory postsynaptic potentials (EPSPs) evoked by intrastriatal stimulation in DA-depleted slices (Calabresi et al., 1988a; Calabresi et al., 1988b). The reduction of the EPSP by D2 agonists is not voltage-dependent and probably involves a presynaptic mechanism. Although the cholinergic and/or the glutamatergic nature of the intrastriatally evoked EPSP is still debated (Cherubini et al., 1988; Misgeld et al., 1980), our data clearly show that the efficacy of excitatory inputs is strongly decreased by D2 agonists probably acting at presynaptic level. Thus, the DA-induced inhibition of excitatory synaptic transmission can depend on a presynaptic decrease of ACh release, via D2 receptors, as suggested also by biochemical evidence (Lehmann and Langer, 1983). In addition, DA may modulate postsynaptically the cholinergic transmission by altering the receptor affinity for nicotinic and muscarinic agonists (Ehlert et al., 1981), or by interfering with the intracellular transduction of the signals via a second-messenger system (Kelly and Nakorsky, 1986).

Morphological and pharmacological observations on intrinsic cholinergic system

The striatum contains a dense plexus of cholinergic axons and terminals, whose absolute majority derives from local interneurons (Takagy et al., 1984). Cholinergic synaptic specializations with local neuronal elements are always of the symmetrical type (DiFiglia, 1987), independent of the nature of the postsynaptic target. A direct demonstration of cholinergic input to the medium size spiny neurons has been recently obtained by combining, in the same section, Golgi impregnation and enzyme histochemistry for ACh-esterase (Bolam et al., 1984): these neurons receive cholinergic input not only to their proximal regions (i.e. perikarya and proximal dendrites), but also to their more distal dendrites and dendritic spines (Izzo et al., 1987).

The muscarinic ACh receptors within the neostriatum as well as in other brain areas can be classified in M1 (sites showing high affinity for pirenzepine) and M2 (sites showing low affinity for pirenzepine) (Hammer et al., 1980). The autoradiographic distribution of M1 and M2 muscarinic cholinergic binding sites has been studied in the striatum of the cat, monkey and human (Nastuk and Graybiel, 1988). The two subtypes of binding sites have distinct striatal distribution. M2 sites are virtually homogeneous. Striatal M1 sites are generally more abundant than M2 sites and, in the dorsal striatum, are densely located in patches corresponding to ACh-E-lacking striosomes.

Such a discrepancy between mosaical distribution of cholinergic cell bodies and processes (Graybiel et al., 1986) and the peculiar concentration of cholinoceptive elements inside striosomes (Izzo and Bolam, 1988) lead to a new approach to the intrinsic spatial segregation of muscarinic receptors and related functions.

Numerous biochemical and electrophysiological studies have attempted to relate muscarinic subtypes to specific second-messenger systems (phosphatidylinositol, protein kinase). For instance, some reports have proposed that M1 receptors are linked to phosphatidylinositol (PI) turnover instead of adenylate cyclase (Gil and Wolfe, 1985). However, since other investigations have proposed that both M1 and

M2 receptors modulate PI breakdown (Fisher and Agranoff, 1987), no conclusive evidence has emerged.

In the striatum, three main electrophysiological effects have been described following the application of exogenous ACh or the increase of endogenous ACh levels by ACh-esterase inhibitors: i) decrease of the intrastriatally evoked EPSP (Bernardi et al., 1976; Misgeld et al., 1980; Misgeld et al., 1986); ii) decrease of postsynaptic calcium influx (Misgeld et al., 1986a; Misgeld et al., 1986b); iii) membrane depolarization coupled to an increase of the firing rate and to a decrease of potassium conductances (slow EPSP) (Bernardi et al., 1976; Misgeld et al., 1986a). None of these responses has been pharmacologically characterized; however, it is possible that different receptor subtypes mediate these physiological responses and that the specific location of these receptors on the neuronal elements may account for the complex action of ACh in neostriatum.

Functional organization of neostriatum and new perspectives on ACh-DA interactions

Many lines of evidence indicate that the nigrostriatal DA system obeys striosomal ordering. Differences have been described in the site of origin of mesostriatal fibers innervating striosomes and matrix (Graybiel et al., 1987; Jimenez-Castellanes and Graybiel, 1987). There is now also evidence that striatal DA binding sites of the D2 type are distributed more densely in the extrastriosomal matrix (Joyce and Marshall, 1985), whereas D1 binding sites are more concentrated in striosomes (Besson et al., 1988). The topography of D2 receptor sites determined from autoradiographs of the rat neostriatum has been compared with previously published values for choline acetyltransferase activity and high-affinity choline uptake within subregions of the striatum. The density of D2 sites in the caudate-putamen correlates strikingly with these parameters of cholinergic neuron distribution (Joyce and Marshall, 1985). This relative location of D2 receptors may be important in view of the opposing effects of D1 and D2 receptor antagonists on ACh levels in the rat striatum. In fact, it is widely accepted that the increased striatal ACh tone caused by classical neuroleptics (D2 and mixed D1/D2 antagonists) plays a major role in the expression of DA-related motor patterns (Fage and Scatton, 1986; MacKenzie and Zigmond, 1985; Scatton, 1982).

Since SCH 23390, a selective D1 antagonist, decreases striatal cholinergic neuronal activity, the changes in extrapyramidal motor function elicited by this drug (e.g. catalepsy, antagonism of apomorphine-induced motor manifestations) (Christensen et al., 1984) are unlikely to be affected via this neuronal type. Indeed, in contrast to the effects of neuroleptics, the anti-stereotypic effect of SCH 23390 is not antagonized by anticholinergic drugs (Christensen et al., 1984), suggesting that D1 receptors directly influence striatonigral neurons rather than interact with cholinergic interneurons. Otherwise, the rather selective location of D1 receptors in the striosomes, coupled with the evidence that M1 receptors are mainly concentrated in the same region, suggest a possible interaction between these two receptors. We can speculate that, in the striosomal compartment, M1 receptors may either increase the release of DA from nigrostriatal terminals (Raiteri et al., 1984) and/or directly influence the postsynaptic electroresponsiveness of striatofugal neurons.

In the data reported above there are clear suggestions of functional and structural segregation between striosomes and matrix compartments regarding both DA and ACh systems. Recently, however, it has been proposed that striatal cholinergic interneurons act as parallel trans-striatal pathways between striosomes and the surrounding matrix, providing a substrate for 'cross-talk' between these two compartments (Izzo and Bolam, 1988).

Conclusions

The presented data show that the balance between DA and ACh within the striatum is still an important target for the therapy of the extrapyramidal disorders (i.e. Parkinson's disease). However, the results recently obtained by utilizing different techniques show that several complex interactions regulate this balance.

The up-regulation of both D1 and D2 receptors following the depletion of DA stores (observed in human and experimental Parkinsonism) (Schultz, 1982) can alter the physiological effects of ACh by influencing, at a presynaptic level, the ACh release (D2 receptors) and, at postsynaptic sites (D1 receptors), neuronal membrane properties. For this reason, new advances in the therapy of Parkinson's disease will be provided not only by the understanding of DA receptor functions, but also by the knowledge of the exact role of the selective cholinergic receptors and their changes following DA depletion and DA receptor supersensitivity.

Arnt, J. (1985) Behavioural stimulation is induced by separate dopamine D1 and D2 receptor sites in reserpine pretreatment, but not in normal rats. Eur. J. Pharmac. 113: 79–88.

Arnt, J., and Hyttel, J. (1986) Differential involvment of dopamine D1 and D2 antagonists of cycling behavior induced by apomorphine, SKF 38393, pergolide and LY 171555 in 6-hydroxydopamine lesioned rats. Psychopharmacology 85: 349–352.

Bernardi, G., Floris, V., Marciani, M. G., Morocutti, C., and Stanzione, P. (1976) The action of acetylcholine and l-glutamic acid on rat caudate neurons. Brain Res. 114: 134–138.

Besson, M. J., Graybiel, A. M., and Nastuk, M. A. (1988) 3H-SCH 23390 binding to D1 dopamine receptors in the basal ganglia of the cat and primate: delineation of striosomal compartments and pallidal and nigral subdivision. Neuroscience, in press.

Bolam, J. P., Ingham, C. A., and Smith, A. D. (1984) The section Golgi impregnation procedure. 3. Combination of Golgi impregnation with enzyme histochemistry and electron microscopy to characterize acetylcholinesterase-containing neurons in the rat neostriatum. Neuroscience 12: 687–709.

Bolam, J. P., Izzo, P. N., and Graybiel, A. M. (1988) Cellular substrate of the histochemically defined striosome/matrix system of the caudate nucleus: a combined Golgi and immunocytochemical study in cat and ferret. Neuroscience 24: 853–875.

Braun, A. R., Barone, P., and Chase, T. N. (1986) Interaction of D1 and D2 receptors in the expression of dopamine agonist induced behaviors. In: Creese, I., and Breese, G. R. (eds), Neurobiology of Central D1 Dopamine Receptors. Plenum Press, New York, pp. 151–166.

Calabresi, P., Benedetti, M., Mercuri, N. B., and Bernardi, G. (1988a) Depletion of catecholamines reveals inhibitory effects of bromocryptine and lysuride on neostriatal neurones recorded intracellularly in vitro. Neuropharmacology 27: 579–587.

Calabresi, P., Benedetti, M., Mercuri, N. B., and Bernardi, G. (1988b) Endogenous dopamine and dopaminergic agonists modulate synaptic excitation in neostriatum: intracellular studies from naive and catecholamine-depleted rats. Neuroscience 27: 145–157.

Calabresi, P., Mercuri, N., Stanzione, P., Stefani A., and Bernardi, G. (1987) Intracellular studies on the dopamine-induced firing inhibition of neostriatal neurons in vitro: evidence for D1 receptor involvement. Neuroscience 20: 757–771.

Cherubini, E., Herrling, P. L., Lanfumey, L., and Stanzione, P. (1988) Excitatory amino acids in synaptic excitation of rat striatal neurons in vitro. J. Physiol. 400: 677–690.

Chesselet, M. F. (1984) Presynaptic regulation of neurotransmitter release in the brain. Neuroscience 12: 347–378.

Christensen, A. V., Arnt, J., Hyttel, J., Larsen, J. J., and Svenson, O. (1984) Pharmacological effects of a specific dopamine D-1 antagonist SCH 23390 in comparison with neuroleptics. Life Sci. 34: 1529.

DiFiglia, M. (1987) Synaptic organization of cholinergic neurons in the monkey neostriatum. J. comp. Neurol. 255: 245–258.

Ehlert, F. J., Roeske, W. R., and Yamamura, H. Y. (1981) Striatal muscarinic receptors regulation by dopaminergic agonists. Life Sci. 28: 441–448.

Fage, D., and Scatton, B. (1986) Opposing effects of D1 and D2 receptor antagonists on acetylcholine levels in the rat striatum. Eur. J. Pharmac. 129: 359–362.

Fisher, S. K., and Agranoff, B. W. (1987) Receptor activation and inositol lipid hydrolisis in the neural tissues. J. Neurochem. 48: 999–1017.

Freund, T. F., Powell, J. F., and Smith, A. D. (1984) Tyrosine hydroxylase immunoreactive synaptic boutons in contact with identified striatonigral neurons with particular reference to dendritic spine. Neuroscience 13: 1189–1215.

Gerfen, C. R. (1984) The neostriatal mosaic: compartmentalization of corticostriatal input and striatonigral output systems. Nature 311: 461–464.

Gerfen, C. R. (1985) The neostriatal mosaic. I. Compartmental organization of projections from the striatum to the substantia nigra in the rat. J. comp. Neurol. 237: 176–194.

Gerfen, C. R., Herkenham, M., and Thibault, I. (1987) The neostriatal mosaic. II. Patch- and matrix-directed mesostriatal dopaminergic and non-dopaminergic system. J. Neurosci. 7: 3915–3934.

Gil, D. W., and Wolfe, B. B. (1985) Pirenzepine distinguishes between muscarinic receptors mediated phosphoinositole breakdown and inhibition of adenylate cyclase. J. Pharmac. exp. Ther. 232: 608–616.

Graybiel, A. M., Baughman, R. W., and Eckstein, F. (1986) Cholinergic neuropil of the striatum observes striosomal boundaries. Nature 323: 625–627.

Graybiel, A. M., Hirsch, E. C., and Agid, Y. A. (1987) Differences in tyrosine hydroxylase-like immunoreactivity characterize the mesostriatal innervation of striosomes and extrastriosomial matrix at maturity. Proc. natl Acad. Sci. USA 84: 303–307.

Graybiel, A. M., and Ragsdale, C. W. (1978) Histochemically distinct compartments in the striatum of human, monkey and cat demonstrated by acetylthiocholinesterase staining. Proc. natl Acad. Sci. USA 75: 5723–5726.

Grigoriadis, D., and Seeman, P. (1984) The dopamine-neuroleptic receptor. Can. J. neurol. Sci. 11: 108–113.

Hammer, R., Berrie, C. P., Birdsall, M. J. M., Burgen, A. S., and Hulme, E. C. (1980)

Pirenzepine distinguishes between different subclasses of muscarinic receptors. Nature 283: 90–92.

Izzo, P. N., Graybiel, A. M., and Bolam, J. P. (1987) Characterization of substance P and Met-Enk-immunoreactive neurons in the caudate nucleus of cat and ferret by a single section Golgi procedure. Neuroscience 20: 577–587.

Izzo, P. N., and Bolam, J. P. (1988) Cholinergic synaptic input to different parts of spiny striatonigral neurons. J. comp. Neurol. 269: 219–234.

Jimenez-Castellanos, J., and Graybiel, A. M. (1987) Subdivisions of the dopamine containing A8-A9-A10 complex identified by their differential mesostriatal innervation of striosomes and extrasomial matrix. Neuroscience 23: 223–242.

Joyce, J. N., and Marshall, J. F. (1985) Striatal topography of D2 receptors correlates with indexes of cholinergic neuron localization. Neurosci. Lett. 53: 127–131.

Kelly, E., and Nakorsky, S. K. (1986) Specific inhibition of dopamine D1-mediated cyclic AMP formation by dopamine D2, muscarinic cholinergic and opiate receptor in dopaminergic control of striatal cholinergic transmission. Life Sci. 31: 2883.

Lehmann, J., and Langer, S. Z. (1983) The striatal cholinergic interneuron: synaptic target of dopaminergic terminals. Neuroscience 10: 1105–1120.

Mackenzie, R. G., and Zigmond, M. S. (1985) Chronic neuroleptic treatment increased D2 but not D1 receptors in rat striatum. Eur. J. Pharmac. 113: 159–165.

Misgeld, U., Calabresi, P., and Dodt, U. (1986a) Muscarinic modulation in the neostriatum: possible involvement of calcium. Exp. Brain Res. S14: 176–184.

Misgeld, U., Calabresi, P., and Dodt, U. (1986b) Muscarinic modulation of calcium-dependent plateau potentials in rat neostriatal neurons. Pflügers Arch. Physiol. 407: 482–487.

Misgeld, U., Weiler, M. H., and Bak, J. J. (1980) Intrinsic cholinergic excitation in the rat neostriatum: nicotinic and muscarinic receptors. Exp. Brain Res. 39: 401–409.

Nastuk, M. A., and Graybiel, A. M. (1985) Patterns of muscarinic cholinergic binding in the striatum and their relation to dopamine islands and striosomes. J. comp. Neurol. 237: 176–194.

Nastuk, M. A., and Graybiel, A. M. (1988) Autoradiographic localization and biochemical characteristics of M1 and M2 muscarinic binding sites in the striatum of the cat, monkey and human. J. Neurosci. 83: 1052–1068.

Penny, G. R., Wilson, C. J., and Kitai, S. T. (1988) Relationship of the axonal and dendritic geometry of spiny projection neurons to the compartmental organization of the neostriatum. J. comp. Neurol. 269: 275–289.

Raiteri, M., Leardi, R., and Marchi, M. (1984) Heterogeneity of presynaptic muscarinic receptors regulating neurotransmitter release in the rat brain. J. Pharmac. exp. Ther. 228: 209–214.

Saller, C. E., and Salama, A. I. (1986) D1 and D2 dopamine receptor blockade: interactive effects in vitro and in vivo. J. Pharmac. exp. Ther. 236: 714–720.

Scatton, B. (1982) Further evidence for the involvement of D2 but not D1 dopamine receptors in dopaminergic control of striatal cholinergic transmission. Life Sci. 31: 2883.

Schultz, W. (1982) Depletion of dopamine in the striatum as an experimental model of parkinsonism: direct effects and adaptive mechanisms. Prog. Neurobiol. 18: 121–166.

Stoof, J. C., and Kebabian, J. W. (1984) Two dopamine receptors: biochemistry, physiology and pharmacology. Life Sci. 35: 2281.

Stoof, J. C., and Kebabian, J. W. (1981) Opposing roles for D1 and D2 dopamine receptors in efflux of cyclic AMP in rat neostriatum. Nature 294: 366–368.

Takagy, H., Somogyi, P., and Smith, A. D. (1984) Aspiny neurons and their local axons in the neostriatum of the rat: A correlated light and electron microscopic study of Golgi-impregnated material. J. Neurocytol. 13: 239–265.

Subject Index